Phylogenetic Comparative Methods in R

Phylogenetic Comparative Methods in R

Liam J. Revell and Luke J. Harmon

PRINCETON UNIVERSITY PRESS

Princeton & Oxford

Published by Princeton University Press
41 William Street, Princeton, New Jersey 08540
99 Banbury Road, Oxford OX2 6JX

press.princeton.edu

Library of Congress Control Number: 2022932452

ISBN 9780691219028
ISBN (pbk.) 9780691219035
ISBN (e-book) 9780691219042

British Library Cataloging-in-Publication Data is available

Editorial: Alison Kalett, Hallie Schaeffer
Jacket/Cover Design: Wanda España
Production: Jacqueline Poirier
Publicity: Matthew Taylor, Charlotte Coyne

Cover Image : mauritius images GmbH/Alamy Stock Photo

This book has been composed in MinionPro and MyriadPro

10 9 8 7 6 5 4 3 2 1

Contents

Phylogenetic Comparative Methods in R

A brief introduction to phylogenetics in R

1.1 Introduction

This book is about carrying out phylogenetic comparative analyses in the R statistical computing environment.

In this chapter, we will:

1. Introduce the general field of evolutionary research called *phylogenetic comparative biology* and discuss how the R scientific computing environment can be used in the analysis of phylogenetic data.
2. Present and explain the general structure of this volume, including how we expect it to be read and used.
3. Introduce the major R function libraries (called "packages") used to analyze phylogenetic data in the R environment.
4. Examine the `"phylo"` object: an important data structure that is used by most phylogenetic R packages to encode a tree.
5. Finally, illustrate a number of important R functions for phylogenetic analysis, including how to read and write trees and phylogenetic data, how to plot trees in various styles, how to manage phylogenetic trees and data, and how to conduct a simple phylogenetic comparative analysis in R.

1.1.1 What is phylogenetic comparative analysis?

Phylogenetic comparative analysis[1] is the general endeavor of using a phylogenetic tree, frequently combined with phenotypic trait data for the species in the tree, to learn something about evolution (Harvey and Pagel 1991; Pennell and Harmon 2013).

Although the modern field of phylogenetic comparative analysis is relatively young (tracing back largely to Felsenstein 1985), phylogenetic comparative methods have diversified in scope, number, and importance in recent decades (reviewed in Harmon 2019).

---

[1] Often called *phylogenetic comparative methods*, *PCMs*, or sometimes just *the comparative method*.

Contemporary phylogenetic comparative methods now encompass an enormous range of topics. For instance, phylogenetic comparative analyses have been employed to measure the relationships between characters while taking the phylogeny into account (Martins and Hansen 1997), to infer the rates of species proliferation and extinction through time (Nee 2006), and to fit sophisticated mathematical models to phylogenies and comparative data in an attempt to explain the diversity of life that we see around us on this planet (Maddison et al. 2007). Comparative methods have also been used to track the spread of diseases (Stadler and Bonhoeffer 2013), to understand contemporary threats to species (Greenberg and Mooers 2017), and to describe the dynamics of evolution over thousands or millions of years (Uyeda et al. 2016). Phylogenetics comparative methods have even been used to study the global SARS-CoV-2 pandemic that started in 2019 (e.g., Wang et al. 2020; Sjaarda et al. 2021).

1.1.2 Phylogenetic comparative analysis in R

Over the past decade, the scientific computing environment R (R Development Core Team 2020) has grown to play a key role in phylogenetic comparative methods. Many developers of PCMs tend to work in R, and many PCM users prefer to conduct their analyses in R. This synergy between users and developers means that R has become an essential tool for scientists interested in employing the comparative method in their research.

The purpose of this book to teach users how to carry out phylogenetic comparative methods using R. We only briefly cover R basics, so readers completely new to the R environment might think about complementing this volume with a simpler book focused on introducing the R computing environment.

This book is designed to complement, not replace, a more complete theoretical treatment of phylogenetic comparative methods. As such, we do not fully explain the mathematics and conceptual basis of all comparative methods covered herein. For a more comprehensive review of the theoretical basis of phylogenetic comparative analysis, we recommend Harmon (2019), Garamszegi (2014), Nunn (2011), or one of several excellent books that cover parts of phylogenetic comparative biology as part of a larger treatment of phylogeny inference (e.g., Felsenstein 2004; Yang 2006). We expect that this book will be used either in parallel with a full course of study on phylogenies and the comparative method (using, for instance, Harmon 2019) or by scientists already familiar with much of the theoretical basis of phylogenetic inference, phylogenetic comparative methods, or both—and ready to immerse themselves in the phylogenetic comparative biology in the R computing environment. Our goal in this book is thus highly practical: to give scientists the tools they need to start analyzing their own comparative data.

The book is not designed to cover all phylogenetic comparative methods. First, we focus exclusively on phylogenetic comparative methods implemented in the R computing environment. Several important phylogenetic analyses (e.g., Pagel and Meade 2013; Rabosky 2014) are implemented in software that run outside of R. As such, we largely consider these methods to be out of scope for the book.[2] Second, we focus especially on methods implemented in our own packages *phytools* (Revell 2012) and *geiger* (Harmon et al. 2008; Pennell et al. 2014), as well as in the core phylogenetics R package *ape* (Paradis et al. 2004). In part, this is due to our own intrinsic biases; however, it's also motivated in equal measure by a desire to ensure that this book remains useful over the medium to long term. As package authors and maintainers, it's much easier for us to guarantee that updates and extensions of the *phytools* and *geiger* R libraries will always remain compatible with the code presented in this book. As such, when a method is implemented in both *geiger* or *phytools* and another R package, we will generally prefer to use

---

[2] Although we promise to discuss the relationship of some of these important methodologies to those that we are covering.

our packages—unless functionality is vastly different between the different implementations. On the other hand, in this book we do cover many important methods implemented in R packages other than our own, and in those cases (obviously), we show how to use these function libraries.

One quick note on controversies in the field. Phylogenetic comparative methods grew organically, with new methods being added rapidly—and sometimes with very little testing or evaluation. Occasionally, methods are shown to have undesirable properties. In other cases, statistical approaches that are commonly used have nuances that can only be appreciated with extensive simulation studies. We subscribe to the philosophy that comparative methods are "normal statistical methods." Consequently, we tend to describe these critiques in terms of standard statistical concepts such as statistical error, model adequacy, identifiability, and so on.

This book is largely based on the content that we developed for a series of classes and workshops that we've taught over the past half-dozen years or so across at least eleven countries and four continents. These workshops were not developed or taught in a vacuum and owe their existence to a long list of collaborators, including (but not restricted to) M. Alfaro, R. Betancur, A. Crawford, S. De Esteban-Trivigno, A. Gonzalez-Voyer, R. Zenil-Feruguson, and J. Tavera, among others.

1.1.3 How to use this book

As noted in the previous section, this book is designed to *complement*, not replace, more comprehensive theoretical treatments of phylogenetic comparative methods, such as Harmon (2019).

We anticipate that some readers will progress through this book from start to finish in a "self-study" course, while others will leap from one chapter to another, depending on their prior R phylogenetics experience, specific questions, or particular topics of interest. As such, we have designed each chapter to "stand apart," in that we reiterate reading input data from file, checking data for completeness, and so on, even if it duplicates computational steps of a prior chapter with the same files. The chapters of the book still do build from beginning to end, so background *explanations* of each R computation or analysis step are not constantly repeated in the text.

We envision that most readers will use this book as a manual or guidebook to undertaking real phylogenetic analyses in an interactive session of R. We picture readers with the book propped open next to their laptop or desktop computer, transcribing (or adapting) our scripts from the book into R. All files that we use in this book are available from download through the book's website,[3,4] so there should be no limit in the reader's ability to follow along.

R is at the same time a statistical software, a scientific computing environment, and a programming language.

---

[3] The book website is https://press.princeton.edu/books/phylogenetic-comparative-methods-in-r; however, for quick access to the files we'll use throughout this volume, readers can refer to http://www.phytools.org/Rbook/.

[4] Readers might notice in frustration that throughout this book, we have used data files that contain phylogenetic trees in different formats, or with mismatched taxa labels, or with data that need to be reorganized or subsampled before analysis. This was an intentional decision, with the aim of helping our readers become more comfortable in working with realistic (and thus sometimes a little *messy*) data sets in the R environment. Please forgive us!

R is distributed free and open source. This means that it is not only free to download and use, but any user or developer can also see the entire source code of the project—and even potentially modify it as they see fit!

1.2.1 The R command line

Although R can be intimidating, most users of R are not doing R programming and can find relatively simple ways to carry out their analyses. However, use of R *does* typically require that you enter commands into a text-based command-line interface, which may be slightly disorienting at first.[5]

One goal of this book in general, and the current chapter in particular, is to help users get comfortable with the commands and language of R.

1.2.2 Packages and resources

The rich functionality of R is built almost entirely on what are called *contributed packages* created by members of the R community of users and developers.

Contributed R packages are best thought of as small libraries of new, usually thematically related R programs known as *functions*. The majority of contributed packages are stored in a public repository called CRAN,[6] an acronym for the *Comprehensive R Archive Network*.

In this chapter, we'll review some of the basics of working with phylogenies in the R environment. We'll assume that the reader has *some* prior experience with R and already knows a little to a lot about phylogenies and the phylogenetic comparative method. Many excellent introductions to R are available both in book form and on the web. Felsenstein (2004) remains an incredible reference for all things phylogeny. Harmon (2019) is (in the humble opinion of the authors) the most comprehensive resource developed to date for phylogenetic comparative analyses.

1.2.3 Code chunks and R output

This chapter and all other chapters in this book have been written by the authors but were assembled using R. As such, all the *gray boxes* consist of what we'll refer to as "code chunks": one or various lines of R script meant to be run in an interactive R session.

All the intervening `courier text` sections and all of the figures are the expected output from R.

That means that to follow along with the R activities of this book, it is possible to simply enter the scripts from the gray boxes into your R session and run them. In fact, this is what we'd recommend!

1.2.4 Entering R commands using a GUI

Rather than typing the commands directly into your R interactive session command prompt, we always suggest entering your R commands first into a scripting window and then executing the code in R.[7]

---

[5]Particularly for computer users raised in a post-MS-DOS world!

[6]https://cran.r-project.org/.

[7]This is easiest to do from within an R *graphical user interface* or GUI, such as *RGui* for Windows or *RStudio* (*RStudio* Team 2020) on pretty much any platform.

This is a good habit to get into not only when learning how to use R for the first time but also down the road when you begin to apply R to analyze your own data. That is because doing so will permit you to easily save all the commands that we've run in R so that you can readily review them, modify them, and rerun them later if necessary. It also permits you to easily publish all the steps of your data analysis alongside your scientific papers or reports, facilitating reproducibility of research by your peers.

Once you have entered your commands into a scripting window, you do not need to copy and paste your code from one window to the other. Instead, most R graphical user interfaces (GUIs) permit us to directly execute lines from our script in our R session with a simple shortcut. In the R Windows GUI (*RGui*), this can be done by typing CTRL-R with the cursor located on the line you want to execute or with various lines selected and highlighted. In *RStudio* for Windows, the shortcut is CTRL-ENTER, whereas in *RStudio* on a Mac, it is Command-ENTER.

1.3 R phylogenetics

R phylogenetics is built on the contributed packages for phylogenetics in R, and there are many such packages. A partial list of the R packages that contain phylogeny-related functionality is available on a website called the *CRAN phylogenetics task view*.[8]

In this book, we'll only be working with a subset of these packages.

1.3.1 Installing R packages and checking version numbers

We can begin by installing a few of the most critical of R phylogenetics packages: *ape* (Paradis et al. 2004; Paradis and Schliep 2019), *phangorn* (Schliep 2011), *phytools* (Revell 2012), and *geiger* (Harmon et al. 2008; Pennell et al. 2014). To ensure that we get the most recent CRAN package versions, we need to have the most up-to-date R version installed on our computer!

In an interactive R session, it's pretty straightforward to see which R version you have installed.

At the time of writing, the most recent version of R was version 4.1.1.[9]

```
R.version

##              _
## platform     x86_64-w64-mingw32
## arch         x86_64
## os           mingw32
## system       x86_64, mingw32
## status
## major        4
## minor        1.1
## year         2021
## month        08
## day          10
```

---

[8]https://cran.r-project.org/web/views/Phylogenetics.html.

[9]Although it will certainly be long out of date by the time this book arrives to your shelf!

```
## svn rev        80725
## language       R
## version.string R version 4.1.1 (2021-08-10)
## nickname       Kick Things
```

Next, let's proceed to install the various packages that we intend to use in this chapter. This can be done easily using the R function `install.packages` as follows.[10]

```
install.packages("ape")
install.packages("phangorn")
install.packages("phytools")
install.packages("geiger")
```

We can proceed to verify the package versions that we've installed by using the *base* R function `packageVersion`:

```
packageVersion("ape")

## [1] '5.5'

packageVersion("phangorn")

## [1] '2.7.1'

packageVersion("phytools")

## [1] '0.7.96'

packageVersion("geiger")

## [1] '2.0.7'
```

Some packages are updated frequently, others less often, but you shouldn't be surprised to see a mismatch between the versions shown above and the package versions you have installed on your computer. Just be aware that sometimes errors can result from using packages that are out of date and thus incompatible with one another.

Installing automatically from CRAN using `install.packages` installs not only your target package but also any libraries on which that package *depends*, if that package has not yet been installed.

For instance, a package dependency of R package *B* on R package *A* means that package *B* uses functions of *A* "internally" (that is, inside of its own functions). Consequently, use of package *B* requires that *A* be installed and loaded. Fortunately, R takes care of these details for us. If a dependent package can't be found or loaded, R will give an error warning us that the missing package needs to be installed.

---

[10] This works for packages that are on CRAN, which covers most common R packages for comparative methods. Some other packages we use in this book must be installed from GitHub using the R package *devtools*.

Now we've installed some critical R phylogenetics packages (*ape, phangorn, phytools,* and *geiger*).

The most important "core" package for phylogenies in R is called *ape* (Paradis et al. 2004; Paradis and Schliep 2019), which stands for **A**nalysis of **P**hylogenetics and **E**volution in R.[11]

1.4.1 Loading the *ape* package

Although we *installed* our main R phylogenetics packages, to make best use of a contributed package, we must proceed to *load* it in our current R session.

Here, we'll do this using the *base* function library as follows.[12]

```
library(ape)
```

1.4.2 Reading a phylogenetic tree file

ape does many different things. To get started, let's read a "toy" phylogenetic tree of vertebrates from a relatively simple Newick text string.[13]

```
text.string<-
    "(((((Robin,Iguana),((((Cow,Whale),Pig),Bat),
    (Lemur,Human))),Coelacanth),Goldfish),Shark);"
vert.tree<-read.tree(text=text.string)
```

1.4.3 Plotting a phylogenetic tree

We can plot this tree in our R session using the *ape* package "phylo" S3 plot method as follows.[14] We see the result in figure 1.1.

---

[11] A good way to think of what makes *ape* a core package in phylogenetics also has to do with dependency relationships between packages. *Many* other R phylogenetics packages depend on *ape*, or depend on packages that depend on *ape*, while *ape* does not itself depend on other phylogenetics packages.

[12] Note that a highly similar function called require will do pretty much exactly the same thing. library and require are subtly different, but for our purposes, they are interchangeable, and you should feel free to use whichever one you prefer!

[13] A *Newick* string—named, believe it or not, after a lobster restaurant in New Hampshire—is a simple way to encode a phylogenetic tree using a series of nested parentheses. More closely nested species are more closely related. For instance, the simple Newick tree *((chimp,human),gorilla);* tells us that the operational taxa *chimp* and *human* are more closely related to each other than either is to *gorilla*. There are other ways that phylogenetic trees can be represented in machine-readable text, but the Newick string is by far the most common.

[14] The terminology *S3 method* refers to a way that R uses to assign a generic function to an object class. This is helpful, because if our object is a set of points in two dimensions and we send this object to plot, R knows—unless we tell it otherwise—to draw a scatterplot. Likewise, if our object is a phylogeny, R knows to draw a tree. Commonly used methods are plot, print, summary, and predict, but there are many others, and it's even possible for R programmers to develop their own new generic methods! One tricky aspect of S3 generic methods is that lazy R programmers can develop methods for new object classes without documenting them—so long as the arguments are nominally equivalent.

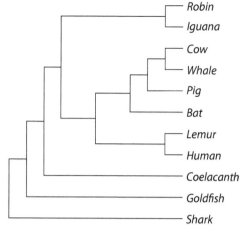

Figure 1.1
A simple phylogenetic plot of vertebrate species drawn with the *ape* method `plot.phylo`.

```
plot(vert.tree,no.margin=TRUE)
```

1.4.4 Getting help for an R function

It's easy to identify ways in which this plot might be improved. For instance, perhaps the lines could be thicker, the font size larger, the margins smaller, and so on. In fact, all of these options are available in the function.

In general, to see the help page for a function, you can call the function `help`[15] on the name of the function you need help with: in our case, `plot`.

```
help(plot)
```

As we are using an S3 method to plot, however, we have to do something different. If we want to see the help page for the `plot` function applied to *phylo* objects, we must run[16]

```
help(plot.phylo)
```

1.4.5 Function arguments and values

Help pages in R are very useful for novice and experienced R users alike. They have a standardized format that details what *arguments* the function takes as input, what the function does, and what *value* the function returns to the user.

Function arguments are best thought of as the *options* and *inputs* of the function. These might include our data, as well as any specifications that the function needs to run our analysis or to generate a plot.

---

[15]Just entering `?function_name` at the command prompt in R will have the same effect—and is quicker too.

[16]This is generally true for S3 methods. That is to say, if the method has been documented for a particular object class, this documentation will be found at `nameOfMethod.classOfObject`.

The function value is what is returned by the function. For some functions, all results are printed to screen or used to make a graph. Many functions, however, return the results of their execution to the user in the form of one or more numerical values or a special object.

Once we've familiarized ourselves with a function via its help page, it is often useful to use the helper base R function `args` in interactive sessions to obtain a list of the arguments that the function accepts:

```
args(plot.phylo)

## function (x, type = "phylogram",
## use.edge.length = TRUE, node.pos = NULL,
## show.tip.label = TRUE, show.node.label =
## FALSE, edge.color = "black",
## edge.width = 1, edge.lty = 1, font = 3,
## cex = par("cex"),
## adj = NULL, srt = 0, no.margin = FALSE,
## root.edge = FALSE,
## label.offset = 0, underscore = FALSE,
## x.lim = NULL, y.lim = NULL,
## direction = "rightwards", lab4ut = NULL,
## tip.color = "black",
## plot = TRUE, rotate.tree = 0, open.angle
## = 0, node.depth = 1,
## align.tip.label = FALSE, ...)
## NULL
```

1.4.6 Different ways to plot a phylogenetic tree

Reviewing the help page for `plot.phylo`, as well as the long list of function arguments listed above, suggests that we can visualize our phylogenies in R in a remarkably large number of different ways, even just using this function (and thus not considering all the other various contributed package functions designed to plot phylogenies).

To see this, let's plot our phylogeny in three different styles (figure 1.2).

```
par(mfrow=c(2,2),mar=c(1.1,1.1,3.1,1.1))
plot(vert.tree)
mtext("(a)",line=1,adj=0)
plot(vert.tree,type="cladogram")
mtext("(b)",line=1,adj=0)
plot(unroot(vert.tree),type="unrooted",
     lab4ut="axial",x.lim=c(-2,6.5),
     y.lim=c(-3,7.5))
mtext("(c)",line=1,adj=0)
```

Now why don't we have a look at what this code does line by line.[17]

---

[17] We'll try to do this as much as possible throughout the book.

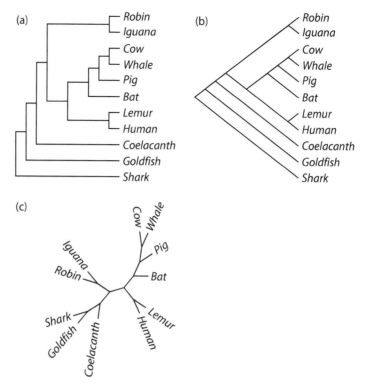

Figure 1.2
A phylogeny plotted in three different styles. (a) A right-facing square cladogram/phylogram. (b) A slanted cladogram/phylogram. (c) An unrooted style. All three graphs were drawn using the *ape* plotting method, `plot.phylo`.

The first line, `par(mfrow=c(2,2),mar=c(1.1,1.1,3.1,1.1))`, tells R to divide our plotting device into four subplots for a 2 × 2 grid—done via the argument `mfrow`. By way of the argument `mar`, it also tells R to set the margins to custom values. The order of this vector is *bottom, left, top*, and *right*—so we see that we are setting all the margins to 1.1 units, except the upper margin, which we set to 3.1.[18]

The second line, `plot(vert.tree)`, plots the tree using the S3 `plot` method with most of its default arguments—but changed `font=1` to print the tip labels in regular (instead of *italic*) font.

The third line, `mtext("(a)",line=1,adj=0)`, adds a subplot label ("(a)").[19] The `mtext` argument `line=1` tells R to put the text one line above the figure margin, while the argument `adj=0` tells R to align the text to the left of the plot area.

Finally the fourth, fifth, sixth, and seventh lines repeat the same pattern for each of the subplots but with different plotting styles: first a slanted cladogram (`type="cladogram"`) in figure 1.2b and then an unrooted tree (`type="unrooted"`) in figure 1.2c.[20] We also adjusted

---

[18]We'll see a lot more of `par` throughout the book.

[19]The function name `mtext` is short for *margin text*.

[20]In the lattermost of these, the argument `lab4ut="axial"`—which stands for *labels for unrooted tree*—tells R to orientate the tip labels in the same direction as the terminal branches of the phylogeny. Who would've guessed? This is what help pages are for!

the *x* and *y* limits of the plot (using the arguments `x.lim` and `y.lim`, respectively)—but this was simply because we discovered that for `type="unrooted"` with `lab4ut` turned on, R was cutting off some of our taxon labels. The specific values that we used are idiosyncratic to the particular tree we're plotting—but these arguments are nonetheless useful to remember should the readers find themselves in a similar situation or want to leave space around their plotted tree for any other reason.[21]

1.5 The internal structure of a tree in R

When we read any phylogeny from file or from a text string (as we did in the previous section), we create an object in the R workspace.

Normally, it won't be necessary to interact directly with this object's internal structure. Instead, we usually pass the object unchanged to other functions—such as when we plotted our phylogeny in different styles to create figure 1.2.

Nonetheless, we believe that for users who commonly work with phylogenies in the R environment, it can be extremely useful to develop a basic working understanding of the structure of phylogenetic objects in memory during an interactive R session.[22]

1.5.1 Trees as lists

This object—that is, the one created in memory when we simulate or estimate a phylogeny or read one from an input file—is a *list* of class `"phylo"`.

In R, a list is just a customizable object type that can combine two or various objects of different types.

For instance, a list might contain a vector of real numbers (with mode `"numeric"`) as its first element, then a vector of strings (with mode `"character"`) as its second element, and so on.

Lists are virtually endlessly flexible, because they can also include other lists[23] (and even functions) among their elements.

Assigning our phylogenetic list with a special class, `"phylo"`, is just a convenient way to tell other functions in R, particularly S3 methods, how to treat that object.

1.5.2 Elements of the "phylo" list

An object of class `"phylo"` always consists of at least three elements.

These components of the object are normally "hidden" from view. That is to say, just typing the name of your `"phylo"` object does not reveal the structure of the object in memory, as it would for a standard list in R.

```
vert.tree

##
## Phylogenetic tree with 11 tips and 10 internal nodes.
```

---

[21] Such as to add additional graphical elements or features to the plot later; see chapter 13.

[22] In fact, we estimate that if we had a penny for every *geiger* or *phytools* user issue that could have been resolved through knowledge of the structure of the `"phylo"` object, we'd have at least two dollars!

[23] Or lists of list, or lists of lists of lists, and so on.

```
##
## Tip labels:
##    Shark, Goldfish, Coelacanth, Human, Lemur, Bat, ...
##
## Rooted; no branch lengths.
```

What's happened here? Why do we see a summary of the object instead of its structure?

What has occurred is that something called an S3 `print` method has been activated to (guess what?) print a *summary* of some of the important attributes of that object.

In the case of a `"phylo"` object, this summary is designed to give us a printout of the number of terminal taxa in the tree and a list of some of their labels.

R lets us, however, reveal the internal structure of this (and, in fact, virtually any) R object using the handy function str[24] as follows.

```
str(vert.tree)

## List of 3
## $ edge    : int [1:20, 1:2] 12 12 13 13 14 14 15
17 21 21 ...
## $ Nnode   : int 10
## $ tip.label: chr [1:11] "Shark" "Gold_fish"
"Coelacanth" "Human" ...
## - attr(*, "class")= chr "phylo"
## - attr(*, "order")= chr "cladewise"
```

This tells us that our `vert.tree` object is a list composed of (in this case) three different elements, along with a couple of different attributes.

More specifically, the different parts of our object include

1. `edge`: a 20×2 (in this case) matrix containing starting and ending indices for the nodes subtending each branch of the phylogeny. By convention, tip nodes (that is, those corresponding to species or operational taxa) are numbered 1 through N for N species, while internal nodes are numbered $N + 1$ (at the root) through $N+$ the number of internal nodes.[25]
2. `Nnode`: an integer value giving the total number of internal nodes in the tree.
3. `tip.label`: a character vector of length N containing the labels for all the tips or terminal taxa in the phylogeny.

1.5.3 Node indices

Now let's see how these different components relate to the structure of the tree by replotting our phylogeny, but this time overlaying the numerical indices from the matrix `edge` onto the nodes and terminals of the tree (figure 1.3). We can do that in R as follows.

---

[24]Short for *structure*.

[25]There will be $N - 1$ of these if our tree is both rooted and perfectly bifurcating. An unrooted, bifurcating tree has $N - 2$ internal nodes. Trees with polytomies can have fewer nodes still, while trees with unbranching nodes can have more.

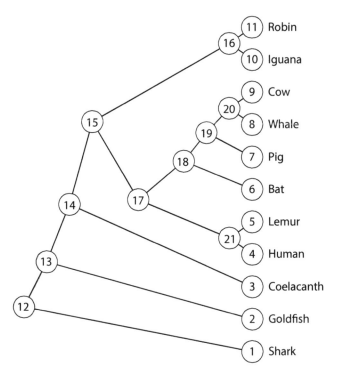

Figure 1.3
A simple phylogeny of vertebrate species with nodes labeled by their indices in the "phylo" edge matrix. We created the plot using the *phytools* function plotTree and added node labels using labelnodes (although the latter could have also been done just as easily with the *ape* function nodelabels).

```
library(phytools)
plotTree(vert.tree,offset=1,type="cladogram")
labelnodes(1:(Ntip(vert.tree)+vert.tree$Nnode),
    1:(Ntip(vert.tree)+vert.tree$Nnode),
    interactive=FALSE,cex=0.8)
```

Reviewing our code line by line, we first loaded an additional R package called *phytools* (library(phytools)).

We then plotted our tree, but instead of using the S3 method, we elected[26] to use the *phytools* function plotTree.

Finally, we used the *ape* function labelnodes to add numerical labels to all the internal and "external" (that is, tip) nodes of the phylogeny.[27]

Just to reiterate, here what we have done is simply *plotted* our tree, and then we've *overlain* the "node numbers" onto the plotted tree. The node numbers are simply the indices from the "phylo" object element edge, which is itself a matrix containing the starting and ending indices for each branch of the phylogeny!

---

[26] For no particular reason.

[27] We could have also used the *ape* functions nodelabels() and tiplabels() without any arguments—but that doesn't look quite as nice, in our opinion. Try it and see if you agree!

```
vert.tree$edge
```

```
##         [,1]  [,2]
##  [1,]    12     1
##  [2,]    12    13
##  [3,]    13     2
##  [4,]    13    14
##  [5,]    14     3
##  [6,]    14    15
##  [7,]    15    17
##  [8,]    17    21
##  [9,]    21     4
## [10,]    21     5
## [11,]    17    18
## [12,]    18     6
## [13,]    18    19
## [14,]    19     7
## [15,]    19    20
## [16,]    20     8
## [17,]    20     9
## [18,]    15    16
## [19,]    16    10
## [20,]    16    11
```

If we now go ahead and compare vert.tree$edge to our plot in figure 1.3, we should see that each *row* of the matrix corresponds to one and only one branch in the tree. In other words, the edge matrix completely represents the topology of our tree using a simple table!

We should also notice

1. that edge has a number of rows that are *equal* to the number of branches (20) in this phylogeny, and
2. that each branch starts and ends with a unique pair of indices, just as we learned above.

1.5.4 Tip labels and node counts of a phylogeny

As we already saw, the other components of our "phylo" object include the vector tip.label and an integer Nnode, which gives the number of interior nodes in the tree. Let's take a look at these two elements now as well.

```
vert.tree$tip.label
```

```
##  [1] "Shark"      "Goldfish"   "Coelacanth"
##  [4] "Human"      "Lemur"      "Bat"
##  [7] "Pig"        "Whale"      "Cow"
## [10] "Iguana"     "Robin"
```

```
vert.tree$Nnode
```

```
## [1] 10
```

By convention, the *order* of the tip labels in `tip.label` corresponds to the numerical order of the tip indices (scored from 1 through *N*, remember) in our phylogeny.

The component `Nnode` has an even more straightforward interpretation—which we think doesn't require any additional explanation.

1.5.5 The "phylo" class

An object of class `"phylo"` also (by definition) has at least one attribute—its class. This is just a value to tell various functions—and, particularly, S3 methods—in R what to do with an object of this type.

For instance, if we call the generic method `plot`, the object class attribute is what instructs R to use the method `plot.phylo` that has been exported by the R package *ape*.

An object of class `"phylo"` can have other components too. The most common of these is `edge.length`: a vector of class `"numeric"` containing all the branch lengths or our tree. Although our object `vert.tree` does not include branch lengths, if it did, we would see that the numeric vector `edge.length` contained the branch lengths of the phylogeny in precisely the order of the rows of `edge`.

In addition, other elements and attributes can be added for special types of phylogenetic trees. Some R functions will behave differently if these additional elements or attributes are present in our `"phylo"` object. We'll see more about this in later chapters of the book!

1.6 Reading and writing phylogenetic trees

Naturally, R can easily read and write trees to and from files.

1.6.1 Reading a tree from a file

For example, let's download the tree file `Anolis.tre` (Mahler et al. 2010, available from the book's website[28]) and read it into R.

For this task, we'll use the *ape* function `read.tree`.[29]

As soon as you have the tree file in your current working directory in R,[30] you can read it in

```
anolis.tree<-read.tree(file="Anolis.tre")
anolis.tree

##
## Phylogenetic tree with 100 tips and 99 internal nodes.
##
## Tip labels:
##   ahli, allogus, rubribarbus, imias, sagrei, bremeri, ...
```

---

[28] The site is http://www.phytools.org/Rbook/, as indicated earlier. Henceforward, we'll only provide the URL of the book website on the first instance that it's referenced in each chapter.

[29] `read.tree` and, likewise, `read.newick` in the *phytools* package read phylogenies in *simple Newick format*. Different functions of *ape*, *phytools*, and other packages can be used to read trees that have been written to file in other formats.

[30] To see your current working directory in R, type `getwd()` at the command prompt. To change your working directory, use the function `setwd`.

```
##
## Rooted; includes branch lengths.
```

```
plotTree(anolis.tree,ftype="i",fsize=0.4,lwd=1)
```

This is a tree containing

```
Ntip(anolis.tree)
```

```
## [1] 100
```

100 species of lizards in the neotropical lizard genus *Anolis* (figure 1.4).

1.6.2 Writing a tree to a file

In addition to *reading* a tree from file, we can also write them. For instance, we can easily write our vertebrate tree from earlier in the chapter to a simple text file in Newick format.

```
write.tree(vert.tree,file="example.tre")
```

This is what the resultant text file example.tre should look like.[31]

```
cat(readLines("example.tre"))
```

```
## (Shark,(Gold_fish,(Coelacanth,(((Hu ...
```

1.7 Plotting and manipulating trees

We've already seen a few in this chapter, but there are a wide range of ways in which we can plot and manipulate trees in R.

Next, let's take a look at a few more of the most common ways that phylogenies are plotted in R.[32]

Meanwhile, we can also see how R can be used to (for lack of a better word) manipulate phylogenies.

Common types of manipulation of phylogenies in R include dropping or "pruning" species from a tree, extracting subtrees, and shrinking or stretching trees to have a particular total length. We'll focus on the former two types of manipulation here.

1.7.1 Pruning taxa from the phylogeny

A convenient and popular plotting method for large *rooted* trees is a circular or "fan" tree. We can start by plotting our *Anolis* tree in this way and then go from there.

---

[31] You can also open the file on your computer using any text editor to check if you'd like.

[32] R is an extremely flexible plotting environment, so there are many plotting options that we are not seeing here; however, some of these will be visited in subsequent chapters.

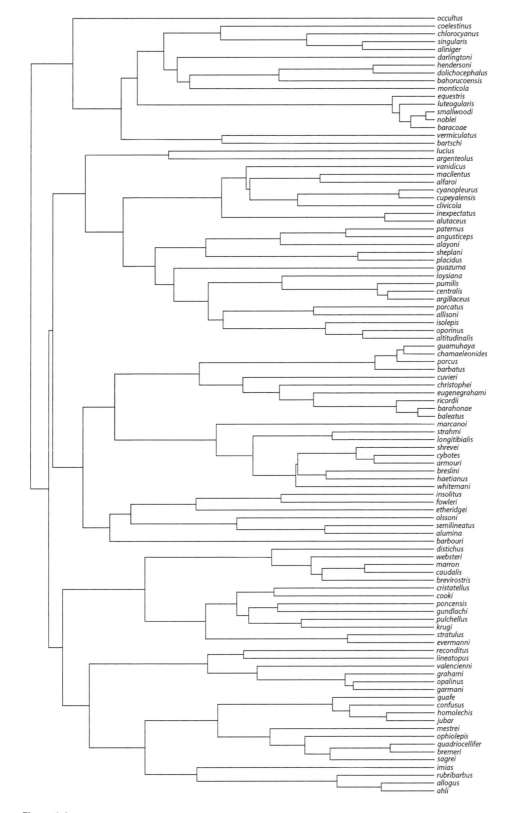

Figure 1.4
A phylogenetic tree of *Anolis* lizards plotted in a right square phylogram style using the *phytools* function `plotTree`.

In comparative analyses with phylogenetic data, we are often called upon to prune species out of the tree or to extract one clade or another.

This might be the case, for instance, when we have phylogenetic data for one set of taxa and morphological, phenotypic, or biogeographic data for a different, but nonetheless overlapping, set. Fortunately, pruning taxa and extracting clades are relatively straightforward operations in an interactive R session.

Let's imagine, for instance, that instead of working with the 100-taxon *Anolis* tree, we would like to analyze a phylogeny that contained only a subset of these taxa.[33] We can focus our attention on the anoles from Puerto Rico, which (in this phylogeny) consist of *A. cristatellus*, *A. cooki*, *A. poncensis*, *A. gundlachi*, *A. pulchellus*, *A. stratulus*, and *A. evermanni* (which form a clade), as well as *A. occultus* and *A. cuvieri*.

As a first step, let's find these Puerto Rican anoles on our complete phylogeny.

The following script uses a *phytools* function called add.arrow to add red[34] arrows pointing to particular tips on the phylogeny that we are interested in.

We have to plot the arrows by indicating the tip *numbers* (not the labels) of the terminal taxa that we want to mark. To find these, we will use the R *base* function grep.

grep matches a character pattern to a vector and returns the positions of the elements of the vector in which that pattern is found. Here, we're going to use it to match the specific epithets[35] of the Puerto Rican anoles to the vector comprising all the tip labels of the tree.

Perhaps the reader is beginning to see how useful it can be to know something about the structure of the "phylo" object—because otherwise, we might not know that these labels can be found in the vector anolis.tree$tip.label!

```
pr.species<-c("cooki","poncensis",
    "gundlachi","pulchellus","stratulus",
    "krugi","evermanni","occultus","cuvieri",
    "cristatellus")
nodes<-sapply(pr.species,grep,x=anolis.tree$tip.label)
nodes

##          cooki    poncensis    gundlachi
##            26           25           24
##      pulchellus    stratulus        krugi
##            23           21           22
##       evermanni     occultus      cuvieri
##            20          100           54
## cristatellus
##            27
```

---

[33] Although *Anolis* is a clade with over 400 described species across the tropical and subtropical Americas, our phylogeny includes only representatives from the Greater Antillean region of the Caribbean.

[34] Here and throughout the volume, we'll refer to the colors that would be plotted if you reproduced our code in R. The figures that you'll *actually* see in the print version of the book, however, have been recolored in grayscale by the publisher to help ensure that the book can be printed, and sold, at a reasonable price. Hopefully this isn't too confusing!

[35] The *specific epithet* is the second part of the Latin binomial name of a species, so for the species *Homo sapiens*, the specific epithet is *sapiens*. In our *Anolis* tree, all tips belong to the same genus, so they've been labeled using *only* the epithet.

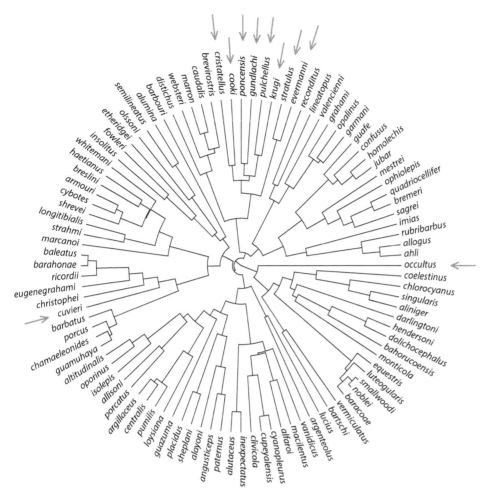

Figure 1.5

Phylogenetic tree of *Anolis* lizards. We plotted the tree using the *phytools* function `plotTree` and then highlighted the species from Puerto Rico using the function `add.arrow`.

Note that we've used a function belonging to the so-called `apply` family of functions—this one called `sapply`. `apply` family functions are designed to iterate operations over the elements of a matrix, vector, or list, without the necessity of writing a loop.[36]

The easiest way to interpret our `sapply` call, `nodes <- sapply(pr.species, grep, x=anolis.tree$tip.label)`, is as "apply to the elements of `pr.species` the function `grep` with the argument `x` of `grep` set to `anolis.tree$tip.label`."

We'll see more uses of various `apply` family functions throughout this chapter and the rest of the book.

Now that we have identified the tip node numbers of all the Puerto Rican *Anolis* lizards in our tree, we can plot our tree and label these species using arrows just as we planned. The result is seen in figure 1.5.

---

[36]The most common loop programming structure in R, as well as in many other programming languages, is called a *for* loop. *for* loops can be very useful in R. We'll see examples of *for* loops later on the book.

```
plotTree(anolis.tree,type="fan",fsize=0.6,lwd=1,
    ftype="i")
add.arrow(anolis.tree,tip=nodes,arrl=0.15,col="red",
    offset=2)
```

The orientation of the arrows in figure 1.5 should exactly match the orientation of the tip branch.[37]

Now let's prune the species that we marked with arrows out of the tree.

```
anolis.noPR<-drop.tip(anolis.tree,pr.species)
plotTree(anolis.noPR,type="fan",fsize=0.6,lwd=1,
    ftype="i")
```

We should see that the function we used here, drop.tip, cuts not only the terminal branch but any branch that leads exclusively to tips that are being pruned. We've plotted the pruned phylogeny in figure 1.6.

1.7.2 Extracting a clade

Alternatively, let's imagine that we want to *extract* the main clade of Puerto Rican *Anolis* species. In our example, this is the clade that includes all but two of the species found on the island.

To extract a clade, we need to identify the node index of the *most recent common ancestor* (MRCA) of the members of the clade we want to prune.

In our case, this corresponds to the MRCA of all the species *except* for *Anolis cuvieri* and *Anolis occultus*.[38] We can find the node number of the MRCA of a set of taxa using getMRCA from the *ape* package.

```
node<-getMRCA(anolis.tree,pr.species[
    -which(pr.species%in%c("cuvieri","occultus"))])
node
```

```
## [1] 123
```

Just for fun, before we pull it out, let's go ahead and visualize the clade that we plan to extract.

To do so, we'll use the *phytools* function paintSubTree.[39] We can also combine this with the function arc.cladelabels to add a nice clade delimiter to our plot. The result is shown in figure 1.7.

---

[37] This is harder to guarantee when we make this kind of figure using a point-and-click image editor!

[38] Here we used *negative indexing* to pull out undesired elements from our vector of taxon names. Negative indexing returns all the elements of a vector, matrix, or list with the exception of those that were indexed. Does that make sense?

[39] This function will also reappear later in the book!

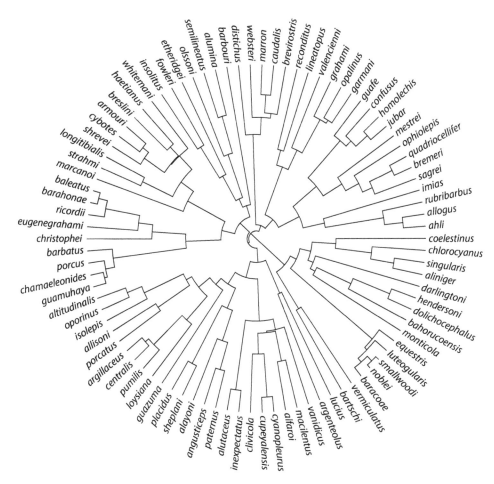

Figure 1.6
Phylogeny of *Anolis* in which we first pruned all Puerto Rican species from the tree using the *ape* function `drop.tip`.

```
plot(paintSubTree(anolis.tree,node,"b","a"),
    type="fan",fsize=0.6,lwd=2,
    colors=setNames(c("gray","blue"),c("a","b")),
    ftype="i")
arc.cladelabels(anolis.tree,"clade to extract",node,
    1.35,1.4,mark.node=FALSE,cex=0.6)
```

The numbers 1.35 and 1.4 have no special significance—in this case, they merely set the relative offset of the clade line and the clade label from the tips of the tree. Readers who want to duplicate this plot with their own phylogeny will probably have to use different values (although sometimes the defaults can work fairly well).

Now, we can proceed to extract the clade of interest using the *ape* function `extract.clade` as follows.

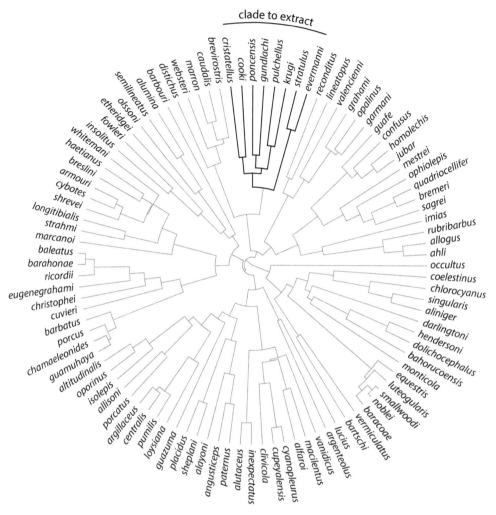

Figure 1.7
Tree of *Anolis* lizards. Here we marked the part of the tree we plan to extract by mapping this clade onto our "phylo" object with the *phytools* function `paintSubTree` and then by drawing the tree using `plot.simmap`.

```
pr.clade<-extract.clade(anolis.tree,node)
pr.clade

##
## Phylogenetic tree with 8 tips and 7 internal
nodes.
##
## Tip labels:
## evermanni, stratulus, krugi, pulchellus,
gundlachi, poncensis, ...
##
## Rooted; includes branch lengths.
```

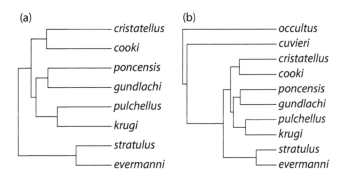

Figure 1.8
(a) Largest clade from Puerto Rico extracted from the full tree of Caribbean *Anolis* lizards using the *ape* function `extract.clade`. (b) Phylogeny containing all of the *Anolis* the species from Puerto Rico obtained from the full tree using the function `keep.tip`.

Likewise, just as we extracted the clade, we can also perform the converse operation—which would be to prune everything in the tree except for these species.

To do this, we will use a *different ape* function called `keep.tip`.

```
pr.tree<-keep.tip(anolis.tree,pr.species)
pr.tree

##
## Phylogenetic tree with 10 tips and 9 internal
nodes.
##
## Tip labels:
## evermanni, stratulus, krugi, pulchellus,
gundlachi, poncensis, ...
##
## Rooted; includes branch lengths.
```

Here are the two resultant phylogenies (figure 1.8).

```
par(mfrow=c(1,2))
plotTree(pr.clade,ftype="i",mar=c(1.1,1.1,3.1,1.1),
    cex=1.1)
mtext("(a)",line=0,adj=0)
plotTree(pr.tree,ftype="i",mar=c(1.1,1.1,3.1,1.1),
    cex=1.1)
mtext("(b)",line=0,adj=0)
```

1.7.3 Interactive tree manipulation using collapseTree

Finally, sometimes it's fun to prune our tree *interactively*—by clicking on nodes or tips of the phylogeny after it has been plotted. In R, this can be done using the *ape* function `drop.tip`, which has an interactive mode, or by using the animated *phytools* function `collapseTree` as follows.

```
anolis.pruned<-collapseTree(anolis.tree)
```

Obviously, this cannot be demonstrated on the pages of a book, but please try it out! As you click on or near nodes of the tree, you should see clades collapse or reexpand.[40] When you're done, just right-click and select *stop*.

1.8 Multiple trees in a single object

It's often useful to store multiple phylogenies in a single object. This could be true, for instance, when we have a set of trees in a posterior sample from Bayesian phylogeny inference, if we're working with a bootstrap distribution of phylogenies, or when we want to replicate a simulation analysis across a large number of trees.

1.8.1 The "multiPhylo" object

In R, multiple phylogenetic trees are usually stored in the form of an object of class `"multiPhylo"`.

This sounds fancy, but it is really nothing more than a list of objects of class `"phylo"` but with the class attribute `"multiPhylo"` assigned!

Many, but not all, functions in *ape*, *phytools*, and other R packages are "vectorized" so that they can be applied to both `"phylo"` and `"multiPhylo"` objects. For instance:

```
anolis.trees<-c(anolis.tree,anolis.noPR,pr.clade,
    pr.tree)
print(anolis.trees,details=TRUE)

## 4 phylogenetic trees
## tree 1 : 100 tips
## tree 2 : 90 tips
## tree 3 : 8 tips
## tree 4 : 10 tips


write.tree(anolis.trees,file="example.trees")
```

Here, we first combined all of our individual trees into a single `"multiPhylo"` object using the function (and S3 method) `c` (short for *combine*).

Next, we printed a summary of our object. Meanwhile, we also turned on the `print` method optional argument `details` so that we could see a bit more information about each tree in the object—in this case, the number of terminal taxa ("tips") in each tree.

Finally, we wrote all the trees to a single text file using the *ape* function `write.tree`.

This output file is merely just a simple text file, with each of our phylogenies written in Newick format onto separate lines:

---

[40]The animation works better in some R GUIs than others!

```
cat(readLines("example.trees"),sep="\n")

## (((((((((ahli:0.131,allogus:0.131):0 ...
## (((((((((ahli:0.131,allogus:0.131):0 ...
## ((evermanni:0.214,stratulus:0.214): ...
## ((((evermanni:0.214,stratulus:0.214 ...
```

1.9 Managing trees and comparative data

Throughout this book, we'll often be called upon to manage not only phylogenies but also comparative phenotypic trait data for species.

To see an example of how to do this with real data, let's use two different data files from the book webpage: `anole.data.csv` and `ecomorph.csv` (Mahler et al. 2010).

We'll combine the data of these files with the phylogeny from our *Anolis* tree file (`Anolis.tre`) that we read into R earlier in the chapter.

1.9.1 The CSV file format

Both of our two data files (`anole.data.csv` and `ecomorph.csv`) are written in a common data file format called *CSV*[41] format.

R can read data in lots of different formats; however, CSV format is pretty reliable and widely used. Precisely as you might expect, CSV format is a simple text format for tabular data, but in which the elements in different rows are separated by a hard return, while the elements in different columns within a row are separated by the comma character: , .

Let's read our CSV files into R using the function `read.csv` as follows:

```
anole.data<-read.csv(file="anole.data.csv",row.names=1,
    header=TRUE)
ecomorph<-read.csv(file="ecomorph.csv",row.names=1,
    header=TRUE,stringsAsFactors=TRUE)
```

Calls to `read.csv`, like those we've executed here, generate data frames in R.[42]

The argument `row.names=1` tells R to look for row names in the first column of our data file, while `header=TRUE` tells R that the first row of our data file contains the column or variable names.

An astute reader might also notice that in our second `read.csv` call, we were careful to set the argument `stringsAsFactors` to TRUE. This is to ensure that the discrete character trait contained in this file was read into R as a multilevel *factor* rather than as a simple set of character strings.[43]

Although we find the CSV format to be a very reliable way to store tabular data, one complication is that in South America and in some parts of continental Europe, the comma (,) is used

---

[41] CSV stands for *comma-separated values*.

[42] A data frame looks like a matrix, but it's technically a *list* arranged in a tabular way such that all the columns of the data frame are vectors with the same number of elements. We'll encounter and work with a lot of data frames in this book!

[43] Prior to R version 4.0, `stringsAsFactors` defaulted to TRUE, and this would not have been necessary.

as a decimal separator in place of the period (.). As such, it would be impractical to demarcate columns in tabular data using commas. Frequently, then, in these places, the columns of a CSV-formatted text file will have been demarcated using the semicolon character (;).

This is no problem at all for R as we can modify the separator and decimal characters using the `read.csv` function arguments `sep=";"` and `dec=","`, respectively, *or* simply by substituting the function `read.csv2` (which uses these arguments by default).

As an aside, one phenomenon that we have often observed in teaching R phylogenetics is that for many users, their spreadsheet software will be set to open CSV format files automatically by default. As a consequence, students often mistakenly open CSV files in their spreadsheet program and then proceed to resave them in a different format instead of as a genuine CSV file. This should obviously be avoided.[44]

Now that we've read our data into R, let's proceed and use the function `head` to inspect the first few rows of the data frames that we've created and then `dim` to review the dimensions[45] of our two data frames.

```
head(anole.data)

##                 SVL      HL     HLL     FLL     LAM
## ahli       4.03913 2.88266 3.96202 3.34498 2.86620
## alayoni    3.81570 2.70212 3.27950 2.80245 3.07527
## alfaroi    3.52665 2.37816 3.30542 2.48366 2.73387
## aliniger   4.03656 2.89884 3.64623 3.15908 3.15677
## allisoni   4.37539 3.35896 3.96069 3.44620 3.23921
## allogus    4.04014 2.86103 3.94018 3.33829 2.80827
##                  TL
## ahli        4.50400
## alayoni     4.07265
## alfaroi     4.41601
## aliniger    4.54173
## allisoni    5.05911
## allogus     4.52189
```

```
dim(anole.data)

## [1] 100   6
```

```
head(ecomorph)

##              ecomorph
## ahli               TG
## allogus            TG
## rubribarbus        TG
## imias              TG
## sagrei             TG
## bremeri            TG
```

---

[44]Some spreadsheet software files can be read directly by R, but this is less reliable and as such we don't really recommend it.

[45]That is, the number of rows and columns, respectively.

```
dim(ecomorph)
```

```
## [1] 82   1
```

Doing this, we can see that our first data frame (anole.data) has 100 rows and contains six different numeric variables, with the names SVL (snout-to-vent length, incidentally—on a log scale), HL (head length), HLL (hindlimb length), and so on. Our data frame ecomorph, by contrast, has only 82 rows and contains one factor variable (also denominated ecomorph). The row names of both data frames contain the taxon labels: in this case, the specific epithets of species of lizard in the genus *Anolis*, just as in our tree.

1.9.2 Comparing a character data set and tree

Although it seems likely that the first data set has the same set of species as our 100-taxon *Anolis* tree from earlier in the exercise, we can (and should!) verify this.

Let's do so using the *geiger* function name.check.[46]

```
library(geiger)
name.check(anolis.tree,anole.data)
```

```
## [1] "OK"
```

This result ("OK") tells us that the taxon names in the phylogeny exactly match those of the data frame.

name.check is useful not only for identifying incongruencies between the phylogeny and data but also instances in which a taxon label may have been misspelled, mistranscribed, or misread by R in either our data set or the tree.

In the case of ecomorph, however, there are obviously fewer observations in the data than in the tree. That suggests that there are at least *some* differences between the data and phylogeny. Let's use name.check again to see how they differ.

```
chk<-name.check(anolis.tree,ecomorph)
chk
```

```
## $tree_not_data
##  [1] "argenteolus"     "argillaceus"
##  [3] "barbatus"        "barbouri"
##  [5] "bartschi"        "centralis"
##  [7] "chamaeleonides"  "christophei"
##  [9] "etheridgei"      "eugenegrahami"
## [11] "fowleri"         "guamuhaya"
## [13] "lucius"          "monticola"
## [15] "porcus"          "pumilis"
## [17] "reconditus"      "vermiculatus"
```

---

[46]We could have also used the *geiger* function comparative.data, which serves some of the same purposes but works a little differently than name.check.

```
##
## $data_not_tree
## character(0)
```

Now we can see that when there are differences between our data and our tree, name.check returns a handy list indicating which taxa are in the tree but not the data, as well as vice versa.

For examples with larger discrepancies between data and tree, we can also print an abbreviated summary of our result as follows:

```
summary(chk)

## 18 taxa are present in the tree but not the data:
##      argenteolus,
##      argillaceus,
##      barbatus,
##      barbouri,
##      bartschi,
##      centralis,
##      ....
##
## To see complete list of mis-matched taxa, print object.
```

1.9.3 Pruning a tree to match your data set, and vice versa

Now, precisely as we learned earlier in the chapter, let's go ahead and prune[47] all the taxa that are present in our phylogeny, but not in our ecomorph data frame. This can be done using *ape*'s drop.tip function as follows:

```
ecomorph.tree<-drop.tip(anolis.tree,chk$tree_not_data)
ecomorph.tree

##
## Phylogenetic tree with 82 tips and 81 internal nodes.
##
## Tip labels:
##    ahli, allogus, rubribarbus, imias, sagrei, ...
##
## Rooted; includes branch lengths.
```

We can similarly subsample our data to include only those taxa present in a phylogeny.

Let's do it for our data frame anole.data so that it contains only data for the species in our new, pruned phylogeny that we've denominated ecomorph.tree.[48]

---

[47] That is, remove from the phylogeny.

[48] This trick subsamples the data frame to include only rows whose names match ecomorph.tree$tip.label—the taxon labels of our tree. One odd behavior of R is that if our data frame has only one column, this operation will return a vector rather than a data frame with one column! This can

```
ecomorph.data<-anole.data[ecomorph.tree$tip.label,]
head(ecomorph.data)
```

```
##                 SVL      HL     HLL     FLL
## ahli        4.03913 2.88266 3.96202 3.34498
## allogus     4.04014 2.86103 3.94018 3.33829
## rubribarbus 4.07847 2.89425 3.96135 3.35641
## imias       4.09969 2.85293 3.98565 3.41402
## sagrei      4.06716 2.83515 3.85786 3.24267
## bremeri     4.11337 2.86044 3.90039 3.30585
##                 LAM      TL
## ahli        2.86620 4.50400
## allogus     2.80827 4.52189
## rubribarbus 2.86751 4.56108
## imias       2.94375 4.65242
## sagrei      2.91872 4.77603
## bremeri     2.97009 4.72996
```

Our new trait data frame (ecomorph.data) should now match our pruned phylogeny exactly—but let's make sure, once again using the function name.check:

```
name.check(ecomorph.tree,ecomorph.data)
```

```
## [1] "OK"
```

This result ("OK") tells us that name.check now thinks that our tree and data match exactly!

1.10 A simple comparative analysis: Phylogenetic principal components analysis

Now that our *Anolis* lizard tree and data sets match, let's go ahead and do a very simple analysis called a "phylogenetic principal components analysis" or phylogenetic PCA (Revell 2009) using our morphological character data.

A phylogenetic PCA is exactly the same as a regular PCA except that we're going to take the nonindependence of species into account when we compute the covariances (or correlations) between different traits.

Whereas in regular (nonphylogenetic) PCA, principal components are *orthogonal*,[49] in phylogenetic PCA, components are evolutionarily orthogonal, meaning that the evolutionary correlations[50] between principal components are all zero. Likewise, whereas principal components describe successive orthogonal dimensions of maximum variance in the original multidimensional trait space, *phylogenetic* principal components correspond to successive evolutionarily orthogonal dimensions of maximum evolutionary variance.

---

be circumvented by using the argument drop=FALSE as follows: ecomorph.data <- anole.data [ecomorph.tree$tip.label„drop=FALSE]. Isn't that weird?

[49] That is to say, *uncorrelated*.

[50] The evolutionary correlation will be discussed in much greater detail in chapters 2 and 3.

The interpretation of the first phylogenetic principal component is thus that it is the axis of greatest, multivariate *evolution*[51] of our traits. Subsequent axes are successive orthogonal dimensions of maximum evolution.

To undertake a phylogenetic principal component analysis in R, we can use the function phyl.pca in the *phytools* package as follows.[52]

```
ecomorph.pca<-phyl.pca(ecomorph.tree,ecomorph.data)
ecomorph.pca

## Phylogenetic pca
## Standard deviations:
##         PC1         PC2         PC3         PC4
## 0.81375257 0.22561158 0.12277034 0.10577996
##         PC5         PC6
## 0.04926765 0.03692593
## Loads:
##             PC1         PC2         PC3
## SVL  -0.9712073  0.16073225  0.01979472
## HL   -0.9644970  0.16959751 -0.01199377
## HLL  -0.9814007 -0.02674374  0.10309671
## FLL  -0.9712156  0.17590524  0.10692548
## LAM  -0.7809539  0.37434869 -0.47406978
## TL   -0.9013706 -0.42546037 -0.07612345
##             PC4         PC5         PC6
## SVL   0.14785037 -0.06199108 -0.069477241
## HL    0.17994467  0.08065005  0.044203206
## HLL  -0.13799438  0.06907952 -0.041160294
## FLL  -0.09104262 -0.06097041  0.048562708
## LAM  -0.15858923  0.00217263 -0.008754817
## TL    0.01713199 -0.01755709  0.010858471
```

```
par(mar=c(4.1,4.1,2.1,1.1),las=1) ## set margins
plot(ecomorph.pca,main="")
```

From this printout, we can see that phylogenetic PC1 loads strongly, and *negatively*, for all of the traits in our data set. This principal component represents evolutionary variation in overall size. Remember that the *sign* of each principal component is arbitrary, so let's flip it. This is easy enough to do in R as follows.

```
ecomorph.pca$Evec[,1]<--ecomorph.pca$Evec[,1]
ecomorph.pca$L[,1]<--ecomorph.pca$L[,1]
ecomorph.pca$S<-scores(ecomorph.pca,
    newdata=ecomorph.data)
```

---

[51] Under our chosen evolutionary model: more in chapters 4 and 5.

[52] The plotting argument las=1, which we can also often set using par(las=1) and we use in many places throughout the book, merely sets the axis tick labels to plot horizontally rather than parallel to the axis (las=0, the default in R) or vertically (las=2).

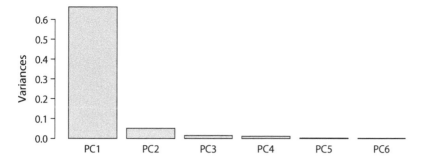

Figure 1.9
Screeplot for phylogenetic PCA of *Anolis* morphological data.

The printout also tells us that PC2 loads primarily for increasing lamellae[53] and decreasing tail length.

This simple screeplot of figure 1.9 shows that these first two principal components (PCs) explain most of the variation in our data, so let's visualize the two components in a bivariate space.

Since the data from phylogenetic PCA are phylogenetic,[54] let's do this while projecting our tree into the same space.

While we're at it, we'll also go ahead and color the tips by the discrete ecomorphological categorization of each species in our `ecomorph` data frame and add a legend.

Our `ecomorph` data frame contains the values for a discrete character that encodes the ecomorphological category of each species in our data set (Mahler et al. 2010). The result is shown in figure 1.10.

```
par(cex.axis=0.8,mar=c(5.1,5.1,1.1,1.1))
phylomorphospace(ecomorph.tree,
    scores(ecomorph.pca)[,1:2],
    ftype="off",node.size=c(0,1),bty="n",las=1,
    xlab="PC1 (overall size)",
    ylab=expression(paste("PC2 ("%up%"lamellae number, "
    %down%"tail length)")))
eco<-setNames(ecomorph[,1],rownames(ecomorph))
ECO<-to.matrix(eco,levels(eco))
tiplabels(pie=ECO[ecomorph.tree$tip.label,],cex=0.5)
legend(x="bottomright",legend=levels(eco),cex=0.8,pch=21,
    pt.bg=rainbow(n=length(levels(eco))),pt.cex=1.5)
```

The first line of the code chunk sets some plotting parameters for our graph.[55]

---

[53] Subdigital scales characteristic of *Anolis* lizards that allow them to adhere to smooth surfaces.

[54] We took the phylogeny into account when computing the rotation, but then to obtain principal component scores, we performed a rigid rotation of the original space. For more information about this, see Revell (2009).

[55] The argument `cex.axis=0.8` controls the font size of the axis numeration, while `mar` sets the margin widths around our plot clockwise from the bottom—just as we learned earlier in the chapter.

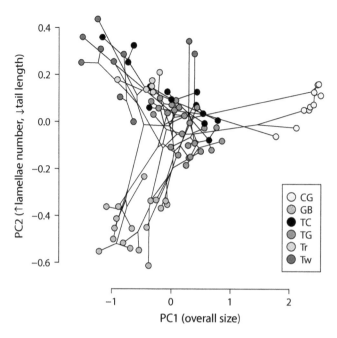

Figure 1.10

Phylomorphospace of PC1 and PC2 from a phylogenetic PCA using *Anolis* morphological data. Tip colors indicate ecomorphological category.

The next line plots a projection of the tree into morphospace for PC1 and PC2 from our phylogenetic PCA.[56] Here we used many of the plotting defaults but added the function `expression` to encode up and down arrow characters (↑ and ↓) to help the reader interpret each principal component.

Then, the second, third, and fourth lines add colored tip labels to the tree, in which the colors are determined by the state of a discrete character (ecomorphological class from our data file `ecomorph.csv`).

Finally, the last line adds a legend to the plot using the flexible base R *graphics* function `legend`.

Awesome. We're now well under way to becoming experts at phylogenetic comparative methods in the R scientific computing environment!

1.11 Practice problems

1.1 Download the following two files for *Phelsuma* geckos from the book website: `phel.csv` and `phel.phy` (Harmon et al. 2010). `phel.csv` is a CSV file containing trait values for ten different morphological traits. `phel.phy` is a phylogeny of thirty-three species. Read both data and tree in from file and use `name.check` to identify any differences between the two data sets. If you find differences, prune the phylogeny and subsample the trait data to include only the species present in both the data file and the tree. Plot the tree.

---

[56]This type of visualization has been dubbed a "phylomorphospace" projection, hence the name of the function.

1.2 Use `phyl.pca` to run a phylogenetic principal component analysis (PCA) of the morphological data set and tree from practice problem 1.1. When data for different variables in a PCA have different orders of magnitude,[57] it often makes sense to transform by the natural logarithm and conduct our analysis on the log-transformed values instead of on the original traits. Inspect your data to see if this applies and then decide whether or not to log-transform before undertaking your phylogenetic PCA. After you've obtained a result for the PCA, create a screeplot to visualize the distribution of variation between the different principal component axes.

1.3 Use `phylomorphospace` to create a single projection of the phylogeny into morphospace for the first two PC axes from practice problem 1.2. Can you think of a way to project the tree into a space defined by more than two principal component dimensions? Hint: look up the help pages for `phylomorphospace3d` and `phyloScattergram` for ideas, or consider simply subdividing your plotting device using `par(mfrow)`.

---

[57] As well as for other reasons, such as when variables are measured in different units.

Phylogenetically independent contrasts

2

Many hypotheses in comparative biology can be investigated by measuring the evolutionary correlation[1] between two characters—that is, by testing to see whether or not two characters have evolved in a coordinated fashion.

For example, one might wonder why some species of mammals have such large home range areas, while others have home ranges that are quite small (Garland et al. 1992). One reasonable hypothesis is that range size is driven by body size, and if so, then we might expect range size and body size to evolve together. When body size increases, so should range size. If body size evolves to be smaller, range size should concordantly shrink.

Hypotheses about evolutionary correlations can be tested using a method known as *phylogenetic independent contrasts* (Felsenstein 1985).

Although other methods for measuring the evolutionary correlation between characters have since been described, the contrasts method is so important to the field of phylogenetic comparative biology that we thought it would be useful to dedicate a full chapter to this topic.

Developing a detailed understanding of phylogenetic contrasts, as well as how to use this method in R, will help us build a stronger general comprehension of comparative methods and of phylogenetic comparative analysis in the R environment.

In this chapter, we will:

1. Introduce the method of phylogenetic independent contrasts.
2. Illustrate the importance of comparative methods by using "Felsenstein's worst-case scenario" (Felsenstein 1985).
3. Use independent contrasts to fit a linear model to a data set, testing for an evolutionary correlation between two characters.

---

[1] An evolutionary correlation is defined as the tendency for two variables to evolve in concert. If two traits are evolutionarily correlated, a large evolutionary change in x should usually tend to be associated with a large—positive for a positive correlation or negative for a negative correlation—change in y and vice versa. See Harmon (2019) for further explanation.

4. Evaluate the statistical properties of linear regression with phylogenetically independent contrasts using simulations.

2.2 Phylogenetic nonindependence

The independent contrasts method was invented, and the modern field of phylogenetic comparative methods born, thanks to a transformational paper published in the mid-1980s by Joseph Felsenstein (1985).

In this *American Naturalist* article, entitled "Phylogenies and the Comparative Method," Felsenstein (1985) fundamentally and irreversibly changed the way in which the field of comparative biology collectively viewed the analysis of species data.

He did this in large part by illustrating why data points obtained from species related to one another by a phylogeny could not and should not be treated as independent from the point of view of traditional statistical analysis.[2]

2.2.1 Felsenstein's worst-case scenario

In particular, Felsenstein (1985) used a kind of *worst case of sorts*[3] to cleverly illustrate how phylogenetic nonindependence of species could lead to vast overconfidence in the inference of an evolutionary correlation between traits—if this correlation were to be estimated without taking the phylogeny properly into consideration.

The following code[4] (and Figure 2.1) illustrates Felsenstein's worst-case scenario using simulated data.

```
## load packages
library(phytools)
## read in tree
tree<-read.tree(
    text="((A,B,C,D,E,F,G,H,I,J,K,L,M),
    (N,O,P,Q,R,S,T,U,V,W,X,Y,Z));")
## set branch lengths on the tree
tree<-compute.brlen(tree,power=1.8)
## simulate data, independently for x & y
x<-fastBM(tree)
y<-fastBM(tree)
## plot the results with clades A & B labeled
## split plotting area
```

---

[2]It's definitely worth checking out the recent historical commentary on the importance of Felsenstein's paper by Huey et al. (2019). We recommend it!

[3]"A worst case of sorts for the naive analysis is shown ... where the phylogeny shows that a large number of species actually consist of two groups of moderately close relatives.... There appears to be a significant regression of Y on X. If the points are distinguished according to which monophyletic group they came from ... we can see that there are two clusters. Within each of these groups there is no significant regression of one character on the other. The means of the two groups differ, but since there are only two group means they must perforce lie on a straight line, so that the between-group regression has no degrees of freedom and cannot be significant. Yet a regression assuming independence of the species finds a significant slope (P < .05). It can be shown that there are more nearly 3 than 40 independent points in the diagram" (Felsenstein 1985, p. 4).

[4]Since it's being included for illustrative purposes only, in this case we won't go through the code line by line. Based on many of the things we've learned already, though, many of our readers can probably figure it out!

```
par(mfrow=c(1,2))
## graph tree
plotTree(tree,type="cladogram",ftype="off",
    mar=c(5.1,4.1,3.1,2.1),color="darkgray",
    xlim=c(0,1.3),ylim=c(1,Ntip(tree)))
## add points at the tips of the tree to match those
## on our scatterplot
points(rep(1,13),1:13,pch=21,bg="lightgray",
    cex=1.2)
points(rep(1,13),14:26,pch=22,bg="black",cex=1.2)
## add clade labels to the tree
cladelabels(tree,"A",node=28,offset=2)
cladelabels(tree,"B",node=29,offset=2)
mtext("(a)",line=1,adj=0,cex=1.5)
## create scatterplot of x & y
par(mar=c(5.1,4.1,3.1,2.1))
plot(x,y,bty="n",las=1)
points(x[1:13],y[1:13],pch=21,bg="lightgray",
    cex=1.2)
points(x[14:26],y[14:26],pch=22,bg="black",cex=1.2)
mtext("(b)",line=1,adj=0,cex=1.5)
```

Here (following Felsenstein 1985), what we see are two clades of equal size (*A* and *B*), each consisting of taxa closely related to one another but separated by a large distance from the root (figure 2.1a).

When we simulate data *independently* for *x* and *y* (figure 2.1b), at first glance, these data appear to be correlated.

Closer inspection, however, reveals that the apparent correlation stems entirely from a chance divergence in both *x* and *y* along the long branches separating the two clades. Values

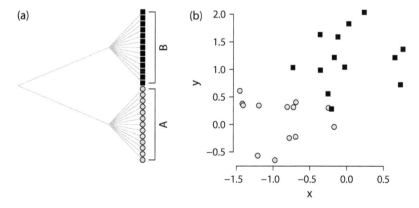

Figure 2.1
Felsenstein's worst-case scenario for phylogenetic nonindependence in regression or correlation analyses of two traits. (a) A hypothetical phylogeny with two clades. (b) The distribution of points for two traits, *x* and *y*, showing an overall relationship but little evidence within clades of an evolutionary correlation between the variables.

within each clade show no evidence of the correlation one might expect to see if x and y consistently tended to coevolve over evolutionary time.

Prior to the foundational contribution of Felsenstein (1985), it would not have been uncommon to interpret a simple correlation between two variables as evidence that they had coevolved. What Felsenstein's paper served to illustrate was that just such a pattern can easily arise in phylogenetic data even absent a genuine evolutionary relationship between traits.

Felsenstein's (1985) article, however, was much more than a critique of the existing paradigm. It also presented a solution for correlation or regression analysis of species data. This was a specific type of data transformation that has come to be known as *phylogenetically independent contrasts* or PICs.

Felsenstein's PIC method is the subject of the present chapter. This method provides a way to test evolutionary correlations among characters, as we'll illustrate by example, below.

2.3 Phylogenetically independent contrasts

The principle behind Felsenstein's method is both remarkably elegant and incredibly simple.

Felsenstein (1985) merely pointed out that while species data per se are not independent (owing to shared history due to the phylogeny), the differences (or contrasts) between species are.[5]

Likewise, the differences (if properly corrected) between trait values that have been interpolated for internal nodes are also independent.

Thus, for a bifurcating N-taxon species tree consisting of observations for two or more continuous traits, one can compute a total of $N - 1$ *phylogenetically independent contrasts* for each trait. These contrasts, once normalized, can then be used in standard regression or correlation analysis.[6]

In this chapter, we'll learn to apply the independent contrasts method of Felsenstein (1985) for estimating the evolutionary correlation between characters.

2.3.1 Fitting linear models in R

To see how to use the contrasts method to fit a linear model (such as a regression model) in R, we'll first need to review some basics of how to fit linear models in the R computing environment.[7]

We can begin by loading some data for body size and home range size in various species of mammals (Garland et al. 1992).

Here we'll use two files, mammalHR.phy and mammalHR.csv, both of which are available from the book website.[8]

To start, why don't we read our character data from file, just as we did in chapter 1.

---

[5] Technically, any differences among species that do not overlap, or share branches in the tree, are independent under a Brownian motion model of evolution. We'll learn more about this important model of continuous character evolution in subsequent chapters.

[6] With the caveat that by virtue of having computed contrasts, this regression should not include an intercept term. This will be explained in more detail later in the chapter.

[7] Readers with abundant experience in fitting linear models in R can probably skip through this section of the chapter.

[8] http://www.phytools.org/Rbook/.

```
mammalHR<-read.csv("mammalHR.csv",row.names=1)
```

As we learned in the previous chapter, the function `head` is a useful means of printing just the first part of a data object.

Let's apply it to our `mammalHR` data frame to ensure that our file has been read properly without printing the entire object to the screen.

```
head(mammalHR)

##              bodyMass homeRange
## U._maritimus    265.0    115.60
## U._arctos       251.3     82.80
## U._americanus    93.4     56.80
## N._narica         4.4      1.05
## P._lotor          7.0      1.14
## M._mephitis       2.5      2.50
```

We can see from this printout that our data frame, `mammalHR`, contains two columns: `bodyMass` (giving the body mass in kilograms [kg] of each species in the data set) and `homeRange` (the home range size). The row names of our object correspond to the species names of each pair of values.

Next, we can go ahead and fit a model in which mean range size (in km^2) varies as a function of overall body size (in kg).

Without taking phylogeny into account, these characters do indeed seem to be correlated, as we see in figure 2.2.[9],[10]

```
## set margins of the plot
par(mar=c(5.1,5.1,1.1,1.1))
## create scatterplot
plot(homeRange~bodyMass,data=mammalHR,
    xlab="body mass (kg)",
    ylab=expression(paste("home range (km"^"2",")")),
    pch=21,bg="gray",cex=1.2,log="xy",las=1,cex.axis=0.7,
    cex.lab=0.9,bty="n")
```

2.3.2 Ordinary least squares regression

Now let's fit a standard regression model using *ordinary least squares* (OLS) (Neter et al. 1996).

In R, and unlike many point-and-click software packages, model fitting (using OLS or, indeed, most methods) and testing hypotheses about a fitted model are done over two steps.

---

[9]In this plot, we first adjusted the margin widths: `par(mar=c(5.1,5.1,2.1,1.1))`. We then proceeded to use the generic (or "S3") method `plot` to generate a bivariate scatterplot graph.

[10]Users following along might find that the labels on their vertical axis are plotted using scientific notation (i.e., 5e-02, 1e-01, etc.), instead of as shown in the figure. This can be changed by assigning the R system options values `scipen` a positive number, for example, `options(scipen=100)`.

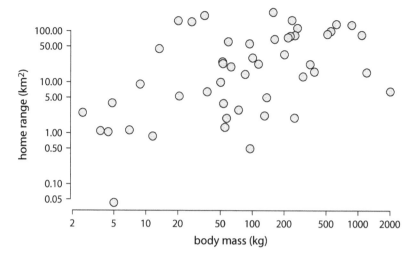

Figure 2.2
Scatterplot of mammal home range size by body size. Note that the values are in their original units, but x- and y-axes have been log-scaled. Data from Garland et al. (1992).

The first of these steps is merely to fit the model to our data, which we'll do here using a powerful function called lm.

The second step is hypothesis testing, and that can be undertaken with one of several functions, including anova and summary,[11] which we'll simply apply to our fitted model object from lm.

Given the distribution of our variables, and as is very common with biological data, we're going to fit our model to the natural logarithm of body size and the natural logarithm of home range size as follows. This is very important for numerical variables (such as these) that vary over orders of magnitude, because transforming to a log-scale will make an evolutionary change of 10 percent in a bear or a mouse equal (Glazier 2013).[12]

We'll do this using the function log.[13]

```
fit.ols<-lm(log(homeRange)~log(bodyMass),data=mammalHR)
fit.ols

##
## Call:
## lm(formula = log(homeRange) ~ log(bodyMass), data = mammalHR)
##
## Coefficients:
##   (Intercept)   log(bodyMass)
##       0.02504         0.60710
```

---

[11] Both of which happen to also be types of the generic S3 methods that we learned about in the previous chapter.

[12] We think this makes a lot more sense compared to, for instance, treating a 0.2-kg change in a rodent and an elephant as equivalent.

[13] By default, log transforms by the natural logarithm—that is, to log base *e*.

Just typing the name of our fitted model into the R interface doesn't do too much. We just get a simple printout of the formula call of our linear model,[14] and the values for the two different fitted coefficients of our model: in this case, our intercept and our slope.

To get a results table that mirrors what we might tend to obtain from other software, we recommend using the S3 method summary.

```
summary(fit.ols)

##
## Call:
## lm(formula = log(homeRange) ~ log(bodyMass), data = mammalHR)
##
## Residuals:
##     Min      1Q  Median      3Q     Max
## -4.1487 -1.0754  0.3137  1.0291  3.2314
##
## Coefficients:
##                Estimate Std. Error t value
## (Intercept)      0.0250     0.6895   0.036
## log(bodyMass)    0.6071     0.1476   4.114
##                Pr(>|t|)
## (Intercept)    0.971183
## log(bodyMass)  0.000155 ***
## ---
## Signif. codes:
## 0 '***' 0.001 '**' 0.01 '*' 0.05 '.' 0.1 ' ' 1
##
## Residual standard error: 1.671 on 47 degrees of freedom
## Multiple R-squared:  0.2648, Adjusted R-squared:  0.2491
## F-statistic:  16.9 on 1 and 47 DF,  p-value: 0.0001553
```

For an object from lm, summary gives us P-values for the model and our fitted coefficients, r^2, F, and many of the other statistics we might obtain using standard statistical software to undertake the same type of analysis.[15]

R also makes it easy to overlay the fitted regression model on a plot of our data, so let's do that too. This result is shown in figure 2.3.

```
## set margins and graph scatterplot
par(mar=c(5.1,5.1,1.1,1.1))
plot(homeRange~bodyMass,data=mammalHR,
    xlab="body mass (kg)",
    ylab=expression(paste("home range (km"^"2",")")),
    pch=21,bg="gray",cex=1.2,log="xy",las=1,
    cex.axis=0.7,cex.lab=0.9,bty="n")
```

---

[14] That is, the structure of the model we fit.

[15] For the F-statistic and ANOVA table, we recommend the method anova.

```
## add the line of best fit from lm
lines(mammalHR$bodyMass,exp(predict(fit.ols)),lwd=2,
    col="darkgray")
```

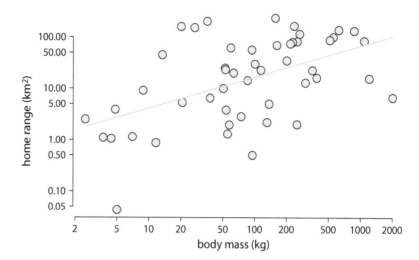

Figure 2.3
Mammal home range size by body size with the fitted regression model overlain (line). Data from Garland et al. (1992).

In this code chunk, the first two lines set the plot margins (using par) and then create a scatterplot (with plot). Then, the last line simply adds a regression to the plot.[16]

Now we've fit a model to our data in which we permitted home range size to vary as a function of body size across a sample of mammal species. We found that they do in fact covary, causing us to reject the null hypothesis that the two traits are independent.

2.3.3 Computing phylogenetically independent contrasts

But wait! What about the phylogeny?

A key assumption of linear regression analysis is that the residual error in the model is *independently and identically distributed*.[17]

Felsenstein (1985) reminded us that phylogenetic data will tend to violate this assumption because closely related species are very often more phenotypically similar than distant ones.

Felsenstein (1985) also gave us the PIC method to take this nonindependence into account by computing phylogenetically independent contrasts (our PICs) and then substituting them for our original data in the linear regression model.

---

[16]Normally, we might have used the very useful function abline instead of calling lines directly; however, in this case, that wouldn't have worked because we fit our regression model on the log-transformed values of body mass and home range size, so our regression slope and intercept are not in the units of our plot! Our solution was to use the function predict to obtain the predicted values for each numerical value of body mass and then exponentiate this predicted value back to our original scale.

[17]Commonly referred to as *i.i.d.*

To do this, we must start by reading our phylogenetic tree from file. This tree is again taken from Garland et al. (1992). Luckily we already learned how to read a phylogeny from file using the `read.tree` function of *ape* in chapter 1.

```
mammal.tree<-read.tree("mammalHR.phy")
```

Now let's plot our tree using the *phytools* function `plotTree`.[18,19]

```
## plot phylogeny of mammals
plotTree(mammal.tree,ftype="i",fsize=0.7,lwd=1)
## add node labels to the plotted tree
nodelabels(bg="white",cex=0.5,frame="circle")
```

The function to compute contrasts is from the *ape* package and is called `pic`. Let's go ahead and calculate contrasts for home range and for body size.

We're going to do this in two steps.

First, we'll pull out separate vectors for each character and assign the vector names that correspond to the row names of the data frame `mammalHR` using the handy function `setNames`.

Next, we'll proceed to compute contrasts for each of these new, named vectors using the *ape* package PIC function, called `pic`.

```
## pull our home range and body mass as
## numeric vectors
homeRange<-setNames(mammalHR[,"homeRange"],
    rownames(mammalHR))
bodyMass<-setNames(mammalHR[,"bodyMass"],
    rownames(mammalHR))
## compute PICs for home range and body size
pic.homerange<-pic(log(homeRange),mammal.tree)
pic.bodymass<-pic(log(bodyMass),mammal.tree)
```

Let's inspect one of our vectors of contrasts.[20]

```
head(pic.homerange,n=20)
##             50             51             52
##    0.025629767  -0.280770953  -0.030984636
##             53             54             55
```

---

[18] We could have also used the generic `plot` method from *ape* as we did in the previous chapter.

[19] We included node labels on our tree plot of figure 2.4 simply so that we'll be able to see that they correspond exactly to our phylogenetically independent contrasts: one contrast per internal node! Examine the tree and decide if you agree that this makes sense.

[20] Using the function `head` with `n=20` tells R to print out just the first twenty elements of our vector. To print the whole thing, either type `print(pic.homerange)` at your R prompt or just type the name of the object, `pic.homerange`, and hit ENTER.

```
##    0.500999613   0.181248561   0.166853947
##            56            57            58
##   -0.042441884  -0.026005970   0.166897552
##            59            60            61
##    0.273228429   0.953184119   0.799393348
##            62            63            64
##   -0.236634270   0.673305909  -0.007568451
##            65            66            67
##    0.572452594  -0.020625533  -0.854572954
##            68            69
##   -0.610533619  -0.016274305
```

We can immediately see that the contrasts take both positive and negative values. This makes sense, because contrasts are (standardized[21]) *differences* in trait values between sister species or nodes. These differences can obviously be *positive* if, say, the right daughter node has a higher value for the trait than the left daughter node or *negative* if the opposite is true.

In addition, if we compare either of our PIC vectors to the tree we plotted in figure 2.4, we should also see that our vector has names that correspond to the node indices of the tree. This too makes sense, because in a bifurcating[22] tree, each contrast subtends one and only one node.

2.3.4 Fitting a linear regression to contrasts

Now we're ready to fit our linear model. We need keep in mind, however, that we'll need to fit this regression model through zero—that is, without an intercept term.

This is because for any node in the phylogeny, the rotation of the right and left daughters of that node is arbitrary—and, as such, so is the direction of subtraction of the contrasts.[23] Our linear model for those contrasts should thus go through the point (0,0) on our plot (Felsenstein 1985).

To do that in R, we can either append +0 or -1 to our formula in lm.[24] Let's use +0.

```
## fit linear model to PICs without intercept
fit.pic<-lm(pic.homerange~pic.bodymass+0)
fit.pic
```

```
##
## Call:
## lm(formula = pic.homerange ~ pic.bodymass + 0)
##
## Coefficients:
```

---

[21] Standardization is done by dividing each contrast by a value proportional to its expected standard deviation under a model following Felsenstein (1985). We'll learn more about this model in chapter 4.

[22] To compute contrasts, our tree must be perfectly bifurcating. Luckily, if it's not, we're allowed to simply resolve each polytomy using internal branches of zero length, and any downstream result will be unchanged. Resolving polytomies this way can be done using the *ape* function multi2di.

[23] A useful way to think about this is that for every pair of contrasts (x, y), there also exists another equivalent pair $(-x, -y)$, rotated 180° around the origin!

[24] Strangely, both of these notations have the same effect for linear models in R!

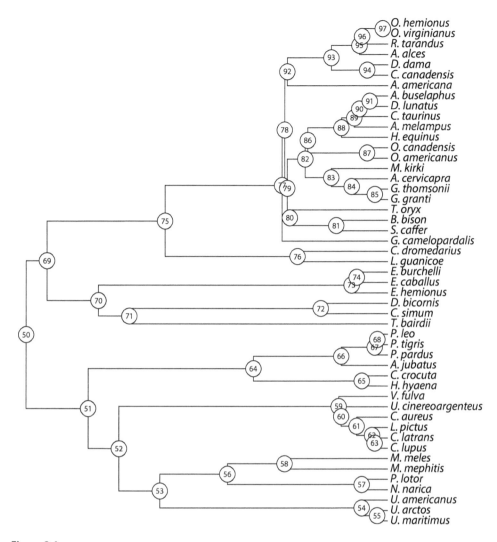

Figure 2.4

A phylogeny of various mammal species from Garland et al. (1992). Numbers indicate node indices in the underlying "phylo" object. Tree from Garland et al. (1992).

```
## pic.bodymass
##      1.262
```

Just as with our linear model fit to the original data, for a summary of the results, we can use the S3 method `summary`:

```
summary(fit.pic)

##
## Call:
```

```
## lm(formula = pic.homerange ~ pic.bodymass + 0)
##
## Residuals:
##       Min      1Q    Median      3Q      Max
## -0.72024 -0.17340  0.01576  0.14859  0.83689
##
## Coefficients:
##                Estimate Std. Error t value Pr(>|t|)
## pic.bodymass     1.2616     0.1768    7.14 5.07e-09
##
## pic.bodymass ***
## ---
## Signif. codes:
## 0 '***' 0.001 '**' 0.01 '*' 0.05 '.' 0.1 ' ' 1
##
## Residual standard error: 0.3456 on 47 degrees of freedom
## Multiple R-squared:  0.5201, Adjusted R-squared:  0.5099
## F-statistic: 50.93 on 1 and 47 DF,  p-value: 5.072e-09
```

The output looks very similar to the summary result from our OLS regression, with the most significant difference being that this model has only one fitted coefficient—the slope.[25] This makes sense since we fixed the intercept term of the model to be zero.

Finally, let's plot independent contrasts and our fitted regression model.

```
## set margins
par(mar=c(5.1,5.1,1.1,1.1))
## graph scatterplot of contrasts
plot(pic.homerange~pic.bodymass,
    xlab="PICs for log(body mass)",
    ylab="PICs for log(range size)",
    pch=21,bg="gray",cex=1.2,las=1,
    cex.axis=0.7,cex.lab=0.9,bty="n")
## add gridlines to the plot
abline(h=0,lty="dotted")
abline(v=0,lty="dotted")
## reset graphing limits of the plot to the
## x/y range of our PICs
clip(min(pic.bodymass),max(pic.bodymass),
    min(pic.homerange),max(pic.homerange))
## graph our fitted line
abline(fit.pic,lwd=2,col="darkgray")
```

---

[25] Here reported as pic.bodymass because, remember, it measures the effect of the contrasts in body mass on our response variable—mammal home range size.

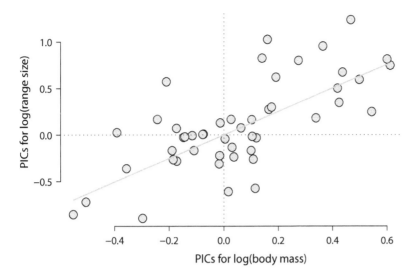

Figure 2.5
The regression of mammal home range size on body size from phylogenetically independent contrasts. Dotted lines show *x* and *y* axes, and the regression has been fit without an intercept term (i.e., through the origin). Data from Garland et al. (1992).

When we make our plot, it's always very important[26] that we indicate that the data are from contrasts in our axis labels, just as we have done in figure 2.5.[27,28]

2.4 What happens if we ignore the tree?

In our mammal home range size example, taking the phylogeny into account (or not) didn't make a huge impact on the slope of our fitted regression model or on the estimated statistical significance of the result.

Nonetheless, it's quite simple to imagine (and almost as easy to simulate) fairly realistic circumstances in which taking the phylogeny into account when fitting a regression model can be much more important.

For fun, let's do just that.

Here, since we're using stochastic simulation, we're going to go ahead and first specify the computer *seed*.[29]

---

[26] Although frequently forgotten in practice.

[27] We also included horizontal and vertical lines with the function `abline` so that we can verify that our regression line passes through the point (0,0) as planned. It does!

[28] We used the function `clip` to limit the range of our regression line to the minimum and maximum values of our contrasts. Otherwise, it would've extended to the edges of the plot area.

[29] We use a seed because no computer-generated random number is ever truly random. To the contrary, random-number generators use complicated algorithms to obtain sequences of numbers that look random but are actually completely deterministic—that is, given that the *seed*, the initial state of the system, is known. Some software that involves random-number generation uses a stimulus external to the program, such as the computer clock, to set the seed. R allows the user to control the seed to ensure that even analyses involving random-number generation can be made entirely reproducible.

To get the same result as in the book, you should set your seed to have the same value as we did. On the other hand, if you'd like to see what happens if you choose a *different* seed, by all means go ahead![30]

```
set.seed(1001)
```

Next we're going to simulate a phylogenetic tree.

To do that, we'll use a stochastic model called the *birth–death model* (Nee 2006). The birth–death model is just a stochastic process for growing phylogenies with constant random speciation (births) and constant random extinction (deaths).[31]

We want to simulate our tree with a high extinction rate, so we'll combine our function call with a `while` loop[32] designed to repeat our simulation until we get a phylogeny that doesn't go completely extinct before the present day.

In so doing, what we hope to obtain is a stochastic phylogenetic tree that was simulated under a relatively high extinction rate compared to the speciation rate while screening out any phylogeny that goes completely kaput[33] before the end of the simulation.

Our simulated tree is shown in figure 2.6.

```
## set starting tree to NULL
tree<-NULL
## repeat simulation until non-NULL (i.e., non-
## extinct) tree is obtained
while(is.null(tree))
    tree<-pbtree(n=100,b=1,d=0.8,extant.only=TRUE)
## plot the simulated tree
plotTree(tree,ftype="off",color="darkgray",lwd=1)
```

Next, let's simulate evolution on this tree for two variables, *x* and *y*.

Note that we're going to simulate the two characters completely *independently*: first one and then the other—that is, without any real *evolutionary correlation* between them.

After we simulate our data, we can plot them, fit a regression model, and graph that fitted model without points.

In this code block, the *phytools* function `fastBM` is used to simulate independent character evolution for *x* and *y*.[34]

---

[30] Our chosen seed has no particular significance.

[31] Don't worry, we'll learn more about this model later. Suffice it to say, this is a stochastic model that is commonly used for phylogenies.

[32] A `while` loop is a programming structure common to both R and many other programming and scripting languages. It can be generally interpreted as follows: *while some logical statement is true, repeat an operation*. Since an extinct tree returns NULL in pbtree, if we set our starting tree to be NULL and then run `while(is.null(tree)) tree<-pbtree(...)`, we will end up repeating the simulation until a nonextinct tree is returned.

[33] That's the technical term.

[34] This simulation is done using a stochastic model called *Brownian motion* that is probably already familiar to many readers. Don't worry. We'll learn a lot more about Brownian motion in chapter 4.

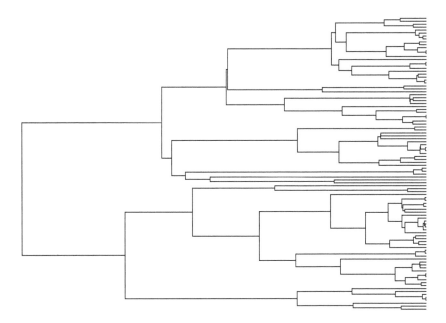

Figure 2.6
A simulated 100-taxon birth–death phylogeny generated using the *phytools* function `pbtree`.

```
x<-fastBM(tree)
y<-fastBM(tree)
```

It's easy to tell that the two traits were simulated independently, because we did it via two separate calls of the simulator!

Next, we'll create a scatterplot of our two simulated variables and then go on to fit an ordinary least squares regression using the function `lm` (ignoring phylogeny). We can add a line showing this fitted regression to our plot using `abline`.

This analysis is illustrated in figure 2.7.

```
## set figure margins
par(mar=c(5.1,4.1,1.1,1.1))
## create scatterplot of x & y
plot(x,y,cex=1.2,pch=21,bg="gray",las=1,
    cex.axis=0.7,cex.lab=0.9,bty="n")
## add gridlines to the plot
grid()
## abbreviate the plotting area to match
## the range of our variables
clip(min(x),max(x),min(y),max(y))
## fit our linear model using OLS
fit.ols<-lm(y~x)
## add our fitted regression line to the plot
abline(fit.ols,lwd=2,col="darkgray")
```

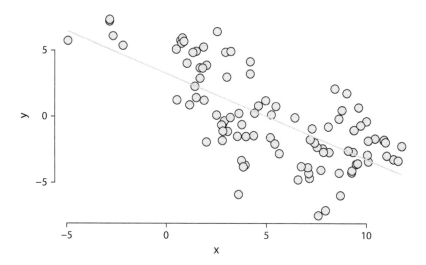

Figure 2.7
Stochastic data for two variables simulated *without* an evolutionary correlation.

Finally, we can print and summarize our fitted OLS model.

```
fit.ols

##
## Call:
## lm(formula = y ~ x)
##
## Coefficients:
## (Intercept)              x
##       3.253         -0.649
```

```
summary(fit.ols)

##
## Call:
## lm(formula = y ~ x)
##
## Residuals:
##     Min      1Q   Median       3Q      Max
## -6.8190 -1.7275   0.0882   1.9320   4.8112
##
## Coefficients:
##              Estimate Std. Error t value Pr(>|t|)
## (Intercept)    3.2529     0.4246   7.662 1.33e-11
## x             -0.6490     0.0653  -9.939  < 2e-16
##
## (Intercept) ***
## x           ***
```

```
## ---
## Signif. codes:
## 0 '***' 0.001 '**' 0.01 '*' 0.05 '.' 0.1 ' ' 1
##
## Residual standard error: 2.477 on 98 degrees of freedom
## Multiple R-squared:  0.502,  Adjusted R-squared:  0.4969
## F-statistic: 98.79 on 1 and 98 DF,  p-value: < 2.2e-16
```

The r^2 of our linear model is not very high (only around 0.5), but the correlation nonetheless comes out as highly significant.

This should be a bit surprising because, remember, these data were simulated in the *absence* of an evolutionary correlation between the two variables! What we should take from this example is that it is not difficult for the phylogeny to induce what might be characterized as a *type I error*.[35]

One way to understand how this comes about is that in our data, there are clusters of closely related (and thus statistically nonindependent) taxa that have highly similar phenotypes, for both x and y—even though x and y are *evolutionarily* independent: that is, the *evolution* of x had no effect on the evolution of y or vice versa.

This can be visualized to some extent using the plotting method (phylomorphospace[36]) that we learned at the end of chapter 1. This illustration is shown in figure 2.8.

```
## set plotting margins
par(mar=c(5.1,5.1,1.1,1.1),
    cex.axis=0.7,cex.lab=0.9)
## graph phylomorphospace projection
phylomorphospace(tree,cbind(x,y),label="off",
    node.size=c(0,0),bty="n",las=1)
## overlay points onto the phylomorphospace plot
points(x,y,pch=21,bg="gray",cex=1.2)
## add gridlines
grid()
## clip plot
clip(min(x),max(x),min(y),max(y))
## add fitted regression line
abline(fit.ols,lwd=2)
```

In this code chunk, we used the *phytools* function phylomorphospace to project the phylogeny into our x and y space, but then we also added back in the original points (using point type and color options[37] that are not available in phylomorphospace). Finally, we graphed a line showing our regression model.

---

[35] A type I error is a kind of statistical error in which we reject the null hypothesis even though it is correct. For our simulated data, we know that x and y evolved absent an evolutionary correlation between them. If we used OLS—that is, linear regression without taking the phylogeny into account—to test the null hypothesis that x and y evolved independently, and then we proceeded to *reject* this null hypothesis and conclude instead that they evolved in a correlated fashion, we would have committed a type I error.

[36] A *projection* of our phylogeny into a two-dimensional phenotype space.

[37] Here, pch=21 sets the point type to be a open circle, while bg="gray" sets the fill of that circle to be gray in color. For more information about plotting points in R, see help(points).

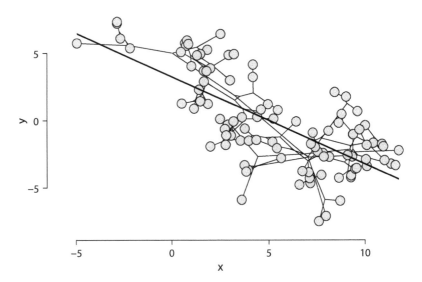

Figure 2.8
A projection of the phylogeny into two-dimensional phenotype space (a.k.a. a phylomorphospace) for two simulated quantitative traits.

Inspecting this projection, we might start to notice that among closely related species, there appears to be very little correlation between the two variables in the model—suggesting that the apparent correlation could be (as described by Felsenstein 1985, and as we reproduced in figure 2.1) due to chance divergences between a relatively small number of nodes in the tree.

Now let's see if by substituting the contrasts between species (and nodes) for the original values in our regression, we resolve this type I error. We'll plot our contrasts and fitted regression line in figure 2.9.

```
## compute PICs for x and y
ix<-pic(x,tree)
iy<-pic(y,tree)
## fit PIC regression through the origin
fit.pic<-lm(iy~ix+0)
fit.pic

##
## Call:
## lm(formula = iy ~ ix + 0)
##
## Coefficients:
##       ix
## -0.07094
```

```
## set plotting margins
par(mar=c(5.1,4.1,1.1,1.1))
## graph scatterplot of PICs
```

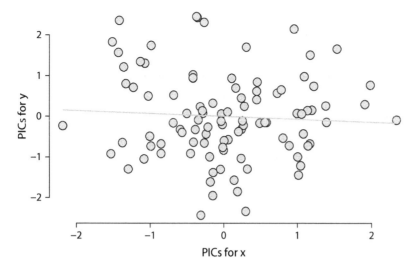

Figure 2.9
Phylogenetically independent contrasts regression for two simulated traits that did not evolve with an evolutionary correlation.

```
plot(ix,iy,cex=1.2,pch=21,bg="gray",las=1,
    xlab="PICs for x",
    ylab="PICs for y",
    cex.axis=0.7,cex.lab=0.9,bty="n")
## add gridlines to plot
grid()
## clip plotting area
clip(min(ix),max(ix),min(iy),max(iy))
## add fitted regression line
abline(fit.pic,lwd=2,col="darkgray")
```

Likewise, if we look at that the statistical significance of the relationship, we should see that it has disappeared.[38]

```
summary(fit.pic)

##
## Call:
## lm(formula = iy ~ ix + 0)
##
## Residuals:
##     Min     1Q  Median     3Q    Max
## -2.4538 -0.6708 -0.1080  0.6791  2.4256
```

---

[38] Of course, readers who reproduce this analysis using different seeds will find a significant relationship between the contrasts for x and y about 5 percent of the time.

```
## 
## Coefficients:
##     Estimate Std. Error t value Pr(>|t|)
## ix -0.07094    0.12032   -0.59     0.557
## 
## Residual standard error: 1.049 on 98 degrees of freedom
## Multiple R-squared:  0.003534,   Adjusted R-squared:   -0.006634
## F-statistic: 0.3476 on 1 and 98 DF,  p-value: 0.5568
```

Remember, this most definitely is not an example of a "real" relationship that has been removed with contrasts. To the contrary, in this case, we know beyond the shadow of a doubt that our data evolved absent a genuine evolutionary relationship between x and y—because we made them that way!

The highly significant correlation that we measure in the OLS regression analysis was a spurious relationship driven by the phylogeny.

A useful way to think about what we found is the x and y seem to be *correlated* but not *evolutionarily correlated*. That is to say, the OLS result tells us that we can use x to predict y—but only because (by knowing its value for x) we know something about where that species occurs in the tree, and so we can predict something about y.[39]

On the other hand, the nonsignificant linear regression of the contrasts tells us that[40] we find *no* evidence of an evolutionary tendency for x and y to coevolve.

2.4.1 An experiment to measure the effect of ignoring phylogeny

Last, let's do a small experiment to see the general effect of ignoring phylogeny when carrying out regression analysis.

For a real experiment, we need replication—so instead of generating just one phylogeny, as we did in the previous section, let's simulate 500.

For each of these 500 trees, we'll simulate two characters—x and y—that have evolved in an uncorrelated fashion on the tree. We'll do this just as we did before by using two separate calls of the *phytools* function fastBM.

Finally, let's fit an OLS regression and a phylogenetic independent contrasts regression to each pair of variables. We'll plot the distribution of *P*-values for OLS (panel a) and contrasts regression (panel b) in figure 2.10.

In our code, we'll use several different R functions of the family apply: lapply, mapply, and sapply, as well as the related function replicate.

As we mentioned in chapter 1, apply functions allow the R user to vectorize lots of different kinds of operations in R without having to write loops.[41] We'll see apply family functions again later on in this book.

---

[39] This is very different from saying that the correlation between x and y is *caused* by the phylogeny, in any genuinely mechanistic way. To the contrary, we may be able to *predict* y from x and vice versa—but only because species with similar values for x tend to be those species that are closely related, and as such, they often share similar values for y as well!

[40] Despite their correlation in the original space.

[41] For the especially curious reader: lapply applies a function to all the elements of a list; mapply applies a function to the corresponding elements in multiple lists or vectors; finally, sapply is like lapply but simplifies the result into a vector or matrix if possible. replicate just repeats the operation of a function the number of times specified by the user and can return its results in the form of a list or vector.

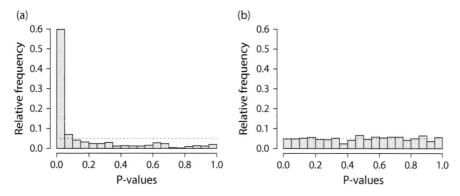

Figure 2.10
Distribution of *P*-values for (a) OLS regression and (b) PIC regression. Data were simulated on stochastic birth–death phylogenies absent an evolutionary correlation between *x* and *y*.

Our first step is to simulate a set of trees.

Actually, pbtree has an nsim for simulating multiple birth–death phylogenies under the same condition. In our case, though, we want to repeat any simulation in which the tree goes extinct before the end of the simulation. As such, we'll write our own custom function, foo,[42] that does exactly that using the same while loop that we learned in the previous example.

We'll use a replicate function call to create 500 trees in a list.

```
## custom function to simulate a birth-death tree,
## under particular fixed conditions, and discard any
## tree that goes extinct before the end of the
## simulation
foo<-function(){
    tree<-NULL
    while(is.null(tree))
        tree<-pbtree(n=100,b=1,d=0.8,
            extant.only=TRUE,quiet=TRUE)
    tree
}
## simulate 500 trees in a list using replicate
trees<-replicate(500,foo(),simplify=FALSE)
## assign "multiPhylo" class attribute
class(trees)<-"multiPhylo"
```

Next, we'll simulate data for *x* and *y* independently using a lapply operation on our list of trees. This, in turn, will generate two lists (denominated x and y), each containing 500 independent *x* or *y* vectors—one for each tree.

---

[42] The name *foo* is a popular placeholder name in computer programming for a variable or function that is being used to demonstrate a concept or that will be used once and then discarded. In our case, the latter applies—we intend to use the function foo only immediately after it has been created and then never again.

Independent contrasts

```
## simulate a list of x & y vectors, one for
## each tree in our "multiPhylo" object
x<-lapply(trees,fastBM)
y<-lapply(trees,fastBM)
```

Now, we can use nonphylogenetic (OLS) regression to fit y~x for every pair of simulated trait vectors from our 500 trees.

Here we first create a new custom function[43] and then use mapply to iterate over the elements in our x and y lists. mapply works a lot like lapply—except that it can repeat an operation (the first argument of the function call) over all the elements of two, three, or more lists, which are supplied as subsequent arguments in the function.

```
## custom function to fit y~x
foo<-function(x,y)  lm(y~x)
## mapply function call to fit y~x to each pair
## of vectors in our two lists
fits.ols<-mapply(foo,x,y,SIMPLIFY=FALSE)
```

For the purposes of this *particular* experiment, we only care about the *P*-value of each fitted model. The easiest way for us to pull these values out is by calling anova on *each* of the fitted models in our long list, fit.ols, from the previous step. We then use sapply to iterate the function call across our list of fitted models!

```
## custom function to get P-value of fitted model
foo<-function(fit)  anova(fit)[["Pr(>F)"]][1]
## sapply call to iterate foo over list of fits
pval.ols<-sapply(fits.ols,foo)
```

OK. Now let's repeat all of these steps—but after computing phylogenetically independent contrasts first using pic.

```
## compute contrasts on all x and y
pic.x<-mapply(pic,x,trees,SIMPLIFY=FALSE)
pic.y<-mapply(pic,y,trees,SIMPLIFY=FALSE)
## custom function to fit linear model without
## intercept
foo<-function(x,y)  lm(y~x+0)
## iterate over contrasts vectors
fits.pic<-mapply(foo,pic.x,pic.y,SIMPLIFY=FALSE)
## custom function to pull out P-value of fitted
## model
```

_____

[43]Likewise called foo, for the same reason as before.

```
foo<-function(fit) anova(fit)[["Pr(>F)"]][1]
## iterate over fitted models
pval.pic<-sapply(fits.pic,foo)
```

As a last step, let's compare the distribution of *P*-values from these two different exercises. We'll do that by generating two different histograms of *P*-values: one from our OLS regressions and then a second from our contrasts regressions (figure 2.10).

```
## compute histograms for OLS without plotting
h1<-hist(pval.ols,breaks=20,plot=FALSE)
## convert counts to relative frequency
h1$counts<-h1$counts/sum(h1$counts)
## repeat for contrasts regression P-values
h2<-hist(pval.pic,breaks=20,plot=FALSE)
h2$counts<-h2$counts/sum(h2$counts)
## subdivide plotting area and adjust margins
par(mfrow=c(1,2),mar=c(5.1,4.1,2.1,1.1))
## plot first histogram of OLS P-values
plot(h1,ylim=c(0,0.6),col="gray",main="",
    xlab="P-values",ylab="Relative frequency",
    cex.axis=0.5,cex.lab=0.7,las=1)
mtext("(a)",line=1,adj=0,cex=0.8)
## show nominal alpha level of 0.05
lines(c(0,1),rep(0.05,2),lwd=1,lty="dotted")
## plot second histogram of contrasts regression
## P-values
plot(h2,ylim=c(0,0.6),col="gray",main="",
    xlab="P-values",ylab="Relative frequency",
    cex.axis=0.5,cex.lab=0.7,las=1)
mtext("(b)",line=1,adj=0,cex=0.8)
## show nominal alpha level of 0.05
lines(c(0,1),rep(0.05,2),lwd=1,lty="dotted")
```

If a statistical method is performing as designed, we would expect to find a *uniform* distribution of *P*-values under the null hypothesis.[44]

This is indeed what we see in figure 2.10b, which uses PICs—but *not* in figure 2.10a where phylogeny is ignored and OLS is employed instead.

In fact, when we ignore the phylogeny, our chances of rejecting the null hypothesis given an α level of 0.05[45] jump to nearly 60 percent (figure 2.10)!

From our experiment, we can thus safely conclude that we are quite likely to be misled if we ignore phylogeny when carrying out regression analysis. In other words, if species are related

---

[44]That is to say, 5 percent between $P = 0$ and $P = 0.05$, 5 percent between $P = 0.05$ and $P = 0.10$, and so on.

[45]In other words, the chances of committing a type I error, assuming that our regression model is being used to test for an evolutionary correlation.

to one another, we should use analyses that account for those phylogenetic relationships rather than assuming that our data are independent.

To quote Felsenstein (1985):[46] *Phylogenies are fundamental to comparative biology; there is no doing it without taking them into account.*

2.5 Practice problems

2.1 Repeat your PIC analysis comparing body mass and home range but for only one clade in your tree: the artiodactyls.[47] Use `extract.clade` to pull out this taxon from your tree of mammals. Do the results and conclusions of your analysis change? If so, in what way?

2.2 On the book website, we supply some data for a group of Asian barbets (`BarbetTree.nex` and `Barbetdata_mod.csv`): members of the bird family Megalaimidae (Gonzalez-Voyer et al. 2013). Using these data and the method of phylogenetic independent contrasts, test for a relationship between the two variables `Lnalt` and `wing`. What do you find? How does the slope of your relationship change if you fit `Lnalt  wing` versus `wing  Lnalt`? How does the *P*-value of a statistical test of this relationship change?

2.3 If you multiply all the branches of your phylogenetic tree by 100, will your independent contrasts analysis change? Why or why not? Can you confirm this using R code? Don't forget to use what you learned about the internal structure of a `"phylo"` object in chapter 1 to modify your tree.

2.4 Repeat the simulation analysis of type I error for OLS, in the last section of the chapter, but using a "pure-birth" model of diversification (i.e., trees simulated using `pbtree` but with the extinction rate, d, set to 0).[48] How does this change to the way your phylogenies are simulated affect your results? In particular, does it cause the type I error rate that results from ignoring the phylogeny to go up or go down? Can you explain the differences that you find?

---

[46] Why wouldn't you?

[47] Artiodactyla is the phylogenetic group descended from node 75 in figure 2.4.

[48] We'll learn more about models of diversification in chapter 9.

Phylogenetic generalized least squares

3.1 Introduction

How do animals see at night? One might suspect, based on the phenotypes of peculiar large-eyed primate species such as the aye-aye (*Daubentonia madagascariensis*) or the Sunda slow loris (*Nycticebus coucang*), that having large eyes is key.

In this chapter, we'll describe a method called *phylogenetic generalized least squares* (PGLS) that can be used to test complex hypotheses about how traits relate to one another and to their environment.

Last chapter, we learned how to use phylogenetically independent contrasts (PICs) to test for an evolutionary correlation between pairs of characters. Many comparative hypotheses, however, require the simultaneous analysis of more than two traits. For instance, in a single model, you might want to include both body size *and* limb length as predictors of home range size. Or perhaps you'd like to analyze the influence of *both* diel activity pattern and foraging mode on the evolution of eye size.[1]

Likewise, perhaps you're not satisfied with the assumption that the amount of variation between species accumulates as a linear function of the time separating them[2] that is required for PICs.

Or, last, perhaps you'd like to fit a model that combines continuous and discrete (i.e., factor) predictor variables, something that is specifically excluded in PIC regression.

In this chapter, we will thus:

1. Describe a much more flexible technique, known as *phylogenetic generalized least squares* or PGLS.
2. Compare independent contrasts regression with PGLS regression.
3. Show how PGLS can be used to permit greater flexibility in our model assumptions than PICs.

---

[1] Echolocating bats, for instance, seem unlikely to be under natural selection to evolve heightened nocturnal visual acuity.

[2] We'll learn more about the basis of this assumption in chapter 4.

4. Finally, describe generalized analysis of variance (ANOVA) and analysis of covariance (ANCOVA) based on PGLS.

3.2 Statistical nonindependence of phylogenetic data

As introduced in chapter 2, we would normally use *ordinary least squares*, abbreviated OLS, to fit a variety of statistical models (such as linear regression, ANOVA, ANCOVA, etc.) to data that consist of a single response variable, y, and one or more predictors.

OLS is a convenient method because it has been mathematically proven to provide the *best linear unbiased estimates*[3] of the coefficients of linear models, given that certain criteria are met by our data. Key among these criteria is an assumption that the residual error of the model (i.e., the variability in y that is not accounted for by the model) is *independently and identically distributed* (i.i.d.) for different observations of y (Neter et al. 1996).

3.2.1 Phylogenetic residuals tend to be correlated

Unfortunately, due to the resemblance of related living organisms that descend from common ancestors in the past, this assumption will not generally hold for data that originate from different species (Felsenstein 1985).

To the contrary, in fact, more closely related species often tend to have *correlated* residual deviations in their values of y from those predicted by the model.

For example, if one species has a large positive residual deviation in y compared to the predicted value, then other closely related sister species are often more likely to have large positive residual values too. If this is the case, then residual errors in y are *correlated*—implying a violation of the OLS assumption of i.i.d.

For the case of linear regression, the technique of phylogenetically independent contrasts solves this problem by first transforming the data into a set of contrasts that are theoretically independent of the phylogeny and thus do not violate i.i.d. This means that we can proceed to use these contrasts in ordinary least squares regression.[4]

3.2.2 Generalized least squares as an estimation method

However, long before the development of independent contrasts, there already existed an estimation method that allowed us to account for non-i.i.d. in the residual error of a regression model. This technique is called *generalized least squares* or GLS. GLS was first described in a paper by Aitken (1936). A few years after the publication of Felsenstein's (1985) contrasts method, Grafen (1989) pointed out that GLS could be used for similar purposes.

In fact, it has subsequently been shown that contrasts regression is a special case of GLS, as we'll see below (Blomberg et al. 2012).

GLS, however, also lets us be a bit more flexible about the specific correlation structure of the residual error in the model, and it also allows us to include discrete factors as independent variables—neither of which is easily done with contrasts. As such, we think that it's quite important to learn about both methods.

Henceforward, we'll refer to the *phylogenetic* generalized least squares estimation method as PGLS.

---

[3] The estimator from this kind of estimation technique is referred to as a *BLUE* by statisticians. We're not kidding.

[4] Albeit without an intercept term, as we learned in the last chapter.

The first thing that we're going to do in this chapter is "prove" the equivalence of contrasts (i.e., PIC) regression, which we learned all about in chapter 2, and PGLS.

In reality, we won't use a mathematical proof.[5] Instead, what we can do is merely fit a contrasts regression and a PGLS regression to the same data and then show that a numerically identical result is obtained.

To illustrate this, we'll use an analysis of primate orbit morphology.[6] We can relate eye size to the size of the skull and, eventually, to species diel activity patterns.

This example comes from a paper by Kirk and Kay (2004), as extended in a wonderful blog post[7] by Randi Griffin (2017).

The data (`primateEyes.csv`) and tree (`primateEyes.phy`) files are available from the book website,[8] just as are all the data files used in this volume.

3.3.1 Review: Fitting a regression model using PICs

Since we want to compare PIC regression with PGLS, let's commence by reading in our data and tree from file, computing independent contrasts using `pic`, and then fitting a linear regression model to our contrasts using `lm`.[9] We can then repeat our analysis using PGLS.

Since this analysis just duplicates what we've already learned, we have just used R *comments* (indicated using the ## characters) to describe our various steps, rather than more complete narrative text.

```
## load packages
library(phytools)
## read data from file
primate.data<-read.csv("primateEyes.csv",row.names=1,
    stringsAsFactors=TRUE)
## inspect data
head(primate.data,4)
```

```
##                             Group
## Allenopithecus_nigroviridis Anthropoid
## Alouatta_palliata           Anthropoid
## Alouatta_seniculus          Anthropoid
## Aotus_trivirgatus           Anthropoid
##                             Skull_length
## Allenopithecus_nigroviridis     98.5
## Alouatta_palliata              109.8
## Alouatta_seniculus             108.0
## Aotus_trivirgatus               60.5
##                             Optic_foramen_area
## Allenopithecus_nigroviridis          7.0
```

---

[5]See Blomberg et al. (2012) for that!

[6]That is, how big their eyeballs are!

[7]https://www.randigriffin.com/2017/11/17/primate-orbit-size.html.

[8]http://www.phytools.org/Rbook/.

[9]All of this just compactly duplicates what we already learned how to do in chapter 2 but for a new data set.

```
## Alouatta_palliata                                    5.3
## Alouatta_seniculus                                   8.0
## Aotus_trivirgatus                                    3.1
##                                 Orbit_area
## Allenopithecus_nigroviridis      298.7
## Alouatta_palliata                382.3
## Alouatta_seniculus               359.4
## Aotus_trivirgatus                297.4
##                                 Activity_pattern
## Allenopithecus_nigroviridis           Diurnal
## Alouatta_palliata                     Diurnal
## Alouatta_seniculus                    Diurnal
## Aotus_trivirgatus                   Nocturnal
##                                 Activity_pattern_code
## Allenopithecus_nigroviridis                       0
## Alouatta_palliata                                 0
## Alouatta_seniculus                                0
## Aotus_trivirgatus                                 2
```

```r
## read tree from file and inspect
primate.tree<-read.tree("primateEyes.phy")
print(primate.tree,printlen=2)
```

```
##
## Phylogenetic tree with 90 tips and 89 internal nodes.
##
## Tip labels:
##    Allenopithecus_nigroviridis, Cercopithecus_mitis, ...
##
## Rooted; includes branch lengths.
```

```r
## extract orbit area from our data frame and add names
orbit.area<-setNames(primate.data[,"Orbit_area"],
    rownames(primate.data))
## extract skull length from our data frame and add names
skull.length<-setNames(primate.data[,"Skull_length"],
    rownames(primate.data))
## compute PICs on the log-transformed values of both traits
pic.orbit.area<-pic(log(orbit.area),primate.tree)
pic.skull.length<-pic(log(skull.length),
    primate.tree)
## fit a linear regression to orbit area as a function of
## skull length, without an intercept term
pic.primate<-lm(pic.orbit.area~pic.skull.length+0)
summary(pic.primate)
```

```
##
## Call:
```

```
## lm(formula = pic.orbit.area ~ pic.skull.length + 0)
##
## Residuals:
##        Min        1Q     Median        3Q        Max
## -0.103535 -0.023104 -0.004624  0.021021  0.175298
##
## Coefficients:
##                   Estimate Std. Error t value
## pic.skull.length  1.37867     0.07734    17.83
##                   Pr(>|t|)
## pic.skull.length   <2e-16 ***
## ---
## Signif. codes:
## 0 '***' 0.001 '**' 0.01 '*' 0.05 '.' 0.1 ' ' 1
##
## Residual standard error: 0.03738 on 88 degrees of freedom
## Multiple R-squared:  0.7831, Adjusted R-squared:  0.7807
## F-statistic: 317.8 on 1 and 88 DF,  p-value: < 2.2e-16
```

The slope of the fitted model is highly significant and has an estimated value of about 1.38.

Since both our traits are log-transformed, this slope gives the best-fitting allometric relationship between skull length and orbit area across all primates in our data set.

To visualize this, why don't we also plot our original data plus our PICs and PIC regression in a similar way to what we saw last chapter.

```
## set plotting parameters
par(mfrow=c(1,2),
    mar=c(5.1,4.6,2.1,1.1))
## plot our raw data in the original space
plot(orbit.area~skull.length,log="xy",
    pch=21,bg=palette()[4],cex=1.2,
    bty="n",xlab="skull length (cm)",
    ylab=expression(paste("orbit area (",mm^2,")")),
    cex.lab=0.8,cex.axis=0.7,las=1)
mtext("(a)",line=0,adj=0,cex=0.8)
## plot our phylogenetic contrasts
plot(pic.orbit.area~pic.skull.length,pch=21,
    bg=palette()[4],cex=1.2,
    bty="n",xlab="PICs for log(skull length)",
    ylab="PICs for log(orbit area)",
    cex.lab=0.8,cex.axis=0.7,las=1)
mtext("(b)",line=0,adj=0,cex=0.8)
## limit the plotting area to the range of our two traits
clip(min(pic.skull.length),max(pic.skull.length),
    min(pic.orbit.area),max(pic.orbit.area))
## add our fitted contrasts regression line
abline(pic.primate,lwd=2)
```

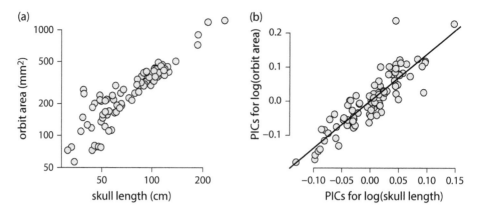

Figure 3.1
Contrasts analysis of the primate orbit and skull length data set. (a) The original data, but with *x* and *y* axes transformed to a log-scale. (b) Independent contrasts. The line of panel (b) shows the fitted contrasts regression line.

Here, we used `par` to split up our plotting device into two panels, and we plotted our data on its original scale, with the *x* and *y* axes both transformed (figure 3.1a). Then, we graphed our independent contrasts with the fitted regression line (figure 3.1b).

3.3.2 Fitting a linear regression model using PGLS

Now let's go ahead and fit the same model to our data—but this time using PGLS instead of contrasts.

To do this, we'll need to load a different R package called *nlme* (Pinheiro et al. 2019; see also Orme et al. 2018).[10],[11]

```
library(nlme)
```

For our next step, we need to take our phylogenetic tree of primates and convert it into a special type of R object called a *correlation structure.*[12]

This is what our GLS model-fitting function will take as input, right alongside our data. The correlation structure will then be used to define the distribution of the residuals from our linear model.

For now, we'll build our `"corStruct"` object using the *ape* function `corBrownian`.

```
spp<-rownames(primate.data)
corBM<-corBrownian(phy=primate.tree,form=~spp)
corBM

## Uninitialized correlation structure of class corBrownian
```

---

[10] Our core phylogenetics package *ape* depends on *nlme*, so we already have it installed. We should still load the package, though: just so that its functions will be more easily available for us to use.

[11] PGLS can also be undertaken using the flexible R package *caper* by Orme et al. (2018).

[12] An object of class `"corStruct"`, to be precise.

Importantly, when we create our correlation structure (corBM), we *must* specify the order of the taxa in our data, which is done using the argument form. Otherwise, it'll be assumed that the order of the rows in our input data frame matches the order of the tip labels of the tree, which can be very dangerous!

Now we're ready to fit our linear model. We'll do this using the *nlme* function gls.

gls works in a very similar way as the function lm that we've already learned but (for our purposes) will take just one additional argument: correlation.[13]

The value of correlation is just our "corStruct" object, corBM, that we generated earlier using corBrownian.

```
pgls.primate<-gls(log(Orbit_area)~log(Skull_length),
    data=primate.data,correlation=corBM)
```

Let's print a summary of our fitted model using (you guessed it) summary.

```
summary(pgls.primate)

## Generalized least squares fit by REML
##   Model: log(Orbit_area) ~ log(Skull_length)
##   Data: primate.data
##        AIC       BIC    logLik
##   -79.58758 -72.15557 42.79379
##
## Correlation Structure: corBrownian
##  Formula: ~spp
##  Parameter estimate(s):
## numeric(0)
##
## Coefficients:
##                      Value Std.Error    t-value
## (Intercept)      -0.2018553 0.3499265 -0.576851
## log(Skull_length) 1.3786743 0.0773418 17.825741
##                    p-value
## (Intercept)         0.5655
## log(Skull_length)   0.0000
##
##  Correlation:
##                    (Intr)
## log(Skull_length) -0.923
##
## Standardized residuals:
##          Min         Q1        Med         Q3
```

---

[13] Things get a bit more complicated if our tree is non-ultrametric; however, we generally recommend working with ultrametric phylogenies except under circumstances in which some of the lineages in the tree are meant to represent extinct taxa. In addition to the "corStruct" object, for a non-ultrametric tree, users should also create an object of class "varFixed" with variances proportional to the total tree height to the end of each tip. Our "varFixed" object is then passed to gls via the argument weights.

```
## -2.5903639 -1.0543078 -0.7395119 -0.2535879
##        Max
##   2.4063079
##
## Residual standard error: 0.3193421
## Degrees of freedom: 90 total; 88 residual
```

As with an object from `lm`, our model summary tells us the significance (*P*-values) of our different model coefficients, as well as other relevant information about model fit (Ives 2019).[14]

3.3.3 Comparing PICs and PGLS

How does this compare to the model we fit to our contrasts earlier?

They are similar at the very least, but are they identical?

To find out, let's use the S3 method `coef` to pull out the estimated coefficient or coefficients of each fitted model and then check to see if they are equal.

We could do this using the logical test `==`; however, in computers, it is usually a good idea to avoid trying to evaluate the equality of real numbers. This is because, for technical reasons,[15] real numbers might differ in the upteenth decimal place even when they are meant to be identical!

Instead, let's compute the absolute value of the difference between the slope coefficients of the two models. If this evaluates to a very small number, then it suggests that[16] their values are the same.

```
coef(pic.primate)

## pic.skull.length
##         1.378674

coef(pgls.primate)

##          (Intercept) log(Skull_length)
##           -0.2018553         1.3786743

abs(coef(pic.primate)[1]-coef(pgls.primate)[2])

## pic.skull.length
##     6.479262e-12
```

Since the absolute value of the difference between our two slope coefficients is *very* small (less than 10^{-11}), we can safely conclude that both methods have returned the same estimated model slopes.

---

[14]Notably, our model summary does not include a value for r^2. In fact, r^2 (as a fraction of variance explained by the model) does not extend well to model fitting with GLS. Some pseudo-r^2 measures have been proposed (Ives 2019), but we feel that a comprehensive discussion of these is beyond the scope of this chapter.

[15]Interested readers can look up *floating point arithmetic* to find out more.

[16]To the numerical precision we have at our disposal.

One difference that you'll have surely noted is that our fitted PGLS model was able to include an intercept term, while our contrasts regression did not.[17] This is simply because in PGLS, we do not transform our data into a new space before analysis, and as such, we're not forced to drop this coefficient from our fitted model.

Not only are the slope coefficients of two fitted models the same, but so are the *F*, *t*, and *P*-values.

The equivalence of these various statistics may not be evident when we merely view the printout to screen—but don't be deceived. This is simply due to the fact that the different print methods of our varying object types display different degrees of numerical precision by default.

Much like the coefficients, we can also compare our *P*-values directly.[18]

```
summary(pic.primate)$coefficients[1,4]

## [1] 5.939575e-31

summary(pgls.primate)$tTable[2,4]

## [1] 5.939575e-31
```

We can see that to at least seven digits of numerical precision, the *P*-values from our PIC regression and the equivalent PGLS regression match precisely.

Collectively, these findings help to bolster our assertion that PIC regression is a special case of linear regression using PGLS (Blomberg et al. 2012).

This special case corresponds to one where the correlation structure of the residual error (as set by corBrownian) is one where the expected correlation between species is directly proportional to their fraction of common ancestry since the root. We'll discuss the ultimate source of this assumption in greater detail in chapter 4.

3.4 Assumptions of PGLS

In the previous section, we used the simplest residual error correlation structure for phylogenetic data, which is called corBrownian. As we noted previously, this structure simply assumes that the correlation between the residual errors of any pair of species in the tree is directly proportional to the height above the root of the common ancestor of that pair.

This expected correlation arises directly from an implicit model[19] that we've assumed regarding how our character traits evolve on the tree. This concept is illustrated in figure 3.2.

The following R code will create that figure—it's not entirely necessary to understand all the code, but it's cool!

---

[17]This is why we compared coef(pic.primate)[1] to coef(pgls.primate)[2]. coef(pgls.primate)[1] contains our PGLS model intercept!

[18]We don't want to get bogged down in the details about how to find where in the object each of these values is stored—let's just say that it involves the use of the handy function str that we learned about in chapter 1.

[19]Called the Brownian motion model, and to be discussed in chapter 4.

```
## set the random number generator seed
set.seed(88)
## simulate a random 5-taxon tree
tree<-pbtree(n=5,scale=10,tip.label=LETTERS[5:1])
## subdivide our plotting area into two panels
par(mfrow=c(2,1))
## plot the tree
plotTree(tree,mar=c(3.1,1.1,4.1,1.1),fsize=1.25,
    ylim=c(0.5,5.4))
## add a horizontal axis
axis(1)
## add edge labels giving the branch lengths
edgelabels(round(tree$edge.length,2),pos=3,
    frame="none",cex=0.9)
mtext("(a)",line=1,adj=0)
## switch to the second panel
plot.new()
## set new plot margins and plot dimensions
par(mar=c(3.1,1.1,4.1,1.1))
plot.window(xlim=c(0,6),ylim=c(0,6))
## add a grid of lines for our correlation matrix
lines(c(0,6,6,0,0),c(0,0,6,6,0))
for(i in 1:5) lines(c(i,i),c(0,6))
for(i in 1:5) lines(c(0,6),c(i,i))
## compute the assumed correlation structure
V<-cov2cor(vcv(tree)[LETTERS[1:5],LETTERS[1:5]])
## print it into the boxes of our grid
for(i in 1:5) text(i+0.5,5.5,LETTERS[i],cex=1.1)
for(i in 1:5) text(0.5,5.5-i,LETTERS[i],cex=1.1)
for(i in 1:5) for(j in 1:5) text(0.5+i,5.5-j,
    round(V[i,j],2),cex=1.1)
mtext("(b)",line=1,adj=0)
```

It's relatively easy to understand how the correlation matrix in figure 3.2b is populated from the phylogeny in figure 3.2a.

Let's take, for instance, the correlation between taxa C and E. To get this, we merely divide the distance from the root of the common ancestor of C and E (4.36) by the total length of the tree (10) and obtain the correlation (0.44).

Likewise, to get the correlation between taxa C and D, we compute the distance from the root to their common ancestor (4.36 + 2.96 = 7.32) and divide it by the total length of the tree (10) to obtain the correlation,[20] which (rounded to two digits) is 0.73.

---

[20] In practice, we actually used the handy base R function cov2cor, which doesn't require that we know the total length of the tree.

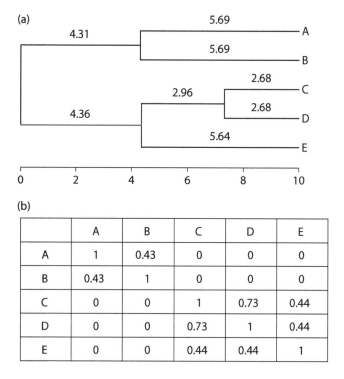

(a)

(b)

	A	B	C	D	E
A	1	0.43	0	0	0
B	0.43	1	0	0	0
C	0	0	1	0.73	0.44
D	0	0	0.73	1	0.44
E	0	0	0.44	0.44	1

Figure 3.2
Brownian motion correlation structure.

Taxa that split at the root of the tree have an expected correlation of 0, and (sensibly) taxa always have an expected correlation with themselves of 1.

3.4.1 Alternative models for the residual error

Now that we understand the form of our model for the correlation of the residual error in PGLS, it's likewise straightforward to envision ways in which the assumptions of this model could be relaxed.

A very simple relaxation of our model might be via the introduction of a single additional parameter, let's say λ, as a multiplier of the off-diagonal elements of the matrix (Pagel 1999a). The neat thing about this model is that it has both OLS (when $\lambda = 0$) and standard PGLS (when $\lambda = 1$) as special cases.

We can figure out which value of λ is best supported by the pattern in our data by estimating it using a procedure called maximum likelihood.[21,22] To apply the λ model in PGLS, we'll use an alternative function to generate the correlation structure of the errors in our linear model. This function is also from the *ape* package and is called `corPagel`.

---

[21] Likelihood will also be explained in a bit more detail in chapter 4.

[22] Technically, by default, `gls` uses *REML*, *restricted* maximum likelihood, for estimation, rather than maximum likelihood. REML uses the likelihood of a transformation of the data, rather than the data themselves. This only becomes important if we want to compare among different models for the correlation structure, in which case we can switch to maximum likelihood by setting `method="ML"`.

```
corLambda<-corPagel(value=1,phy=primate.tree,form=~spp)
corLambda

## Correlation structure of class corPagel representing
## lambda
##      1
```

You may notice that the arguments taken by corPagel are highly similar to those of corBrownian but also include the additional argument value, which we have assigned 1. This is the initial condition for our λ scaling factor that we'll be estimating at the same time as we fit the regression model to our data. The specific starting condition for λ isn't too important, but we should be sure to choose a number that is on the range over which λ is defined.[23] Since λ is always defined between 0 and 1, it's a pretty safe bet to set the initial value of λ to 1.[24]

We can now proceed to fit the same regression model as before to our data but this time with our updated correlation structure. Note again that the particular value we specified for λ is just a starting value. The final value of λ will be estimated jointly with our fitted regression model.

```
pgls.Lambda<-gls(log(Orbit_area)~log(Skull_length),
    data=primate.data,correlation=corLambda)
summary(pgls.Lambda)

## Generalized least squares fit by REML
##   Model: log(Orbit_area) ~ log(Skull_length)
##   Data: primate.data
##          AIC        BIC     logLik
##    -80.88544  -70.97609  44.44272
##
## Correlation Structure: corPagel
##  Formula: ~spp
##  Parameter estimate(s):
##   lambda
## 1.010023
##
## Coefficients:
##                          Value Std.Error    t-value
## (Intercept)         -0.3756847 0.3437745  -1.092823
## log(Skull_length)    1.4204872 0.0749575  18.950578
##                          p-value
## (Intercept)               0.2775
## log(Skull_length)         0.0000
##
##   Correlation:
##                      (Intr)
```

---

[23] For our purposes, this is just the range of values for λ over which it's possible to compute the probability of our data under the model.

[24] We could've also set it to some intermediate value, such as 0.5.

```
## log(Skull_length) -0.911
##
## Standardized residuals:
##        Min          Q1         Med          Q3
## -2.4320899 -1.0518571 -0.7037664 -0.2437600
##        Max
##   2.3518420
##
## Residual standard error: 0.3356994
## Degrees of freedom: 90 total; 88 residual
```

This result shows that the ML estimate of lambda, 1.01, is extremely close to 1 but a tiny bit higher.

That means that under our model, close relatives have correlated residuals and even a bit more so than they would under our original model—and we still conclude that there is an evolutionary correlation between the two traits.

In addition to the λ model, there are a number of other ways in which the correlation structure of the residual error of our model can be made more flexible. The principles that underlie working with the alternative error structures in R are basically the same. We first create an uninitialized object of class `"corStruct"`, and then we optimize the parameters of the error structure jointly with our model. Typically, each error structure has `corBrownian` as a special case, and many also have OLS (for the λ model, λ = 0) as a special case as well. To pick the model that best fits our data, it is a valid exercise to fit alternative models for the error structure to our tree and data set and compare them.[25]

3.5 Phylogenetic ANOVA and ANCOVA

Perhaps the most attractive feature of PGLS when compared to contrasts regression is that it's very straightforward to fit a linear model that includes one or more factors as independent variables (i.e., an ANOVA model) or a combination of continuous and discrete factors (i.e., an ANCOVA).[26]

To see how this works, we can test the hypothesis that diel activity pattern—nocturnal, diurnal, or cathemeral—affects relative eye size. We say relative eye size here because our analysis is going to also control for allometry by including skull size as an additional covariate.

This kind of model is called an ANCOVA model, and we'll be assuming that the correlation structure of the residual error is given by the phylogeny. We'll thus call our model a *phylogenetic generalized ANCOVA*.

Activity pattern is already a column in our data set `primate.data`. We've also already built our correlation structure, `corBM`, which depends only on our tree.[27,28]

---

[25] Although in that case, as we mentioned in a prior footnote, we recommend fitting the model in question with `method="ML"` instead of (`method="REML"`).

[26] Note that this method should not be confused with the related but distinct approach developed by Garland et al. (1993), also called phylogenetic ANCOVA, that uses simulation.

[27] Along with our assumed evolutionary model for the residual error, of course.

[28] We could've also fit an ANCOVA model with an interaction between our discrete factor, diel activity pattern, and the covariate. For our example, this would be written as `log(Orbit_area) ~ log(Skull_length) * Activity_pattern`. For the primate data, the interaction term is non-significant; however, if it had been significant, this could be interpreted as evidence for a difference

```
primate.ancova<-gls(log(Orbit_area)~log(Skull_length)+
    Activity_pattern,data=primate.data,
    correlation=corBM)
anova(primate.ancova)

## Denom. DF: 86
##                         numDF    F-value  p-value
## (Intercept)                 1  2017.1877  <.0001
## log(Skull_length)           1   376.6683  <.0001
## Activity_pattern            2     9.1575  2e-04
```

Our results show us that there is a significant effect of activity pattern on orbit area after controlling for the also significant allometric effect of skull size.

Let's make a plot that helps us to see this pattern.

```
## set the margins of our plot using par
par(mar=c(5.1,5.1,2.1,2.1))
## set the point colors for the different levels
## of our factor
pt.cols<-setNames(c("#87CEEB","#FAC358","black"),
    levels(primate.data$Activity_pattern))
## plot the data
plot(Orbit_area~Skull_length,data=primate.data,pch=21,
    bg=pt.cols[primate.data$Activity_pattern],
    log="xy",bty="n",xlab="skull length (cm)",
    ylab=expression(paste("orbit area (",mm^2,")")),
    cex=1.2,cex.axis=0.7,cex.lab=0.8)
## add a legend
legend("bottomright",names(pt.cols),pch=21,pt.cex=1.2,
    pt.bg=pt.cols,cex=0.8)
## create a common set of x values to plot our
## different lines for each level of the factor
xx<-seq(min(primate.data$Skull_length),
    max(primate.data$Skull_length),length.out=100)
## add lines for each level of the factor
lines(xx,exp(predict(primate.ancova,
    newdata=data.frame(Skull_length=xx,
    Activity_pattern=as.factor(rep("Cathemeral",100))))),
    lwd=2,col=pt.cols["Cathemeral"])
lines(xx,exp(predict(primate.ancova,
    newdata=data.frame(Skull_length=xx,
    Activity_pattern=as.factor(rep("Diurnal",100))))),
    lwd=2,col=pt.cols["Diurnal"])
```

---

in the slope of the relationship between the two continuous variables as a function of the factor: a common use for ANCOVA models.

```
lines(xx,exp(predict(primate.ancova,
    newdata=data.frame(Skull_length=xx,
    Activity_pattern=as.factor(rep("Nocturnal",100))))),
    lwd=2,col=pt.cols["Nocturnal"])
```

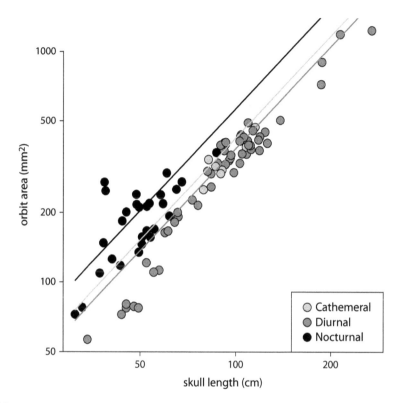

Figure 3.3
Results of our ANCOVA analysis on primate orbit area as a function of skull length and diel activity pattern.

To create figure 3.3, we first set the colors to fill each plotted point as a function of the level of our factor: diel activity pattern. Next, we created our plot and overlaid a legend. Finally, we added lines to show the fitted model for each level of our factor.[29]

Our results show that we can best predict primate orbital area if we account for both their skull length and activity pattern.

We also see from figure 3.3 that nocturnal species have *large* orbital areas relative to their skull size (the black line in figure 3.3), compared to cathemeral or diurnal species. Since we didn't include an interaction term in our final model,[30] we're assuming that the allometric slope

---

[29] We did this using predict with the argument newdata instead of using abline so that we could plot our log-log model in the original space but with the axes transformed to a log-scale. See if you can figure out why!

[30] Because it was nonsignificant, as we mentioned in a prior footnote.

of the orbital–skull relationship[31] is equal across all activity patterns. The eye of the aye-aye doesn't lie!

In summary, PGLS is a more flexible method to test for evolutionary correlations, and it allows for alternative models of the residual error in y given our fitted model, multiple predictor variables, and both continuous and discrete predictors. All assumptions in PGLS are assumptions about the distribution of species' residuals from the linear model.

3.6 Practice problems

3.1 In practice problem 2.2 of the previous chapter, you used Asian barbet data to test whether the variable `Lnalt` varied as a function of `wing`.[32] Again, using these data from Asian barbets (`BarbetTree.nex` and `Barbetdata_mod.csv`), carry out the same analysis using PGLS. Confirm that you get the same results from PGLS and PICs. What happens to your fitted model if you also estimate λ using the `corPagel` function?

3.2 If you multiply all the branches of your phylogenetic tree by 100, will your PGLS analysis change? Why or why not? Can you confirm this using R code?

3.3 Use the data files from chapter 1 to run a phylogenetic ANCOVA for anoles testing for the effect of body size (`"SVL"`) and ecomorphological state (`"ecomorph"`) on forelimb length (`"FLL"`) in anoles, using an *Ornstein–Uhlenbeck* model,[33] as implemented in the *ape* function `corMartins`, as your correlational structure for the residual error of the model. You can use the data files from chapter 1, but you will need to do some work[34] to combine data across files.

---

[31] In other words, the way in which orbital area varies with skull size, which we can think of as an index of body size.

[32] We also tested `Lnalt` ~ `wing`. Let's do *just* `wing` ~ `Lnalt` here.

[33] We'll learn about the Ornstein–Uhlenbeck model in chapter 4.

[34] In R! Please avoid the temptation to use your favorite spreadsheet software instead.

Modeling continuous character evolution on a phylogeny

4

In chapter 2, we learned the method of independent contrasts of Felsenstein (1985), and we said that our contrasts, "once properly normalized," would theoretically be independently and identically distributed.

Later, in chapter 3, we proposed a structure for the residual error of our phylogenetic generalized least squares (PGLS) model in which the correlation between species is assumed to be directly proportional to their shared ancestry (Martins and Hansen 1997).

Many readers probably realize that when we said "properly normalized" and when we asserted this expected correlation between species, we were implicitly assuming a particular model for the evolution of our characters—and that this model is one called *Brownian motion* (sometimes shortened to the abbreviation BM).

In this chapter we intend to:

1. Introduce in slightly greater detail the model of Brownian motion evolution.
2. Show how to fit a Brownian model of trait evolution to data and a phylogeny using R.
3. Discuss the concept of phylogenetic signal of continuously measured traits and show how phylogenetic signal is measured from trait data and a phylogeny.
4. Finally, describe other, closely related models of trait evolution for continuous characters and demonstrate how these models can be fit and compared one to another in R.

4.2 The Brownian motion model

Brownian motion is a stochastic continuous-time random walk model in which changes from one time period to the next are random draws from a normal distribution with a mean of zero and a variance σ^2 (Felsenstein 1973).

Given these conditions, the expected variance under Brownian motion increases linearly through time with an instantaneous rate that's equivalent to the variance of the normal distribution from which the evolutionary changes are drawn (that is, σ^2).

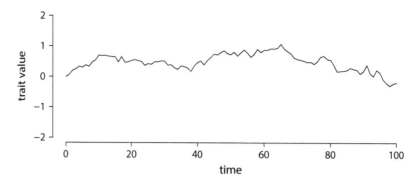

Figure 4.1
A single Brownian motion simulation.

4.2.1 Simulating Brownian motion

Brownian motion is a simple model, and it is very easy to emulate using a computer.

To start off, let's simulate a single instance of Brownian motion evolution for 100 generations of discrete time in which the variance of the diffusion process[1] (σ^2) is equal to 0.01 per iteration of the process (figure 4.1).

In this case, we'll draw our evolutionary changes from a normal distribution; however it's probably worth noting that[2] regardless of the distribution, evolution will proceed by Brownian motion as the width of our time steps decreases toward zero, so long as certain other reasonable assumptions are met.

```
# set values for time steps and sigma squared parameter
t<-0:100
sig2<-0.01
## simulate a set of random changes
x<-rnorm(n=length(t)-1,sd=sqrt(sig2))
## compute their cumulative sum
x<-c(0,cumsum(x))
# create a plot with nice margins
par(mar=c(5.1,4.1,2.1,2.1))
plot(t,x,type="l",ylim=c(-2,2),bty="n",
    xlab="time",ylab="trait value",las=1,
    cex.axis=0.8)
```

You may have noticed that instead of simulating additive Brownian evolution step-by-step, we *began* by simulating all the steps, then we proceeded to compute the state of the process through time by calculating the cumulative sum from time 0 through t for all possible values of t.

This is possible to do because a property of the model is that the distribution of changes under Brownian motion is invariant and does not depend on the state of the chain.

---

[1] A diffusion process is a continuous-time stochastic process, of which Brownian motion is a type.

[2] Due to an important theorem in probability theory called the *central limit theorem*.

We can similarly do a whole bunch of simulations like this at once, using the same conditions.[3]

```
# set number of simulations
nsim<-100
# create matrix of random normal deviates
X<-matrix(rnorm(n=nsim*(length(t)-1),sd=sqrt(sig2)),
    nsim,length(t)-1)
# calculate the cumulative sum of these deviates
# this is now a simulation of Brownian motion
X<-cbind(rep(0,nsim),t(apply(X,1,cumsum)))
# plot the first one
par(mar=c(5.1,4.1,2.1,2.1))
plot(t,X[1,],ylim=c(-2,2),type="l",bty="n",
    xlab="time",ylab="trait value",las=1,
    cex.axis=0.8)
# plot the rest
invisible(apply(X[2:nsim,],1,function(x,t) lines(t,x),
    t=t))
```

Looking across simulations (figure 4.2), we see that some evolve up (that is, toward larger values of the trait x), while others evolve down—but, overall, there is no collective tendency to evolve up or down.

We also see that, in general, the variation among simulations starts off small and increases as time elapses from left to right on our graph.

4.2.2 The rate parameter of Brownian motion, σ^2

To understand how the Brownian process depends on σ^2, the instantaneous rate of the process, let's see what happens when we use a value for the rate that is 1/10 as large as the one we employed for our previous simulation.

The code to do this is as follows.

```
# create matrix of random normal deviates
# but with a smaller sd
X<-matrix(rnorm(n=nsim*(length(t)-1),sd=sqrt(sig2/10)),
    nsim,length(t)-1)
# calculate the cumulative sum of these changes
# this is now a simulation of Brownian motion
X<-cbind(rep(0,nsim),t(apply(X,1,cumsum)))
# plot as above
par(mar=c(5.1,4.1,2.1,2.1))
plot(t,X[1,],ylim=c(-2,2),type="l",bty="n",
```

---

[3] We used generic method, plot, to graph the first simulation and then an apply call of lines to add all the remaining simulations. Note that apply functions like to return a value to the user. Since we're just plotting, there's nothing of interest to return. Wrapping the function call with invisible prevents the function from just printing NULL a bunch of times, which would be kind of annoying.

```
    xlab="time",ylab="trait value",las=1,
    cex.axis=0.8)
invisible(apply(X[2:nsim,],1,function(x,t) lines(t,x),
    t=t))
```

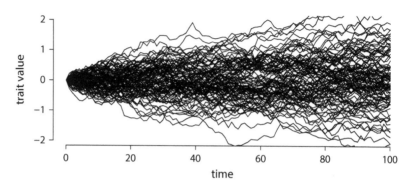

Figure 4.2
One hundred replicate Brownian motion simulations.

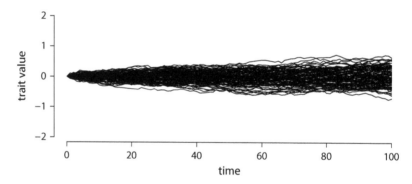

Figure 4.3
One hundred replicate Brownian motion simulations, but for a lower rate of evolution than in figure 4.2.

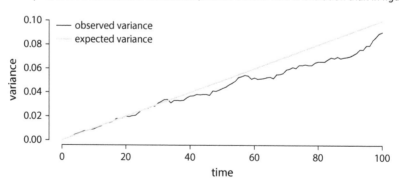

Figure 4.4
Variance among simulations through time under Brownian motion evolution.

Evolution proceeds similarly (figure 4.3), with some replicates of the process evolving toward higher values for the phenotype and others evolving lower. Meanwhile, however, variation among simulations accrues at a much lower rate, just as we probably imagined it would.

In fact, the expected variance under Brownian motion is just σ^2 multiplied by t (Felsenstein 1973).

To see this, we can simply take our last set of 100 replicate simulations and, for each value of t, compute the variance across the entire set of simulations.

We should see that the accumulation of variation *among* simulations more or less follows a straight line in which the slope of the line is approximately equal to the value of σ^2 that we used for the simulations (figure 4.4). Let's check

```
# calculate variance of columns
v<-apply(X,2,var)
# plot the results
par(mar=c(5.1,4.1,2.1,2.1))
plot(t,v,ylim=c(0,0.1),type="l",xlab="time",
    ylab="variance",bty="n",las=1,
    cex.axis=0.8)
lines(t,t*sig2/10,lwd=3,col=rgb(0,0,0,0.1))
legend("topleft",c("observed variance","expected variance"),
    lwd=c(1,3),col=c("black",rgb(0,0,0,0.1)),
    bty="n",cex=0.8)
```

Likewise, the variance at the *end* of the simulation should just be σ^2 (0.001, in our case) multiplied by the total time elapsed (100), or about 0.1.

```
# find variance at the end of the simulations
var(X[,length(t)])
```

```
## [1] 0.09060629
```

Within a certain margin of error,[4] this is indeed what we find.

4.3 Brownian motion on a phylogeny

So far, we've considered only either a single instance of Brownian evolution or numerous, *independent* realizations of the process.

What happens if instead we simulate Brownian motion evolution from the root up the branches of a phylogenetic tree? Let's try it and find out.

To do so, we can use the *phytools* function simBMphylo as follows. The result is shown in figure 4.5.

---

[4]Remember, this result is from a relatively small number of stochastic simulations.

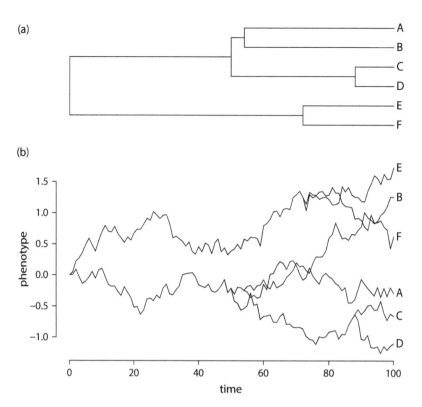

Figure 4.5
Brownian motion on a six-species phylogeny.

```
## load phytools package
library(phytools)
## simulate a tree and Brownian evolution on that
## tree using simBMphylo
object<-simBMphylo(n=6,t=100,sig2=0.01,
    fsize=0.8,cex.axis=0.6,cex.lab=0.8,
    las=1)
```

Here we see that (much like when we simulated replicated, independent Brownian motion), variation between lineages increases through time.

Likewise, however, we also see that since lineages can only begin to differentiate after they've diverged, the amount of variation that tends to accrue between species appears to increase as a function of the amount of evolutionary time since they shared common ancestry.

4.4 Properties of Brownian motion

By now we should be beginning to see many of the properties of Brownian motion emerge.

First, from figures 4.1 and 4.2, we learned that Brownian motion is a stochastic, directionless process. A single realization of Brownian motion might move up or down (figure 4.1), but (averaged across many replicates of the process; e.g., figures 4.2 and 4.3), there is no greater tendency to move up than down.

Second, we observed that Brownian motion is a process in which variation accumulates through time. In fact, we saw that the variance among different, independent simulations of Brownian motion increases exactly linearly with elapsed time, in which the rate of variance accumulation is equal to the instantaneous Brownian rate—a quantity that we've denoted σ^2 (figure 4.4).

Finally, third, we see that when we simulate Brownian evolution on a phylogenetic tree, we find that closely related species (such as C and D in figure 4.5) have more similar phenotypes than do distant taxa (such as A and E).

In fact, if we conducted *many* phylogenetic Brownian simulations on a given phylogeny, instead of just one, we would see that on average, the correlation between related taxa would be exactly proportional to the fraction of shared history they have in common: in other words, the time from the root of the tree to their MRCA,[5] divided by the total tree length.

Remember, this is precisely what we assumed about the correlation structure of the residual error of our PGLS model in chapter 3 (see figure 3.2)!

Let's undertake exactly this simulation.

Now, though, to conduct 1,000 independent simulations of Brownian evolution, we'll use the much faster *phytools* function `fastBM` instead of the `simBMphylo` function that we employed before.

```
## pull the phylogeny out of the object we simulated
## for figure 4.5 using simBMphylo
tree<-object$tree
## simulate 1000 instance of Brownian evolution on that
## tree
X<-fastBM(tree,nsim=1000)
```

Let's graph our results.

In this case, our plot (figure 4.6) will consist of a scatterplot matrix in which each x, y coordinate of each point in each panel represents the *pair* of phenotypic values for a single simulation of the Brownian process in each pair of species corresponding to the panel of our matrix.[6]

If the scatter of points for a hypothetical pair of species i and j is tightly clustered to the 1:1 line, then this indicates that (across many replicates of the evolutionary process) species i and j always tend to evolve similar values for the trait. By contrast, if the scatter of points for species i and j is diffuse and uncorrelated, this indicates that there is no tendency in our simulation for species i and j to evolve similar values of the trait across many instances of Brownian evolution on our tree from figure 4.5.

```
## set the orientation of the axis labels to be
## horizontal
par(las=1)
## create a scatterplot matrix from our simulated
## data using pairs
```

---

[5] *Most recent common ancestor*, for the uninitiated.

[6] The rows and columns of the matrix of scatterplots correspond to different pairs of species in the tree.

```
pairs(t(X)[,tree$tip.label[6:1]],pch=19,
    col=make.transparent("blue",0.05),
    cex.axis=0.9)
```

When we compare this plot to the phylogeny of figure 4.5, it shows us—just as we'd expect—that the trait values of closely related species (such as *C* and *D*) tend to be more highly correlated across simulations than do the trait values of distant taxa.

Furthermore, species whose MRCA is also the global root of the tree have no correlation at all!

4.5 Fitting a Brownian model to data

Just as we can simulate data on a phylogeny under a Brownian motion evolution model, we can likewise fit a Brownian model to our data and tree.

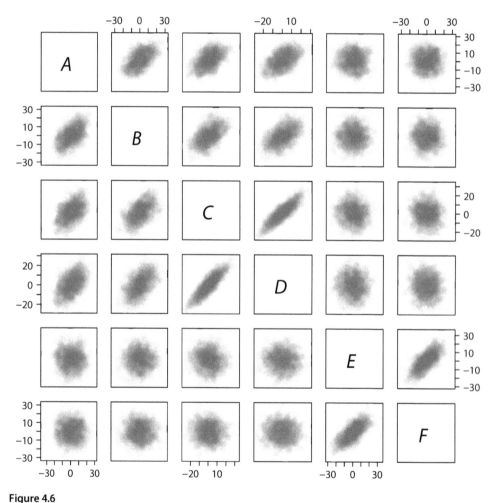

Figure 4.6
The correlation among species under Brownian evolution. These results come from 1,000 simulations of Brownian motion evolution on the phylogenetic tree of figure 4.5a.

The Brownian model has two different parameters that we'll try to estimate.

The first of these is the "instantaneous variance" of the stochastic evolutionary process under Brownian motion, usually referred to as the *evolutionary rate*: σ^2.

The second of these is the initial state of the process. This corresponds to the ancestral condition at the root node of our tree, and here we'll denote this quantity as x_0.

4.5.1 Maximum likelihood estimation

To estimate the two parameters of our model from a set of observations for species and our tree, we need a criterion for choosing values of σ^2 and x_0 that *best fit* our data.

One strategy is to select the values that maximize the probability of obtaining the data that we've observed.

This strategy is called estimation by *maximum likelihood* and was initially developed by Ronald Fisher near the start of the twentieth century (Fisher 1922; see historical review in Aldrich 1997). Maximum likelihood estimators have since been shown to possess many desirable statistical properties, such as consistency, efficiency, and asymptotic unbiasedness (Edwards 1992).[7]

4.5.2 Fitting a Brownian model using likelihood

Fitting a Brownian model to data in R is straightforward. To see how this works, let's begin by reading some data and a tree from file.

For this analysis, we'll use data from a study by Gibson and Eyre-Walker (2019) on the rate at which bacteria accumulate mutations (i.e., the mutation rate) and on the overall sizes of their genomes.

Gibson and Eyre-Walker (2019) obtained data for both genome size and mutation rate across thirty-four different bacterial species, which we'll use here.[8]

The files to use for this analysis, bac_rates.phy and bac_rates.csv, can both be obtained from the book website.[9]

The first thing we can do is read in our data and our phylogeny from file. The data are contained in a CSV file, so we can read them in using the function read.csv that we've seen in prior chapters. As we've done for some prior examples, we'll use the function head to print out the first part of our data frame. This helps us to ensure that our data file has been read into R correctly.

```
## read bacterial data from file
bacteria.data<-read.csv("bac_rates.csv", row.names=1)
head(bacteria.data,3)

##                          Accumulation_Rate
## Acinetobacter_baumannii          1.99e-06
## Bordetella_pertussis             2.24e-07
```

---

[7] Some maximum likelihood estimators are biased, but *asymptotically unbiased*. This means that bias tends to decrease to zero as more and more data points are obtained.

[8] We've constructed an ultrametric tree from their phylogeny and rescaled the tree to an (arbitrary) total length of 1. As such, our results might differ slightly from those of the original paper.

[9] http://www.phytools.org/Rbook/.

```
## Buchnera_aphidicola                              1.10e-07
##                             Genome_Size_Mb
## Acinetobacter_baumannii         4.0369921
## Bordetella_pertussis            4.1151522
## Buchnera_aphidicola             0.5915785
##                             GC_Content_Percent
## Acinetobacter_baumannii            39.02430
## Bordetella_pertussis               67.70278
## Buchnera_aphidicola                25.18965
##                             Lab_Doubline_Time_Hours
## Acinetobacter_baumannii                           NA
## Bordetella_pertussis                             3.8
## Buchnera_aphidicola                               NA
##                              piN.piS
## Acinetobacter_baumannii  0.04851772
## Bordetella_pertussis     0.46038453
## Buchnera_aphidicola      0.05394829
```

Our phylogeny is stored as a simple Newick string, so we can read that in using the *ape* function[10] read.tree.

```
bacteria.tree<-read.tree("bac_rates.phy")
print(bacteria.tree,printlen=2)

##
## Phylogenetic tree with 34 tips and 33 internal nodes.
##
## Tip labels:
##   Mycoplasma_gallisepticum, Mycobacterium_tuberculosis, ...
##
## Rooted; includes branch lengths.
```

To make sure that our tree is read in properly, let's plot it using the *phytools* tree plotter, plotTree (figure 4.7).

```
## graph phylogeny using plotTree
plotTree(bacteria.tree,ftype="i",fsize=0.5,
    lwd=1,mar=c(2.1,2.1,0.1,1.1))
## add a horizontal axis to our plot
axis(1,at=seq(0,1,length.out=5),cex.axis=0.8)
```

Great! So far, so good.

Next, we can load the R package *geiger* that we've used in previous chapters. *geiger* contains the function fitContinuous that we'll be using to fit our Brownian model to these data.

---

[10]We loaded *phytools* earlier, and since *phytools* depends on *ape*, that means *ape* should already be loaded.

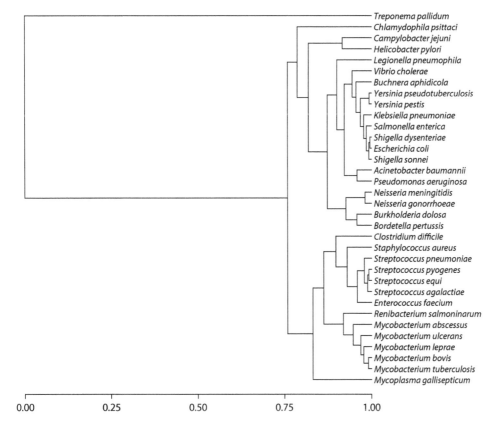

Figure 4.7
Phylogeny of thirty-four bacteria species from Gibson and Eyre-Walker (2019).

```
library(geiger)
```

geiger also contains the handy helper function name.check that we learned about in chapter 1, so let's use name.check to make sure that the labels in our input data and on our phylogeny match.

```
name.check(bacteria.tree,bacteria.data)
```

```
## [1] "OK"
```

This tells us that the names in our phylogeny and data set match exactly, so we're good to go!

Our character data have the form of a data frame; however, fitContinuous takes a single trait vector as input. Let's proceed, then, to extract a single trait to analyze, genome size, from our data frame.

```
genome_size<-bacteria.data[,"Genome_Size_Mb"]
genome_size
```

```
##  [1]  4.0369921  4.1151522  0.5915785  6.4090900
##  [5]  1.6767532  1.1698112  4.2182558  3.0148471
##  [9]  5.0945238  1.6251464  5.6341215  3.4300281
## [13]  5.0295090  4.3600613  3.2681350  4.4043280
## [17]  5.8057600  0.9699609  2.2106470  2.1890708
## [21]  6.6193001  3.1552500  4.8180117  4.5205550
## [25]  5.0991850  2.8536100  2.0675049  2.1404940
## [29]  2.1154909  1.8365169  1.1386050  4.1043307
## [33]  4.7494239  4.7837531
```

One thing you're likely to notice immediately is that even though our data frame had row names corresponding to the species names of our tree, our vector doesn't.

This won't do, so we'll assign the row names of our data frame to our trait vector using the base R function names,[11] as follows.

```
names(genome_size)<-rownames(bacteria.data)
head(genome_size)
```

```
## Acinetobacter_baumannii      Bordetella_pertussis
##               4.0369921                 4.1151522
##      Buchnera_aphidicola      Burkholderia_dolosa
##               0.5915785                 6.4090900
##     Campylobacter_jejuni  Chlamydophila_psittaci
##               1.6767532                 1.1698112
```

Now we're ready to fit a Brownian model to our data.

To do this, we will be using the *geiger* function fitContinuous. As we'll see further along in the chapter, fitContinuous, as its name suggests, fits a variety of different continuous character models to data and a phylogeny.

The default model is Brownian motion, so we can start with that.

```
## fit Brownian motion model using fitContinuous
fitBM_gs<-fitContinuous(bacteria.tree,genome_size)
fitBM_gs
```

```
## GEIGER-fitted comparative model of continuous data
##  fitted 'BM' model parameters:
##  sigsq = 24.990480
##  z0 = 1.984134
##
##  model summary:
##  log-likelihood = -59.279289
##  AIC = 122.558578
##  AICc = 122.945675
```

----

[11]We could also do this in one step using the very useful function setNames that we first learned in chapter 2. This operation would be as follows: genome_size <- setNames(bacteria.data$Genome_Size_Mb, rownames(bacteria.data)).

```
##   free parameters = 2
##
## Convergence diagnostics:
##   optimization iterations = 100
##   failed iterations = 0
##   number of iterations with same best fit = 100
##   frequency of best fit = 1.00
##
##   object summary:
##   'lik' -- likelihood function
##   'bnd' -- bounds for likelihood search
##   'res' -- optimization iteration summary
##   'opt' -- maximum likelihood parameter estimates
```

From the summary of our fitted model, we see that the maximum likelihood estimate (MLE) of σ^2 is about 25.0 and that the MLE of the root state x_0 (here given as z0) is around 1.98.

How do we interpret these values? Well, since our genome sizes have been measured in Megabases[12] (Mb), a σ^2 of 25 indicates that under a Brownian process, we'd expect a large number of independently evolving lineages to accumulate the variance among each other of 25 Mb^2 after one unit of time.[13]

Our MLE of x_0 of 1.98 means that the most likely state at the root of the tree, under our model, is a genome size of 1.98 Mb.

The printout also gives us some information about the model fit to our data: the model likelihood and the Akaike information criterion (AIC and AICc[14]) values.

These quantities are not particularly interpretable in absolute terms—especially since the branch lengths of our tree are not in any especially meaningful units. They will become much more important when we progress to comparing alternative models of evolution for our data later on in this chapter.

In addition to the model parameters and fit, fitContinuous also reports some information about the optimization process.

To find the MLEs, fitContinuous uses numerical optimization, and in this case, it tells us how many independent optimization iterations were used (optimization iterations) and how frequently optimization converged on the same, best solution.

Brownian motion is such a simple model that most often this latter quantity will be 1.00, indicating that the optimizer found the same optimal solution in 100 percent of iterations.

For more complex models, the fraction will often be lower than 100 percent.[15]

Finally, let's do the same analysis—but this time with the mutation accumulation rate. We can start by pulling mutation accumulation rate from our data frame into a named vector. This time, we'll do this in one step using setNames.

---

[12] A Megabase is a million bases—not a stereo system with really good sound amplification, as we thought.

[13] The Brownian rate is like a variance, so its units are the units of the original trait, squared!

[14] AICc is just a small-sample corrected version of AIC.

[15] This should not concern us especially until the *number* of iterations that converge on the same best solution (that is, *optimization iterations* × *frequency of best fit*) falls to a very low value. In that case, we might have good cause to really worry that our optimizer failed to find the true MLEs of our model parameters!

```
## pull our mutation accumulation rate as a named vector
mutation<-setNames(bacteria.data[,"Accumulation_Rate"],
    rownames(bacteria.data))
head(mutation)

##  Acinetobacter_baumannii        Bordetella_pertussis
##                 1.99e-06                    2.24e-07
##        Buchnera_aphidicola        Burkholderia_dolosa
##                 1.10e-07                    3.28e-07
##       Campylobacter_jejuni      Chlamydophila_psittaci
##                 3.23e-05                    1.74e-05
```

If we plot the distribution of mutation accumulation rate across our different species, we'll see that it is highly left-skewed[16] (figure 4.8a). Although we don't necessarily expect a precisely normal distribution of trait values under Brownian evolution,[17] the *very* highly skewed distribution that we see for this trait is nonetheless a strong signal that Brownian evolution is very likely a bad fit to this trait on its original scale. Let's transform mutation accumulation rate to a log-scale and plot that too.

```
## set up for side-by-side plots
par(mfrow=c(1,2),mar=c(6.1,4.1,2.1,1.1))
## histogram of mutation accumulation rates on original scale
hist(mutation,main="",las=2,xlab="",
    cex.axis=0.7,cex.lab=0.9,
    breaks=seq(min(mutation),max(mutation),
    length.out=12))
mtext("(a)",adj=0,line=1)
mtext("rate",side=1,line=4,cex=0.9)
## histogram of mutation accumulation rates on log scale
ln_mutation<-log(mutation)
hist(ln_mutation,main="",las=2,xlab="",
    cex.axis=0.7,cex.lab=0.9,
    breaks=seq(min(ln_mutation),max(ln_mutation),
    length.out=12))
mtext("(b)",adj=0,line=1)
mtext("ln(rate)",side=1,line=4,cex=0.9)
```

It seems clear from figure 4.8 that it's going to be better to work with mutation accumulation rate on a log-scale rather than on its original scale.

Let's take our log-transformed variable and fit the Brownian model again using fitContinuous.

---

[16] Meaning the distribution has a long right tail.

[17] Brownian motion is a Gaussian process—meaning that the outcome is described by a multivariate Gaussian distribution—but it doesn't necessarily produce a normal distribution of trait values at the tips of the tree. This is largely a function of the structure of our phylogeny. For instance, if our tree consists of two deeply divergent clades, each consisting of a large number of closely related species, then the most likely distribution of trait values under Brownian motion is bimodal—one mode for each clade.

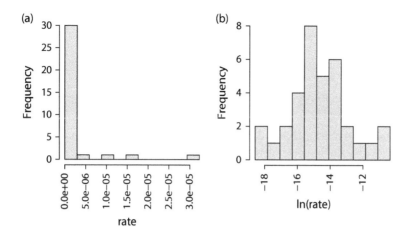

Figure 4.8
Mutation accumulation rate across thirty-four species of bacteria. (a) On the original scale. (b) On a natural logarithm scale.

```
## fit Brownian motion model to log(mutation accumulation)
fitBM_ar<-fitContinuous(bacteria.tree,ln_mutation)
fitBM_ar

## GEIGER-fitted comparative model of continuous data
##  fitted 'BM' model parameters:
##  sigsq = 65.935009
##  z0 = -13.720388
##
##  model summary:
##  log-likelihood = -75.772258
##  AIC = 155.544515
##  AICc = 155.931612
##  free parameters = 2
##
## Convergence diagnostics:
##  optimization iterations = 100
##  failed iterations = 0
##  number of iterations with same best fit = 100
##  frequency of best fit = 1.00
##
##  object summary:
##  'lik' -- likelihood function
##  'bnd' -- bounds for likelihood search
##  'res' -- optimization iteration summary
##  'opt' -- maximum likelihood parameter estimates
```

Our model parameter estimates, model fit, and convergence diagnostics have much the same interpretation as for our previous analysis of the rate of evolution of genome size.

You may be tempted to compare the rate parameter estimate for mutation accumulation rate with our estimate of σ^2 for genome size—but, keep in mind, the former is in units of $\log(\text{rate})^2$, while the latter is in Mb^2, so this comparison is probably not particularly meaningful! We'll return to the topic of rate comparisons in chapter 5.

4.6 Phylogenetic signal

Something that we haven't mentioned so far in this book is the popular concept of *phylogenetic signal*, which we'd define as the tendency for related species to resemble one another more than expected by chance (Revell et al. 2008).[18]

Phylogenetic signal, however, is very closely related to Brownian motion because the most popular ways of *measuring* signal do so with reference to this model. That is, they ask whether species tend to resemble each other less or more than one would expect based on a Brownian model of evolutionary change through time.

Here, we'll examine two different methods for measuring phylogenetic signal of quantitative characters: Blomberg et al.'s (2003) K and Pagel's (1999a) λ.

4.6.1 Blomberg et al.'s K statistic

Blomberg et al.'s (2003) K is best summarized as a normalized ratio comparing the variance among clades on the tree to the variance within clades.

If the variance *among* clades is high (compared to the variance within clades), then phylogenetic signal is said to be *high*. Conversely, if the variance *within* clades is high (compared to the variance among clades), then phylogenetic signal will be low.

This ratio is then normalized by dividing it by its expected value under a Brownian evolutionary process. As such, Blomberg et al.'s (2003) K has an expected value of 1.0 under evolution by Brownian motion.

We can compute Blomberg et al.'s (2003) K using the *phytools* function `phylosig` as follows.

```
phylosig(bacteria.tree, genome_size)

##
## Phylogenetic signal K : 0.349525
```

In this case, we see that our measure of phylogenetic signal, K, for genome size has a value of around 0.35, which is *lower* than we'd expect under evolution by Brownian motion.

4.6.2 Testing hypotheses about Blomberg et al.'s K

What does this mean though?

A quite popular endeavor is to test whether the amount of phylogenetic signal in our data (by whatever measure) exceeds the quantity of signal expected by random chance.

One sensible way to do that is by simply randomizing the data across the tips of the tree a large number[19] of times and then repeatedly recalculating phylogenetic signal by the same measure for each randomized data set.

---

[18]Phylogenetic signal has many definitions—some that relate more directly to the quantitative ways in which signal is measured in practice. We prefer a simpler definition that applies to most if not all measures of signal.

[19]One thousand or 10,000, at least. Fortunately, the calculations are quite rapid.

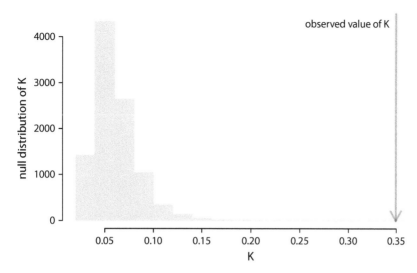

Figure 4.9
Null distribution of *K* for a hypothesis test of phylogenetic signal in bacterial genome size. The observed value of *K* is indicated by the arrow.

The fraction of randomizations with equal or *higher* values of phylogenetic signal than our observed value is our *P*-value for a null hypothesis test of no signal. Let's try this, once again using the `phylosig` function.

```
## test for significant phylogenetic signal using
## Blomberg's K
K_gs<-phylosig(bacteria.tree,genome_size,
    test=TRUE,nsim=10000)
K_gs
```

```
##
## Phylogenetic signal K : 0.349525
## P-value (based on 10000 randomizations) : 1e-04
```

phytools makes it quite easy for us to plot this result—showing the null distribution of *K* that we generated using randomization, along with our observed value of *K* from the original data. Let's see what that looks like.

```
## set plot margins and font size
par(cex=0.8,mar=c(5.1,4.1,2.1,2.1))
## plot null-distribution and observed value of K
plot(K_gs,las=1,cex.axis=0.9)
```

This result, and our plot of figure 4.9, tells us that despite being considerably less than 1.0, our observed value of phylogenetic signal, *K*, is still larger than the value of *K* we'd expect to find if our data for genome size were random with respect to the phylogeny.

Now let's repeat exactly the same exercise, but this time using mutation accumulation rate.

Continuous character models

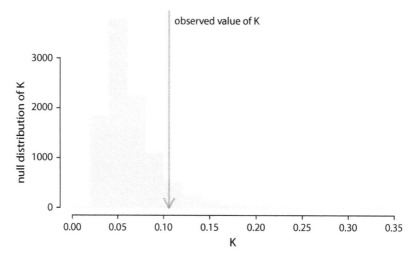

Figure 4.10
Null distribution of *K* for a hypothesis test of phylogenetic signal in bacterial mutation accumulation rate. The observed value of *K* is indicated by the arrow.

```
## test for phylogenetic signal in mutation accumulation
## rate
K_ar<-phylosig(bacteria.tree,ln_mutation,
    test=TRUE,nsim=10000)
K_ar

##
## Phylogenetic signal K : 0.105888
## P-value (based on 10000 randomizations) : 0.0831

## plot the results
par(cex=0.8,mar=c(5.1,4.1,2.1,2.1))
plot(K_ar,las=1,cex.axis=0.9)
```

Our phylogenetic signal is again very low, and this time we *can't* reject the null hypothesis. This means that our observed value of *K* for mutation accumulation rate (on a log-scale) is entirely consistent with what we'd expected to obtain by chance if species mutation accumulation rates were randomly arrayed on the phylogeny (figure 4.10).

Another quite reasonable question is whether our measured values of phylogenetic signal—rather than being greater than expected by chance—are nonetheless *less* than expected under a sensible null model for evolution, such Brownian evolution.

There's no automated routine in R to run this analysis, so to test this null hypothesis, we'll have to generate a null distribution of values for *K* under our null hypothesis: Brownian motion.

Luckily, in R, this is quite easy to do via simulation,[20] so let's give it a try.

---

[20]We'll use the *phytools* function fastBM, just as we did earlier, for our Brownian motion simulation.

```
## simulate 10000 datasets
nullX<-fastBM(bacteria.tree,nsim=10000)
## for each, carry out a test for phylogenetic signal
## and accumulate these into a vector using sapply
nullK<-apply(nullX,2,phylosig,tree=bacteria.tree)
## calculate P-values
Pval_gs<-mean(nullK<=K_gs$K)
Pval_gs
```

```
## [1] 0.2172
```

```
Pval_ar<-mean(nullK<=K_ar$K)
Pval_ar
```

```
## [1] 6e-04
```

In the preceding code chunk, we first generated 10,000 data sets under a Brownian model by simulating evolution on our mammal tree using fastBM. We then computed a value of K for each simulated data vector by iterating across the columns of our simulated data using apply.[21]

Finally, we compute a *P*-value by counting the proportion of times our simulated values of K were *smaller*[22] than our observed values.[23]

What we find is that only the phylogenetic signal for mutation accumulation rate is significantly lower than we'd expect under Brownian evolution. For genome size, we often obtain signal values that are as small as the observed data, even when data are simulated under Brownian motion.

Just as we did using our plot method earlier, we can also visualize these null distributions.

Unfortunately, there's no nifty plot method to do this for us. We've got to do it manually. The results are shown in figure 4.11.

```
## set up for side-by-side plots
par(mfrow=c(1,2))
## plot for Genome size
## null distribution
hist(c(nullK,K_gs$K),breaks=30,col="lightgray",
    border="lightgray",main="",xlab="K",las=1,
    cex.axis=0.7,cex.lab=0.9,ylim=c(0,4000))
```

---

[21] The way to interpret the call apply(nullX,2,phylosig,tree=bacteria.tree) is as follows: apply to the matrix nullX, in the second dimension (that is, the columns), the function phylosig, with the additional argument tree=bacteria.tree. Get it?

[22] We only counted the fraction of times they were *smaller* because our observed K values are both less than 1. If they were above 1, we could count the number of simulated data sets with larger values than our observed value. For a two-tailed test, we would need to multiply this count by 2.

[23] The reason we can compute the mean of a logical vector (that is, the vector of TRUE and FALSE that results from applying a logical test to a series of values) to get a proportion is because, for mathematical operations, R treats logical values as 1s (for TRUEs) and 0s (for FALSEs).

```
## actual value as an arrow
arrows(x0=K_gs$K,y0=par()$usr[4],y1=0,length=0.12,
    col=make.transparent("blue",0.5),lwd=2)
text(K_gs$K,0.96*par()$usr[4],
    paste("observed value of K (P = ",
    round(Pval_gs,4),")",sep=""),
    pos=4,cex=0.8)
mtext("(a)",line=1,adj=0)
## plot for mutation accumulation rate
## null distribution
hist(c(nullK,K_ar$K),breaks=30,col="lightgray",
    border="lightgray",main="",xlab="K",las=1,
    cex.axis=0.7,cex.lab=0.9,ylim=c(0,4000))
## actual value as an arrow
arrows(x0=K_ar$K,y0=par()$usr[4],y1=0,length=0.12,
    col=make.transparent("blue",0.5),lwd=2)
text(K_ar$K,0.96*par()$usr[4],
    paste("observed value of K (P = ",
    round(Pval_ar,4),")",sep=""),
    pos=4,cex=0.8)
mtext("(b)",line=1,adj=0)
```

4.6.3 Pagel's λ

In addition to Blomberg et al.'s K, another popular measure of phylogenetic signal is called Pagel's λ.

λ is a scaling coefficient for the off-diagonal elements in the expected correlations among species that we learned about in chapter 3 (Pagel 1999a).

Thus, values of λ < 1 correspond to *less* phylogenetic signal than expected under a Brownian motion model. Unlike K, however, λ is not generally well defined outside of the range of (0,1). As such, λ is more appropriate for detecting phylogenetic signal that is lower than expected under Brownian motion than the converse.

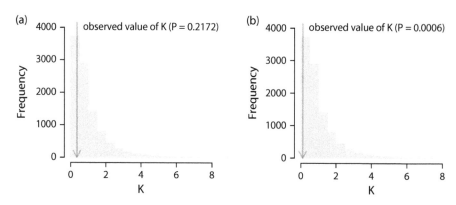

Figure 4.11
Null distributions for K under the null hypothesis of Brownian motion. (a) Genome size. (b) Mutation accumulation rate.

Just as we did for Blomberg et al.'s (2003) *K*, we can calculate Pagel's (1999a) λ using the function `phylosig` from the *phytools* package. Let's do precisely that for the same two variables, genome size and accumulation rate.

```
## compute phylogenetic signal, lambda, for genome size
## and mutation accumulation rate
phylosig(bacteria.tree,genome_size,method="lambda")
```

```
##
## Phylogenetic signal lambda : 1.0017
## logL(lambda) : -59.1365
```

```
phylosig(bacteria.tree,ln_mutation,method="lambda")
```

```
##
## Phylogenetic signal lambda : 0.923697
## logL(lambda) : -64.5216
```

We see immediately that both traits have lambda estimates that are close to 1; however, λ is higher for genome size than for mutation accumulation rate.

In some ways, this is similar to our findings for *K*, in which we showed that genome size had phylogenetic signal similar to the expectation under Brownian motion, while mutation accumulation rate had less.

It's important to note, though, that we *do not* expect a one-to-one correspondence between *K* and λ. To the contrary, we suggest that the two metrics actually measure different aspects of phylogenetic signal and thus we should probably expect them to be different, not the same (also see Boettiger et al. 2012 for additional commentary on λ estimation).

The λ method also allows for a hypothesis test of a null that λ = 0. Since λ is estimated using likelihood, this can be done most easily using a likelihood ratio test.[24]

Here is what that looks like, also plotting the likelihood surface for each trait (figure 4.12).

```
## test for significant phylogenetic signal, lambda,
## in each of our two traits
lambda_gs<-phylosig(bacteria.tree,genome_size,
    method="lambda",test=TRUE)
lambda_gs
```

```
##
## Phylogenetic signal lambda : 1.0017
## logL(lambda) : -59.1365
## LR(lambda=0) : 11.3482
## P-value (based on LR test) : 0.00075521
```

---

[24]The theory of likelihoods tells us that two times the difference in log-likelihood between a simple model and a more complex one that has the simple model as a special case should have a χ^2 distribution with degrees of freedom equal to the difference in the number of parameters estimated in the two models, under the null hypothesis that the simpler model is correct. This theory is used frequently by scientists and statisticians to test hypotheses about models that are fit to data using likelihood.

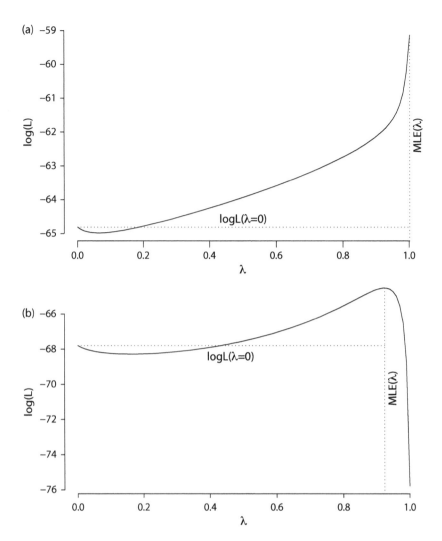

Figure 4.12
Likelihood surfaces for λ for each trait.

```
lambda_ar<-phylosig(bacteria.tree, ln_mutation,
  method="lambda",test=TRUE)
lambda_ar

##
## Phylogenetic signal lambda : 0.923697
## logL(lambda) : -64.5216
## LR(lambda=0) : 6.55541
## P-value (based on LR test) : 0.0104566

## plot the likelihood surfaces
## first set plotting parameters, including subdividing
```

```
## our plot area into 1 column and two rows
par(mfrow=c(2,1),mar=c(5.1,4.1,2.1,2.1),
    cex=0.8)
## plot the likelihood surfaces of lambda for each of our
## two traits
plot(lambda_gs,las=1,cex.axis=0.9,bty="n",
    xlim=c(0,1.1))
mtext("(a)",line=1,adj=0)
plot(lambda_ar,las=1,cex.axis=0.9,bty="n",
    xlim=c(0,1.1))
mtext("(b)",line=1,adj=0)
```

This tells us that we can reject a null hypothesis of $\lambda = 0$ for both phenotypic traits in our data set.

Just as with Blomberg et al.'s K, it's possible to test a null hypothesis of $\lambda = 1$.

In this case, however, we will take advantage of the fact that, in the case of method="lambda", phylosig exports the likelihood function that it used for optimization. As such, we can just compute the likelihood for $\lambda = 1$ and then compare it to our MLE of λ using a likelihood ratio test with 1 degree of freedom.

Let's see this for genome size step-by-step.

First, our likelihood ratio test statistic.

```
LR_gs<--2*(lambda_gs$lik(1)-
    lambda_gs$logL)
LR_gs
```

```
## [1] 0.2855661
```

Now, to compute a P-value for this statistic, we need compare it to a χ^2 distribution with 1 degree of freedom—accounting for the one additional parameter estimated in our λ model. This can be done with the base R function pchisq as follows.

```
Pval_lambda_gs<-pchisq(LR_gs,df=1,
    lower.tail=FALSE)
Pval_lambda_gs
```

```
## [1] 0.593076
```

This tells is that we cannot reject a null hypothesis of $\lambda = 1$ for genome size. Now, let's repeat the same thing for mutation accumulation rate.

```
LR_ar<--2*(lambda_ar$lik(1)-
    lambda_ar$logL)
Pval_lambda_ar<-pchisq(LR_ar,df=1,
    lower.tail=FALSE)
Pval_lambda_ar
```

```
## [1] 2.100025e-06
```

pchisq (for lower.tail=FALSE) gives us the probability of observing an equally or more extreme value of our likelihood ratio than the one we calculated for each trait, assuming that the likelihood ratio is χ^2-distributed under the null!

Interestingly, and consistent with what we saw with Blomberg's *K*, we're only able to reject our null hypothesis of $\lambda = 1$ for mutation accumulation rate.

4.7 Other models of continuous character evolution on phylogenies

In addition to Brownian motion, there exist a number of other popular models for continuous character evolution on phylogenies.

For now, we'll focus on just two of these: the "early burst" (a.k.a. EB) model of Blomberg et al. (2003) and the Ornstein–Uhlenbeck (OU) model of Hansen (1997; see also Butler and King 2004).

In chapter 5, we'll consider other more complex scenarios, such as trait evolution models in which the rate changes in different places in our phylogeny.

4.7.1 The early burst (EB) model

The EB model (Blomberg et al. 2003) is quite straightforward to understand.

Under this model, the rate of evolution, σ^2, starts with some initial value at the root of the tree and then declines monotonically through time according to an exponential decay function.

Figure 4.13 shows the rate of evolution through time under conditions of an initial rate of evolution (σ_0^2) of 1.0 and an exponential decay parameter (a) of -0.04.

```
## set parameters of the EB process
sig2.0<-1.0
a<-0.04
t<-100
## compute sigma^2 as a function of time under this
## process
sig2.t<-sig2.0*exp(-0.04*0:t)
## graph sigma^2 through time
par(mar=c(5.1,4.1,2.1,2.1))
plot(0:t,sig2.t,type="l",xlab="time",
    ylab=expression(sigma^2),bty="l",
    las=1,cex.axis=0.9)
```

The effect of this change in rate through time can be quite profound.

In particular, a declining rate of evolution through time will tend to result in large differences between clades and relatively small differences within them.

Just as we did for Brownian evolution earlier in the chapter, we can also graph evolutionary change through time on a phylogeny under EB using simBMphylo as follows (figure 4.14).[25]

---

[25] Here, we send only the first 100 elements of sig2.t to the function because 0:100 produces 101 values, but our tree only contains 100 units of time from the root to any tip!

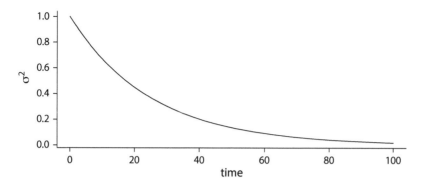

Figure 4.13
The rate of evolution through time under an EB model with $a = -0.04$.

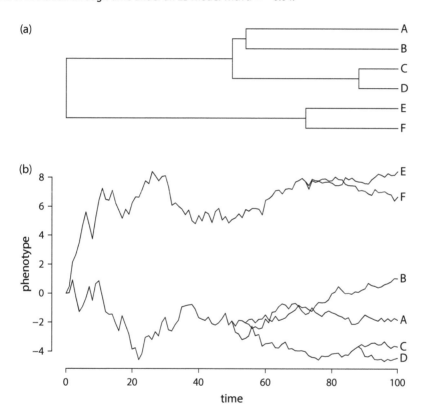

Figure 4.14
EB evolution on a simulated phylogeny.

```
## visualize early-burst evolution using simBMphylo
object<-simBMphylo(6,100,sig2=sig2.t[1:100],
    fsize=0.8,cex.axis=0.6,cex.lab=0.8,
    las=1)
```

The EB model is one in which differences between lineages tend to accumulate rapidly near the beginning of a diversification. As such, it has often been linked to the concept of *adaptive radiation*.

4.7.2 The Ornstein–Uhlenbeck (OU) model

The final model we'll consider for this chapter is the Ornstein–Uhlenbeck model of Hansen (1997).

This model is a relatively simple extension of Brownian motion, with an additional parameter (α) that describes the tendency to return toward a particular central value (θ). Because this model involves evolutionary change toward a particular value, the Ornstein–Uhlenbeck model is most often interpreted as a model for adaptive evolution in which α corresponds to the strength of natural selection and θ to the position of the optimum.

We'll learn more about this model later in the book.

4.8 Fitting and comparative alternative continuous character models

All three of the models that we've seen so far—Brownian motion, early burst, and Ornstein–Uhlenbeck—can be fit and compared to each other using likelihood.

Let's go ahead and do this.

Remember, we already fit a Brownian evolution model to our genome size and accumulation data. We can now proceed to do the same with the EB and OU models.

We'll do this using the same *geiger* function fitContinuous that we tried earlier, but in which we now specify a value for the argument model, starting with model="EB".

```
## fit EB model to genome size
fitEB_gs<-fitContinuous(bacteria.tree,genome_size,
    model="EB")

## Warning in fitContinuous(bacteria.tree, genome_size,
## model = "EB"):
## Parameter estimates appear at bounds:
##  a
```

The first thing that we probably notice is the warning message In fitContinuous (bacteria.tree, genome_size, model = "EB") : Parameter estimates appear at bounds: a.

Let's examine the fitted model to see what this is about:

```
fitEB_gs

## GEIGER-fitted comparative model of continuous data
##  fitted 'EB' model parameters:
##  a = -0.000001
##  sigsq = 24.990508
##  z0 = 1.984134
##
```

```
##   model summary:
##   log-likelihood = -59.279290
##   AIC = 124.558580
##   AICc = 125.358580
##   free parameters = 3
##
## Convergence diagnostics:
##   optimization iterations = 100
##   failed iterations = 0
##   number of iterations with same best fit = 4
##   frequency of best fit = 0.04
##
##   object summary:
##   'lik' -- likelihood function
##   'bnd' -- bounds for likelihood search
##   'res' -- optimization iteration summary
##   'opt' -- maximum likelihood parameter estimates
```

Our printed model summary tells us that our ML estimated value of the decay parameter, a, is almost exactly zero.

When $a = 0.0$, the EB model reduces to a constant rate Brownian motion. This is not an error! It merely indicates that when we impose an EB model on our data, the best-fitting EB model is Brownian evolution!

Next, we can fit an OU model using model="OU".

```
## fit OU model to genome size
fitOU_gs<-fitContinuous(bacteria.tree,genome_size,
    model="OU")

## Warning in fitContinuous(bacteria.tree, genome_size,
## model = "OU"):
## Parameter estimates appear at bounds:
##   alpha
```

We see a highly similar error message, but this time for the parameter α (alpha in the model object) for the OU model. Once again, let's look at our fitted OU model to see what's going on.

```
fitOU_gs

## GEIGER-fitted comparative model of continuous data
##   fitted 'OU' model parameters:
##   alpha = 2.718282
##   sigsq = 30.119232
##   z0 = 2.331269
##
```

```
##  model summary:
##  log-likelihood = -57.129984
##  AIC = 120.259969
##  AICc = 121.059969
##  free parameters = 3
##
## Convergence diagnostics:
##  optimization iterations = 100
##  failed iterations = 0
##  number of iterations with same best fit = 23
##  frequency of best fit = 0.23
##
##  object summary:
##  'lik' -- likelihood function
##  'bnd' -- bounds for likelihood search
##  'res' -- optimization iteration summary
##  'opt' -- maximum likelihood parameter estimates
```

In this case, our best-fitting OU model *is not* a Brownian model (which would correspond to $\alpha = 0.0$); α is quite different from zero, in fact.

In this case, the warning message we received is very important because it indicates not that the MLE of α is near 0 (which would be equivalent to a Brownian motion model) but that it is at the default upper bound for optimization. Let's change these default bounds and see what result we obtain.

```
fitOU_gs<-fitContinuous(bacteria.tree,genome_size,
    model="OU",bounds=list(alpha=c(0,10)))
fitOU_gs
```

```
## GEIGER-fitted comparative model of continuous data
##  fitted 'OU' model parameters:
##  alpha = 6.871261
##  sigsq = 40.534696
##  z0 = 2.750270
##
##  model summary:
##  log-likelihood = -56.380463
##  AIC = 118.760925
##  AICc = 119.560925
##  free parameters = 3
##
## Convergence diagnostics:
##  optimization iterations = 100
##  failed iterations = 0
##  number of iterations with same best fit = 37
##  frequency of best fit = 0.37
##
##  object summary:
```

```
##  'lik' -- likelihood function
##  'bnd' -- bounds for likelihood search
##  'res' -- optimization iteration summary
##  'opt' -- maximum likelihood parameter estimates
```

This time, there's no warning message—and both our optimized value of α and our log-likelihood are higher, which makes sense.

To compare either EB and Brownian or OU and Brownian, we could use a likelihood ratio test.[26] However, to compare all three models simultaneously, our best option is probably to use an information criterion such as the AIC.[27]

Let's do just that.

```
## accumulate AIC scores from our three models into
## a vector
aic_gs<-setNames(c(AIC(fitBM_gs),
    AIC(fitEB_gs),AIC(fitOU_gs)),
    c("BM","EB","OU"))
aic_gs
```

```
##       BM       EB       OU
## 122.5586 124.5586 118.7609
```

Since our preferred model under this criterion should be the one with the *lowest* AIC, this result tells us that the best-supported model (among those tested) for genome size is Ornstein–Uhlenbeck.

We can also compute Akaike weights[28] using the *phytools* function `aic.w`.

```
aic.w(aic_gs)
```

```
##         BM         EB         OU
## 0.12428638 0.04572238 0.82999124
```

Here you can see that the overwhelming majority of weight falls on the OU model—but evidence in support of the different models is somewhat split, and there is also quite a bit of weight on Brownian motion.

Let's now repeat all of this analysis, but for mutation accumulation rate. In this case, we tried it ahead of time and know the MLE of α for our OU model is going to fall well outside of our default bounds, so we set `bounds=list(alpha=c(0,100))` to see if that helps.

```
## fit EB model
fitEB_ar<-fitContinuous(bacteria.tree,ln_mutation,
    model="EB")
```

---

[26]This is because Brownian motion is a special case of both EB (when $a = 0$) and OU (when $\alpha = 0$).

[27]In practice, we'd normally recommend using the sample-size corrected AIC, AICc, but this adjustment is pretty easy to make and does not affect our result here anyway.

[28]Roughly interpretable as the weight of evidence in support of each model.

```
## fit OU model
fitOU_ar<-fitContinuous(bacteria.tree,ln_mutation,
    model="OU",bounds=list(alpha=c(0,100)))
## accumulate AIC scores in a vector
aic_ar<-setNames(c(AIC(fitBM_ar),
    AIC(fitEB_ar),AIC(fitOU_ar)),
    c("BM","EB","OU"))
## compute and print Akaike weights
aic.w(aic_ar)
```

```
##          BM          EB          OU
## 0.00044427 0.00016344 0.99939230
```

This result indicates that almost 100 percent of the weight of support falls on the OU model, with virtually none at all on the Brownian or EB models. Let's just inspect our fitted OU model.

```
fitOU_ar

## GEIGER-fitted comparative model of continuous data
##   fitted 'OU' model parameters:
##   alpha = 22.065869
##   sigsq = 159.923040
##   z0 = -14.279631
##
##   model summary:
##   log-likelihood = -67.053782
##   AIC = 140.107565
##   AICc = 140.907565
##   free parameters = 3
##
## Convergence diagnostics:
##   optimization iterations = 100
##   failed iterations = 0
##   number of iterations with same best fit = 37
##   frequency of best fit = 0.37
##
##   object summary:
##   'lik' -- likelihood function
##   'bnd' -- bounds for likelihood search
##   'res' -- optimization iteration summary
##   'opt' -- maximum likelihood parameter estimates
```

This tells us that a stochastic process with a tendency to revert toward a central value is better at explaining our data for mutation accumulation rates than is a model of constant-rate Brownian motion, or a model of early burst evolution, in which large differences between clades arise early in diversification and then the rate of evolution slows through time. This is actually quite a common pattern across comparative data sets (Harmon et al. 2010).

4.1 The data set used in this chapter also includes estimates of GC content ("GC_Content_Percent"). Fit the three models (BM, OU, and EB) for this character. Also test for phylogenetic signal in GC content. What do you conclude?

4.2 Reanalyze the data for mutation accumulation rate, but this time do not log-transform the data. What happens? Can you explain the discrepancy?

4.3 You might wonder about the relationship between the models considered here and phylogenetic signal. Use a simulation study to find out! Focus for now on the OU model, which can be simulated using fastBM in *phytools*. For simplicity, you can use the bacterial tree, bacteria.tree, that we used in this exercise. On that tree, simulate OU characters with $\sigma^2 = 1$, expected mean $\theta = 0$, and root state $a = 0$. Vary α over a range of values from small $\alpha = 0.1$ to large $\alpha = 10$. For each simulation, determine the amount of phylogenetic signal, and do a significance test. See if you can detect the general pattern!

Multi-rate, multi-regime, and multivariate models for continuous traits

5

In the previous chapter, we learned about the important Brownian motion model of evolutionary change for continuous traits.

We were also introduced to a couple of different relatively simple extensions of Brownian motion: the early burst model (EB), in which the rate of evolution (σ^2) varies as a continuous function of time since the root of the tree (Blomberg et al. 2003), and the Ornstein–Uhlenbeck model (OU; Hansen 1997), in which evolution proceeds by a random walk, but with the tendency to revert toward a particular central value.

We also learned that EB has often been associated with the concept of adaptive radiation and OU with natural selection toward a specific value as captured by the OU parameter θ.

In the current chapter, we'll extend the Brownian motion model in several additional ways. In particular, we will:

1. Consider a model of Brownian motion evolution in which the rate of evolutionary change (σ^2) is allowed to differ between clades or between different, predefined regions of the tree and learn how to fit this model to data in R.
2. Model Ornstein–Uhlenbeck evolution in which θ (the parameter used to describe the central tendency toward which the process reverts) is permitted to vary among different clades and branches across the phylogeny.
3. Examine the simultaneous evolution of multiple traits, in which the evolutionary correlations between different characters have been allowed to assume different values in different parts of the tree.
4. Finally, see a different class of method in which regimes are not specified a priori by the user but, instead, identified from the data using an approach called *reversible-jump Markov chain Monte Carlo*.

This chapter is not intended to be a comprehensive survey of the ways in which Brownian motion has been extended or a review of the entire realm of models for the evolution of continuous characters on trees.

Nonetheless, we'll try to sample across a range of relatively basic approaches that we feel are important for any phylogenetic comparative biologist to know well and to understand.

5.1 Multi-rate Brownian evolution

Almost the simplest imaginable way in which one might consider extending the Brownian motion model of evolution that we learned in the previous chapter is simply by permitting the *rate* of evolution under the model (σ^2) to differ in different clades or in different parts of the phylogeny (O'Meara et al. 2006; Revell et al. 2018).

There are several different R functions that have been developed to fit this exact model to data.

Why don't we start by fitting one in which we'll permit two or more different trees or clades to evolve with different Brownian motion rates ($\sigma_1{}^2$, $\sigma_2{}^2$, $\sigma_3{}^2$, and so on) and then compare this to a simpler model in which all of our clades have evolved with the same rate (Revell et al. 2018)? This approach was originally described as the "censored" method by O'Meara et al. (2006; also see Revell et al. 2018).

This model is implemented in the *phytools* function `ratebytree`.

5.1.1 Comparing the rate between different trees

To try it out, we'll apply the method to some data for two different lizard groups: the North American Phrynosomatidae family and the well-known South American clade Liolaemidae (data from Harmon et al. 2003).

What we intend to do is compare the rate of body size evolution between these two clades.

For this, we'll use three different files (`Phrynosomatidae.phy`, `Liolaemidae.phy`, and `Iguania.csv`), all three of which can be downloaded from the website of this book.[1]

Naturally, our first step is to read our phylogenies from file and to plot them, which we can proceed to do as follows.

```
## load packages
library(phytools)
## read trees from file
phryn.tree<-read.tree("Phrynosomatidae.phy")
liol.tree<-read.tree("Liolaemidae.phy")
## subdivide plotting area
par(mfrow=c(1,2))
## plot the two trees, adding labels using mtext
plotTree(phryn.tree,color="lightblue",
    fsize=0.5,ftype="i",
    mar=c(1.1,1.1,2.1,1.1))
mtext(text="(a)",line=0,adj=0)
plotTree(liol.tree,color="lightgreen",
    fsize=0.5,ftype="i",
    mar=c(1.1,1.1,2.1,1.1))
mtext(text="(b)",line=0,adj=0)
```

---

[1] http://www.phytools.org/Rbook/.

As in previous chapters, we use the function `par` to split our plotting device into two equal-width columns, and then we graphed our two different phylogenies[2] with the *phytools* function `plotTree` (figure 5.1).

From here, we can next go ahead and read our phenotypic trait data from file as well.

```
## read data file
Iguania<-read.csv("Iguania.csv",row.names=1)
head(Iguania)

##               SVL      FIXTL       TAL       TLL
## amcris 192.39000  312.4000 109.49000 146.48000
## anlon   67.82000  205.4700  27.77000  42.84000
## anacut  62.11578  120.7835  27.74346  44.34500
## anaen   70.10541  127.8682  29.51799  44.47824
## anahl   57.64300   91.0000  27.64300  46.78600
## analin  55.90000   92.2500  23.70000  34.80000
```

This data frame contains continuous character data for both Liolaemidae and Phrynosomatidae, as well as for several other iguanian lizard clades.

To subsample the data to match our two trees, we'll once again use the handy *geiger* function `name.check` that we've already used a number of times in prior chapters.

```
## load geiger
library(geiger)
## run name.check on phrynosomatid tree
chk<-name.check(phryn.tree,Iguania)
summary(chk)

## 212 taxa are present in the data but not the tree:
##      amcris,
##      anacut,
##      anaen,
##      anahl,
##      analin,
##      analli,
##      ....
##
## To see complete list of mis-matched taxa, print object.
```

As we can see, all of the species in `phryn.tree` are represented in our `Iguania` data frame but not the converse.

Just as we first learned in chapter 1 and have seen in subsequent exercises, in R, it's a simple thing to subsample the rows of our `Iguania` data frame to match the tip labels of our `phryn.tree` phylogeny.

Since we're going to do the same with our liolaemid tree, we should create a *new* data frame instead of simply modifying and overwriting `Iguania`.

---

[2] Using different colors—just for fun.

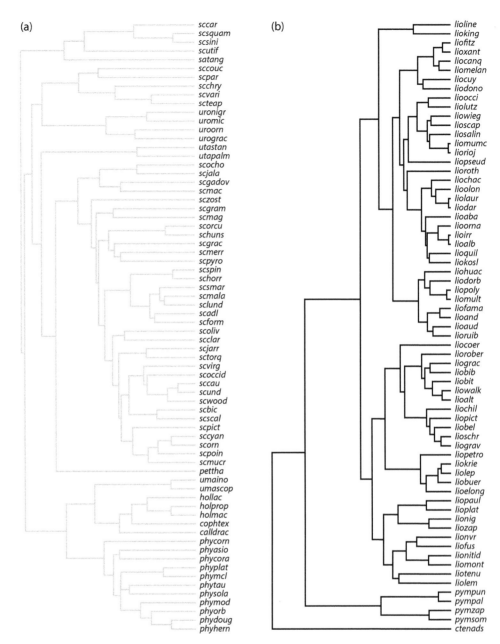

Figure 5.1
Phylogenetic trees of (a) Phrynosomatidae and (b) Liolaemidae.

```
## subsample Iguania to include only phrynosomatids
phryn.data<-Iguania[phryn.tree$tip.label,]
## check to see it matches the tree
name.check(phryn.tree,phryn.data)
```

```
## [1] "OK"
```

Now let's do the same thing with South American Liolaemidae.

```
## run name check on liolaemid tree
chk<-name.check(liol.tree,Iguania)
summary(chk)
```

```
## 215 taxa are present in the data but not the tree:
##      amcris,
##      anacut,
##      anaen,
##      anahl,
##      analin,
##      analli,
##      ....
##
## To see complete list of mis-matched taxa, print object.
```

```
## subsample Iguania to include only liolaemids
liol.data<-Iguania[liol.tree$tip.label,]
name.check(liol.tree,liol.data)
```

```
## [1] "OK"
```

Our data consist of morphological measurements for four traits. However, we'll for now just focus on overall body size—measured here as snout-to-vent length (SVL), as it often is in reptiles and amphibians.

Let's pull out SVL for each data set—in both cases transforming by the natural logarithm as we do so.

As we've seen earlier in the book, we can use the function setNames to make sure that the vectors we create both have names that match the row names of their corresponding data frames.

```
## extract body sizes (SVL) for phrynosomatids and liolaemids
## into separate vectors
phryn.size<-log(setNames(phryn.data$SVL,
    rownames(phryn.data)))
liol.size<-log(setNames(liol.data$SVL,
    rownames(liol.data)))
```

Finally, let's fit our alternative models—as mentioned at the beginning of this section of the exercise. To do that, we'll use the *phytools* function `ratebytree`.

`ratebytree` takes a list of trees[3] and a list of character vectors as input.

We can create the former by just using the combine function, `c`, to concatenate our two iguanian lizard family trees together within our `ratebytree` function call. Likewise, the latter is easy enough to make using the function `list`.

```
## fit our censored multi-rate Brownian model using
## phytools::ratebytree
fit.size<-ratebytree(c(phryn.tree,liol.tree),
    list(phryn.size,liol.size))
fit.size
```

```
## ML common-rate model:
##  s^2  a[1]    a[2]     k    logL
## value   0.2613  4.1772  4.2621  3    -4.8496
## SE      0.0316  0.1478  0.2394
##
## ML multi-rate model:
##   s^2[1] s^2[2]   a[1]    a[2]     k   logL
## value   0.1911  0.3346  4.1772  4.2621  4    -2.1877
## SE      0.0323  0.0578  0.1264  0.2709
##
## Likelihood ratio: 5.3238
## P-value (based on X^2): 0.021
##
## R thinks it has found the ML solution.
```

This result tells us that we can *reject* a null hypothesis of equal rates of body size evolution between these two clades, in favor of an alternative hypothesis that the rates differ. Inspecting our parameter estimates for the two models, we see that estimated rate of body size evolution in Liolaemidae is over 50 percent higher than the rate of body size evolution in Phrynosomatidae.

Although in this example we focused on comparing the rate of evolution (σ^2) between phylogenies, a similar approach can be used to fit a model with different values of σ^2 in different clades of a single tree or even on different branches of the tree.[4]

5.2 Multi-optimum Ornstein–Uhlenbeck evolution

Just as we can fit a Brownian model that permits different values of σ^2 for different trees (as we saw in the previous example) or different parts of a single tree, it's also possible to extend the Ornstein–Uhlenbeck (OU) model by allowing the parameters of that model to have different values in different clades or on different branches of our phylogeny.

Before we do, let's step back and reconsider the OU model.

---

[3] In fact, an object of class "multiPhylo"—see chapter 1 for more details.

[4] This type of analysis is implemented in the *phytools* function `brownie.lite` as well as in the *OUwie* package, as we'll see below.

Remember, OU consists of at least three parameters: θ, the position of the central value toward which the process will tend to revert; α, the strength of the force drawing the evolutionary process toward θ; and, finally, σ^2, the instantaneous variance of stochastic evolution in the model (Hansen 1997).[5]

Although in theory, we could imagine fitting a model in which θ, α, or σ^2 differed in different parts of the phylogeny, most often we tend to be most interested in the θ parameter as this is used to approximate the position or positions of the optimum or various optima under an adaptive evolutionary process through time.

5.2.1 Fitting a multi-optimum OU model to data

To explore this model, why don't we start by reading in a phylogeny of *Anolis* lizards as well as data for gross morphology. Then, we can proceed to fit a multi-optimum (i.e., multiple θ) OU model in which regime shifts are associated with transitions between the ecomorph state of different anole species.

The data files for this part of the chapter are the same as we used in chapter 1: anole.data.csv and ecomorph.csv (Mahler et al. 2010); however, we'll use a different tree file: anolis.mapped.nex.[6]

First, we should read our *Anolis* morphology and ecomorph data from file using read.csv.

```
## read morphological data from file
anole.morphology<-read.csv("anole.data.csv",
    row.names=1)
## read ecological states from file
anole.ecomorph<-read.csv("ecomorph.csv",
    row.names=1,stringsAsFactors=TRUE)
```

The tree file anolis.mapped.nex contains, instead of a regular phylogeny, a phylogeny with a six-state discrete character, *ecomorph*, mapped onto its nodes and branches.

For now, we won't worry about how this kind of mapping is generated. We'll see about that in a later chapter.

To read in the file, though, we need to use the *phytools* function read.simmap as follows.

```
## read in the phylogeny with ecomorph mapped
## using phytools::read.simmap
ecomorph.tree<-read.simmap(file="anolis.mapped.nex",
    format="nexus",version=1.5)
ecomorph.tree
```

```
##
## Phylogenetic tree with 82 tips and 81 internal nodes.
```

---

[5]Sometimes the OU model is also parameterized with a fourth parameter—the state at the root node of the phylogeny. Unfortunately, for most circumstances, the root state and θ are not *separately estimable*, which can make this parameterization of the OU model nonidentifiable.

[6]Also from Mahler et al. (2010), but modified—as we'll see in a sec!

```
##
## Tip labels:
##   ahli, allogus, rubribarbus, imias, sagrei, ...
##
## The tree includes a mapped, 6-state discrete character
## with states:
##   CG, GB, TC, TG, Tr, Tw
##
## Rooted; includes branch lengths.
```

The object printout seems pretty similar to what we're used to from typical "phylo" with one difference:

```
The tree includes a mapped, 6-state discrete character
with states:
        CG, GB, TC, TG, Tr, Tw
```

This tells us that we are working with a special type of phylogenetic object in R: the *phytools* "simmap" object.[7] A "simmap" object is a special type of phylogeny in which the two or more states of a discrete character have been mapped or *painted* onto the edges and nodes of the tree.

Since we're using a different tree file than the one we used in chapter 1, let's check again see if the data and tree coincide.

```
## check to see if data and phylogeny match using
## geiger::name.check
chk<-name.check(ecomorph.tree,anole.morphology)
summary(chk)
```

```
## 18 taxa are present in the data but not the tree:
##      argenteolus,
##      argillaceus,
##      barbatus,
##      barbouri,
##      bartschi,
##      centralis,
##      ....
##
## To see complete list of mis-matched taxa, print object.
```

This time, we see that there are species in the data that are not in the tree, but not the converse.

Let's remove these taxa from our data set by using the technique of negative numerical indexing that we learned in an earlier chapter.

---

[7] "simmap" objects actually have two different class attributes: "simmap" *and* "phylo". In R, this means that if no appropriate method can be identified for the first class, the second class will be used instead.

```
## trim our input data to match the tree using
## negative indexing
ecomorph.data<-anole.morphology[
    -which(rownames(anole.morphology)%in%
    chk$data_not_tree),]
name.check(ecomorph.tree,ecomorph.data)
```

```
## [1] "OK"
```

To fit our model—that is, a multiple-"regime" (i.e., multi-θ) Ornstein–Uhlenbeck model—we'll use the package *OUwie* (Beaulieu et al. 2012; Beaulieu and O'Meara 2020).

Since we haven't used this package yet in the book, you might not have it installed. To follow along, you'll need to install *OUwie* from CRAN, which can be done using the base R function install.packages as we saw for other R phylogenetic packages in chapter 1.

After installing *OUwie*, we just load it in the usual way.

```
library(OUwie)
```

As these data are multivariate and we're interested in how body shape evolves on the tree, we'll use the same phylogenetic principal components analysis (PCA)[8] (phyl.pca) that we saw in chapter 1.

We could analyze any of our principal components (PCs),[9] but here we will focus only on principal component 3.

```
## run phylogenetic PCA and print the results
pca<-phyl.pca(ecomorph.tree,ecomorph.data)
print(pca)
```

```
## Phylogenetic pca
## Standard deviations:
##         PC1        PC2        PC3        PC4
## 0.81378367 0.22553447 0.12277042 0.10577197
##         PC5        PC6
## 0.04920179 0.03691686
## Loads:
##              PC1         PC2         PC3
## SVL -0.9712234  0.16067288  0.01972570
## HL  -0.9645111  0.16955087 -0.01203113
## HLL -0.9814164 -0.02674808  0.10315533
## FLL -0.9712265  0.17585694  0.10697935
## LAM -0.7810052  0.37429334 -0.47398703
## TL  -0.9014509 -0.42528918 -0.07614571
```

---

[8] Remember: phylogenetic PCA is just like a regular PCA, but in which we take the phylogeny into account when estimating the covariances or correlations between traits. We can interpret the results exactly as we would a regular PCA.

[9] Or all of them, if we took steps to control our experiment-wise type I error.

```
##                   PC4          PC5          PC6
## SVL   0.14782215  -0.06211906   0.06935433
## HL    0.17994634   0.08064324  -0.04406887
## HLL  -0.13790763   0.06887922   0.04126248
## FLL  -0.09105747  -0.06075142  -0.04864769
## LAM  -0.15871456   0.00217418   0.00875408
## TL    0.01709649  -0.01750404  -0.01088743
```

PC3 loads positively on forelimb (`FLL`) and hindlimb (`HLL`) lengths and negatively on lamellae number (`LAM`). In *Anolis* lizards, lamellae are expanded scales on their digits that they use to enhance clinging on smooth surfaces (Glossip and Losos 1997; Losos 2009). Lots of previous work suggested that more arboreal lizards will tend to evolve *shorter* limbs and *more* toepad lamellae (Losos 2009), so this PC dimension is of special interest in looking at the evolutionary divergence of microhabitat specialists in anoles.

To work in *OUwie*, we need to make a special data frame for our analysis that contains our species names, the OU regimes at the tips of the tree for each species, and our continuous trait. To pull out PC3 from our `"phyl.pca"` object, we'll use the expression `scores(pca)[,3]`.

Let's see how it works:

```
## create our OUwie data frame
ouwie.data<-data.frame(Genus_species=rownames(scores(pca)),
    Reg=anole.ecomorph[rownames(scores(pca)),],
    X=as.numeric(scores(pca)[,3]))
head(ouwie.data,n=10)

##      Genus_species Reg          X
## 1             ahli  TG  0.26504666
## 2          alayoni  Tw -0.19938619
## 3          alfaroi  GB -0.10584289
## 4         aliniger  TC -0.15296783
## 5         allisoni  TC -0.16227073
## 6          allogus  TG  0.30083764
## 7     altitudinalis  TC -0.31196975
## 8          alumina  GB -0.04393829
## 9         alutaceus  GB -0.14093032
## 10     angusticeps  Tw -0.08274274
```

We should be able to see that our data frame contains our trait of interest (PC3), as well as the OU regimes (in this case, `"ecomorph"`) for the tips.

We also need to either supply the regimes for the nodes of the tree or map our regimes directly onto the branches. For us, the latter is the easier of these two options—because the tree we read in from file already has these regimes encoded (figure 5.2)!

Here we'll plot our mapped regimes. Remember, this is our *hypothesis* for how we think the parameters of our OU model (and, in particular, θ) varies across the tree. Our plot is shown in figure 5.2.

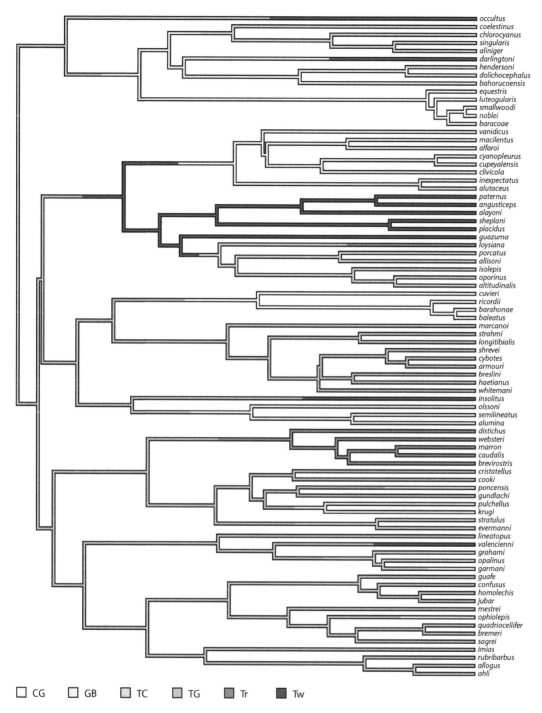

□ CG   □ GB   ■ TC   ■ TG   ■ Tr   ■ Tw

Figure 5.2
Habitat specialty (ecomorph) mapped onto a phylogenetic tree of Caribbean *Anolis*. Ecomorphs are as follows. CG, crown-giant; GB, grass-bush; TC, trunk-crown; TG, trunk-ground; Tr, trunk; Tw, twig.

```
cols<-setNames(rainbow(n=6),
    levels(anole.ecomorph[,1]))
plot(ecomorph.tree,cols,lwd=2,ftype="i",fsize=0.4,
    ylim=c(-4,82),outline=TRUE)
add.simmap.legend(colors=cols,prompt=FALSE,x=0,y=-2,
    vertical=FALSE,fsize=0.9)
```

Now we're finally ready to fit our model.

Instead of fitting just one model, however, we're going to fit three different models.

The first is a single-rate Brownian motion model. This is exactly the same model that we learned how to fit using *geiger*'s flexible fitContinuous function in chapter 4, but we can fit this model using *OUwie* too.

Next, we'll fit a model in which we allow only the rate of evolution (σ^2) to vary as a function of our mapped ecomorph regimes on the tree.

Last, we'll fit our multi-regime OU model. This model is the one that we're really interested in. According to this model, the θ parameter (remember, the value or values toward which the OU process is expected to revert—i.e., our phenotypic optima) is allowed to assume different values according to the regimes we've mapped onto our phylogeny (figure 5.2).

```
## fit standard, one-rate Brownian model
fitBM<-OUwie(ecomorph.tree,ouwie.data,model="BM1",
    simmap.tree=TRUE)

## Warning: An algorithm was not specified.
## Defaulting to computing the determinant and
## inversion of the vcv.

## Initializing...
## Finished. Begin thorough search...
## Finished. Summarizing results.

fitBM

##
## Fit
##        lnL       AIC       AICc       BIC model
##   87.00424 -170.0085 -169.8566 -165.195    BM1
##   ntax
##     82
##
## Rates
##        alpha    sigma.sq
##           NA  0.01488732
##
## Optima
##                           1
```

```
## estimate -5.385946e-08
## se         3.726990e-02
##
## Arrived at a reliable solution
```

We've supplied the tree with mapped regimes (ecomorph.tree) and set the argument simmap.tree to TRUE; however, since we are just fitting a simple, single-rate Brownian model, these regimes are ignored.

Next, we'll fit the more complicated model in which we allow each of the mapped regimes to evolve with a different value of the Brownian rate parameter, σ^2.

In the *OUwie*, this model is called the "BMS" model, and the only additional argument we'll add is root.station=FALSE.[10]

```
## fit multi-rate Brownian model
fitBMS<-OUwie(ecomorph.tree,ouwie.data,model="BMS",
    simmap.tree=TRUE,root.station=FALSE)
```

```
## Warning: An algorithm was not specified.
## Defaulting to computing the determinant and
## inversion of the vcv.

## Initializing...
## Finished. Begin thorough search...
## Finished. Summarizing results.
```

```
fitBMS
```

```
##
## Fit
##      lnL       AIC       AICc       BIC model ntax
##   93.3399 -172.6798 -171.1663 -155.8328   BMS   82
##
## Rates
##                    CG          GB          TC
## alpha              NA          NA          NA
## sigma.sq 0.005041015 0.007821223 0.02815262
##                    TG          Tr          Tw
## alpha              NA          NA          NA
## sigma.sq 0.01933404 0.002776003 0.01021321
##
## Optima
##                    CG          GB          TC
## estimate -0.01453764 -0.01453764 -0.01453764
## se        0.03902636  0.03902636  0.03902636
```

---

[10]This argument tells the OUwie function to fit the noncensored model of O'Meara et al. (2006). This means just that all edges of the tree are assigned to a regime, and thus only one value for the global root is estimated. root.station=TRUE would result in an analysis equivalent to the model we fit using ratebytree earlier in the chapter.

```
##                    TG          Tr          Tw
## estimate -0.01453764 -0.01453764 -0.01453764
## se         0.03902636  0.03902636  0.03902636
##
## Arrived at a reliable solution
```

Finally, we'll fit our multiple-θ OU model. This model is called "OUM" in *OUwie*.

```
## fit multi-regime OU model
fitOUM<-OUwie(ecomorph.tree,ouwie.data,model="OUM",
    simmap.tree=TRUE,root.station=FALSE)
```

```
## Warning: An algorithm was not specified.
## Defaulting to computing the determinant and
## inversion of the vcv.

## Initializing...
## Finished. Begin thorough search...
## Finished. Summarizing results.
```

```
fitOUM
```

```
##
## Fit
##       lnL       AIC       AICc       BIC model ntax
##  112.2198 -208.4396 -206.467 -189.1858   OUM   82
##
##
## Rates
##                 CG         GB         TC
## alpha     1.60477943 1.60477943 1.60477943
## sigma.sq 0.01619871 0.01619871 0.01619871
##                 TG         Tr         Tw
## alpha     1.60477943 1.60477943 1.60477943
## sigma.sq 0.01619871 0.01619871 0.01619871
##
## Optima
##                  CG          GB          TC
## estimate -0.16755924 -0.16497801 -0.38367569
## se         0.06725631  0.03745325  0.04532934
##                  TG          Tr          Tw
## estimate 0.1851189 0.03477840 -0.20700986
## se       0.0217364 0.07293059  0.04469685
##
##
## Half life (another way of reporting alpha)
##        CG        GB        TC        TG        Tr
## 0.4319268 0.4319268 0.4319268 0.4319268 0.4319268
```

```
##          Tw
## 0.4319268
##
## Arrived at a reliable solution
```

We should see that our fitted model gives us estimated values of α (the strength of pull toward the optima), σ^2 (the stochastic parameter), and θ (the position of the optima) for each mapped regime. Because we fit the `"OUM"` model, only θ differs between regimes.

In theory, we can also use the *OUwie* package to fit models in which $\alpha, \sigma^2, and \theta$ vary across the clades and branches of a phylogeny. In practice, we would recommend doing so only with a great deal of caution because the number of parameters to estimate can quickly grow very large, making the most parameter-rich of these models very difficult to fit to data sets typical of interspecific comparative studies.

The `"BM1"` and `"BMS"` (with `root.station=TRUE`) models should be *exactly* the same as the models fit using a function called `brownie.lite`[11] of the *phytools* package. If we want, we can use `brownie.lite` to double-check our results from *OUwie*.

Let's try:

```
## use brownie.lite to double-check OUwie results
brownie.lite(ecomorph.tree,scores(pca)[,3])
```

```
## ML single-rate model:
##              s^2          se a k      logL
## value 0.0148888 0.002302173 0 2 87.00426
##
## ML multi-rate model:
##        s^2(CG)      se(CG)    s^2(GB)      se(GB)
## value 0.0051565 0.002774887 0.0078668 0.003209361
##        s^2(TC)      se(TC)   s^2(TG)     se(TG)
## value 0.0280411 0.01144552 0.019357 0.005282045
##        s^2(Tr)      se(Tr)   s^2(Tw)     se(Tw)
## value 0.0029992 0.00148324 0.0102536 0.005630275
##               a k       logL
## value -0.014162 7 93.33284
##
## P-value (based on X^2): 0.0268123
##
## R thinks it has found the ML solution.
```

Finally, let's compare among the models. First, we'll pull out Akaike information criterion (AIC) scores for each of the three fitted models, and then we'll use these scores to obtain Akaike weights—quantities that represent the weight of evidence that supports each model (Burnham and Anderson 2003).

---

[11] The function `brownie.lite` is an R implementation of Brian O'Meara's software *Brownie*. To some degree, *OUwie* makes both programs obsolete.

```
## extracting AIC scores
aic<-setNames(c(fitBM$AIC,fitBMS$AIC,fitOUM$AIC),
    c("BM1","BMS","OUM"))
aic
```

```
##       BM1       BMS       OUM
## -170.0085 -172.6798 -208.4396
```

```
## compute Akaike weights
aic.w(aic)
```

```
##   BM1    BMS    OUM
## 0e+00  2e-08  1e+00
```

This tells us that (among the three different models we've fit to our data set), the multi-peak (that is, multi-θ) OU model is the one that's best supported by our data.

Our result makes a lot of sense because we hypothesize that lizards using different microhabitats should be subject to different regimes of natural selection and that this natural selection should likewise cause ecologically similar species in different parts of the phylogeny to have more similar phenotypes than expected under Brownian motion (Losos 2009).

In fact, if we show our phenotypic trait adjacent to the tips of the tree using a barplot in which bars are colored using the same scheme as figure 5.2, this pattern of evolution toward similar values for PC3 by ecologically similar lineages in different parts of the tree—that is, the pattern of *convergent evolution* in anoles (Losos 2009)—is quite apparent (figure 5.3).

```
## get the tip state for ecomorph for each species
tips<-getStates(ecomorph.tree,"tips")
## set these tip states to have the colors using the
## color scheme of Figure 5.2
tip.cols<-cols[tips]
## plot tree with adjacent barplot using
## phytools::plotTree.barplot
plotTree.barplot(ecomorph.tree,scores(pca)[,3],
    args.plotTree=list(fsize=0.4),
    args.barplot=list(col=tip.cols,
    xlab=expression(paste("PC3 (",""%up%"","limbs, ",
    ""%down%"","lamellae)",sep="")),
    cex.lab=0.8))
## add an informative legend
legend("topright",levels(anole.ecomorph[,1]),
    pch=22,pt.bg=cols,pt.cex=1.5,cex=0.9)
```

5.3 Multivariate Brownian evolution

So far in both chapter 4 and in the present chapter, we've focused on the analysis of just one trait at a time.

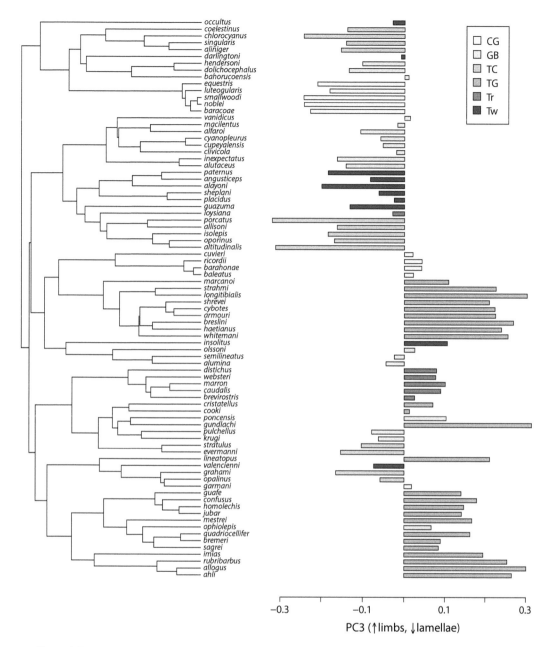

Figure 5.3

Phylogeny of *Anolis* with the value of principal component 3 from a phylogenetic PCA of six morphological traits. Ecomorphological category is as in figure 5.2.

Sometimes, however, it makes sense to fit a model for the evolution of multiple traits at once.[12]

If Brownian motion is occurring in more than one dimension at the same time, then it can no longer be fully described by a single-rate parameter, σ^2. Instead, we need to include a rate for each trait dimension, $\sigma^2(i)$—and an evolutionary covariance between them, $\sigma(i,j)$ (Hohenlohe and Arnold 2008; Revell and Harmon 2008). The evolutionary correlation between traits is just the ratio of the latter divided by the square root of the product of the former for each trait.

The penultimate analysis that we'll do in this chapter is fit a multivariate Brownian model in which the evolutionary covariance (and thus correlation) between characters is allowed to be different in different parts of the phylogeny (Revell and Collar 2009).

This kind of model makes sense to fit to our data under circumstances in which we've hypothesized that the way that different phenotypic traits work together or interact, and thus the way natural selection acts on their relationship, may have changed through time or among lineages of our tree.

5.3.1 Fitting multivariate Brownian evolution with multiple evolutionary correlations

For this example, we'll use data and a phylogeny for fish species in the family Centrarchidae (Collar and Wainwright 2006; Revell and Collar 2009). The data are in the following two files: Centrarchidae.csv and Centrarchidae.nex, which can be downloaded from the book website.

Let's read our tree and data.

The tree has an encoded discrete state, just as did anolis.mapped.nex in the previous exercise. As such, we'll read it using the *phytools* function read.simmap.

```
## read in tree with encoded discrete state using
## phytools::read.simmap
fish.tree<-read.simmap(file="Centrarchidae.nex",format="nexus",
    version=1.5)
print(fish.tree,printlen=2)

##
## Phylogenetic tree with 28 tips and 27 internal nodes.
##
## Tip labels:
##   Acantharchus_pomotis, Lepomis_gibbosus, ...
##
## The tree includes a mapped, 2-state discrete character
## with states:
##   non, pisc
##
## Rooted; includes branch lengths.
```

The data are in a simple CSV file, so we can read it in using read.csv.

---

[12]Our analysis with contrasts regression and PGLS involved more than one trait; however, our evolutionary model is *only* for the residual error—so these were not *really* multivariate trait evolution models.

```
## read in trait data using read.csv
fish.data<-read.csv("Centrarchidae.csv",header=TRUE,
    row.names=1,stringsAsFactors=TRUE)
head(fish.data)
```

```
##                      feeding.mode gape.width
## Acantharchus_pomotis         pisc      0.114
## Lepomis_gibbosus             non      -0.133
## Lepomis_microlophus          non      -0.151
## Lepomis_punctatus            non      -0.103
## Lepomis_miniatus             non      -0.134
## Lepomis_auritus              non      -0.222
##                      buccal.length
## Acantharchus_pomotis         -0.009
## Lepomis_gibbosus             -0.009
## Lepomis_microlophus           0.012
## Lepomis_punctatus            -0.019
## Lepomis_miniatus              0.001
## Lepomis_auritus              -0.039
```

Just as in the previous section, we're using a tree with a mapped discrete character that encodes our different evolutionary regimes. In this case, the regimes that we've mapped onto the nodes and branches of the tree are simply the feeding mode (piscivorous vs. nonpiscivorous) for each terminal taxon. As we've noted previously, later in this volume we'll see how this type of character history can be produced using a statistical method.

For now, let's just go ahead and plot our tree with its mapped regimes (figure 5.4).

```
## set colors to be used for plotting
cols<-setNames(c("white",palette()[2]),c("non","pisc"))
## plot "simmap" object
plot(fish.tree,cols,lwd=3,ftype="i",outline=TRUE,
    fsize=0.6)
## add a legend
legend("topleft",c("non-piscivorous","piscivorous"),
    pch=22,pt.bg=cols,pt.cex=1.5,cex=0.7,bty="n")
```

The hypothesis we want to test is whether feeding mode influences the evolutionary correlation between different aspects of the buccal (i.e., mouth) morphology of centrarchid fishes.[13] Our two continuous traits are gape width (the relative width of the mouth when gaped) and buccal length (the relative length of the mouth cavity).

For this analysis, we'll use evol.vcv in the *phytools* package. The evol.vcv function takes as input a phylogeny with a mapped discrete state (our regimes) and a numeric matrix. Since our data frame fish.data also contains one discrete factor (feeding.mode) in column 1, we'll convert columns 2 and 3 of the data frame to a matrix using as.matrix before continuing to fit our model.

---

[13] The plural of fish is fish. But the plural of fish species is fishes—so this is not a grammatical error!

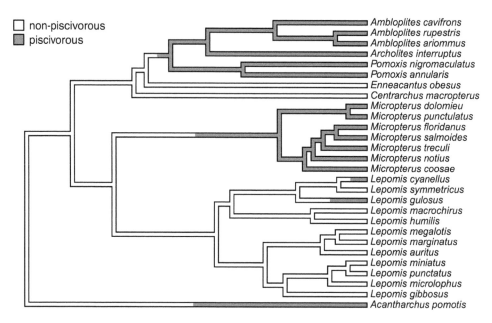

Figure 5.4
Feeding mode mapped on to a phylogeny of centrarchid fishes.

```
## convert numerical trait data to a matrix
fish.buccal<-as.matrix(fish.data[,2:3])
## fit multi-regime multivariate Brownian model
fitMV<-evol.vcv(fish.tree,fish.buccal)
fitMV
```

```
## ML single-matrix model:
##   R[1,1]   R[1,2]   R[2,2]   k    log(L)
## fitted   0.114    0.033    0.0556   5    72.1893
##
## ML multi-matrix model:
##   R[1,1]   R[1,2]   R[2,2]   k    log(L)
## non   0.1387   -0.0022 0.012    8    81.2393
## pisc 0.0758   0.0796   0.1291
##
## P-value (based on X^2): 4e-04
##
## R thinks it has found the ML solution.
```

This shows us that the model with two among-trait covariance matrices (one for piscivorous fish species and a second for their nonpiscivorous relatives) fits significantly better than a model with only one evolutionary covariance matrix.

For cases in which we have only two continuous traits, it's actually possible to fit two different intermediate models between the two extremes of identity and no similarity (Revell and Collar 2009). We'll do this using the *phytools* function evolvcv.lite:

```
fitMV.all<-evolvcv.lite(fish.tree,fish.buccal)
```

```
## Fitting model 1: common rates, common correlation...
## Best log(L) from model 1: 72.1893.
## Fitting model 2: different rates, common correlation...
## Best log(L) from model 2: 77.9869.
## Fitting model 3: common rates, different correlation...
## Best log(L) from model 3: 73.6144.
## Fitting model 4: no common structure...
## Best log(L) from model 4: 81.2393.
```

```
fitMV.all
```

```
## Model 1: common rates, common correlation
##  R[1,1]  R[1,2]  R[2,2]  k   log(L)  AIC
## fitted  0.114   0.033   0.0556  5    72.1893 -134.3786
##
## (R thinks it has found the ML solution for model 1.)
##
## Model 2: different rates, common correlation
##  R[1,1]  R[1,2]  R[2,2]  k   log(L)  AIC
## non  0.183   0.0271  0.0199  7    77.9869 -141.9738
## pisc 0.0489  0.0296  0.0889
##
## (R thinks it has found the ML solution for model 2.)
##
## Model 3: common rates, different correlation
##  R[1,1]  R[1,2]  R[2,2]  k   log(L)  AIC
## non  0.0991  0.0129  0.0634  6    73.6144 -135.2287
## pisc 0.0991  0.0538  0.0634
##
## (R thinks it has found the ML solution for model 3.)
##
## Model 4: no common structure
##  R[1,1]  R[1,2]  R[2,2]  k   log(L)  AIC
## non  0.1386  -0.0022 0.012   8    81.2393 -146.4785
## pisc 0.0758  0.0795  0.129
##
## (R thinks it has found the ML solution for model 4.)
```

We see from the results that evolvcv.lite has now fit two additional models compared to evol.vcv: one (Model 2) in which the two regimes have different evolutionary rates for each trait, but the same evolutionary correlation, and another (Model 3) in which the regimes evolve with the same rate for each trait, but with different evolutionary correlations between the traits, depending on the regime. It turns out that the no common structure model (which is the same as the one we fit using evol.vcv) is still the best supported of the four.

Since correlations tend to be a bit easier to interpret than covariances, we might also like to extract these values for our alternative evolutionary models.

We'll do this using the handy *stats* function cov2cor.

```
cov2cor(fitMV$R.single)

##              gape.width buccal.length
## gape.width    1.000000      0.414274
## buccal.length 0.414274      1.000000

cov2cor(fitMV$R.multiple[,,"non"])

##              gape.width buccal.length
## gape.width    1.00000000   -0.05508253
## buccal.length -0.05508253   1.00000000

cov2cor(fitMV$R.multiple[,,"pisc"])

##              gape.width buccal.length
## gape.width    1.0000000     0.8041757
## buccal.length 0.8041757     1.0000000
```

This shows us that in our best-fitting model, the evolutionary correlation between gape width and buccal length is *much higher* (over 0.8 compared to −0.05) in piscivorous than nonpiscivorous fish species.

This result makes biological sense, because piscivory (which is accomplished via suction feeding in Centrarchidae, as in many percomorph fishes) requires close integration of the different aspects of the feeding apparatus.

Even though this result seems clear-cut, we believe that it's *always* wise to visualize our data.

In this case, why don't we project our tree into phenotype space for the two characters using the *phytools* function phylomorphospace (as we've done in prior chapters)—but this time while retaining our regime mapping on the tree (figure 5.5).

```
## modify the margins of the plot area and adjust axis
## and label font sizes
par(cex.lab=0.7,cex.axis=0.6,mar=c(5.1,4.1,1.1,2.1))
## plot the phylomorphospace without the mapped
## regimes
phylomorphospace(as.phylo(fish.tree),fish.buccal,
    ftype="off",lwd=4,xlab="gape width",
    ylab="buccal length",node.size=c(0,0),
    bty="n")
## add the phylomorphospace projection with the
## mapped regimes
phylomorphospace(fish.tree,fish.buccal,
    ftype="off",colors=cols,lwd=2,
    node.by.map=TRUE,xlab="gape width",
    ylab="buccal length",node.size=c(0,1.3),
    add=TRUE)
## add a legend
legend("topleft",c("non-piscivorous","piscivorous"),
    pch=22,pt.bg=cols,pt.cex=1.5,cex=0.7)
```

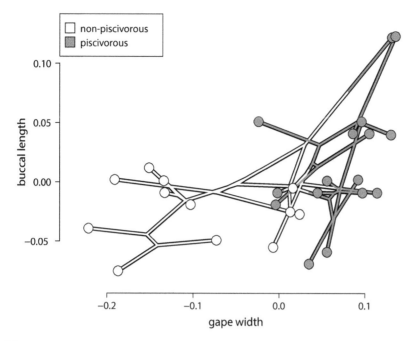

Figure 5.5
Phylomorphospace for centrarchid fish.

One trick we used in figure 5.5 was to first create a phylomorphospace *without* feeding mode mapped and coloring the lines of the projection black. We did this by using the custom S3 method `as.phylo` to *convert* our `"simmap"` object that we read from file into a simpler `"phylo"` object.

Then, we simply replotted our phylomorphospace showing feeding mode mapped onto the edges of the tree but set `add=TRUE` so that it gets plotted *on top* of our previous graph. Because we generated the first plot using slightly thicker edge widths (`lwd=4` instead of `lwd=2`), this overplotting will create the effect of a fine black outline around our phylomorphospace projection![14]

The pattern that we see in this projection (figure 5.5) is also quite clear: piscivorous lineages (in red) evolve with a higher evolutionary correlation between relative gape width and buccal length than do nonpiscivorous taxa (in white), just as indicated by our best-fitting evolutionary model (Revell and Collar 2009).

5.4 Exploring evolutionary heterogeneity

Up to this point in the chapter, all of our analyses have involved testing prespecified hypotheses about how the rate or the process of evolution differs between different trees or different parts of a single tree.

But what if we think the rate of evolution may have varied over time, but we lack an a priori hypothesis about how that variation is structured among the clades or branches of our phylogeny? In that case, it would be nice if we could use the data in our continuous trait to tell

---

[14]The same type of approach can be used to simulate outlines and shadows in other kinds of R plots too.

---

us something about how the rate or process of evolutionary change has varied throughout our phylogeny (Revell et al. 2012; Revell 2021).

5.4.1 Testing for temporal shifts in the rate of evolution

One straightforward approach would be to imagine the relatively simple scenario in which the rate of evolution shifts[15] at one point, or at more than one point, between the root of the tree and the present day.

If we could draw out a hypothesis proposing how the rate of evolution shifted, we could likewise paint this hypothesis onto the edges of the tree and fit our model using *OUwie* or using `brownie.lite` from the *phytools* package.

However, we could also treat the location(s) in which the rate shifts occurred as a free parameter or set of free parameters in our model and then simultaneously optimize the rate shift locations jointly with the rates of evolution ($\sigma_1{}^2$, $\sigma_2{}^2$, and so on) for our trait. Precisely this model is implemented in the *phytools* function, `rateshift`.

To try this method, we'll use a data set from Broeckhoven et al. (2016) in which the authors studied defensive structure evolution in the South African lizard family, Cordylidae. Broeckhoven et al. (2016) found that an early burst (EB) model[16] was best supported for the principal component associated with body armament in cordylids. Let's use `rateshift` to fit a model in which instead of changing as a continuous function of time, as in the EB model, the rate of evolution shifts a discrete time point, as indicated by our data.

We can start off by reading our lizard phylogeny (`cordylid-tree.phy`) and data (`cordylid-data.csv`) into R. Both files can be obtained from the book website.

```
## read cordylid tree from file
cordylid.tree<-read.tree(file="cordylid-tree.phy")
print(cordylid.tree,printlen=4)

##
## Phylogenetic tree with 28 tips and 27 internal nodes.
##
## Tip labels:
##    C._aridus, C._minor, C._imkeae, C._mclachlani, ...
##
## Rooted; includes branch lengths.

## read cordylid data from file
cordylid.data<-read.csv(file="cordylid-data.csv",
    row.names=1)
head(cordylid.data)

##                    pPC1      pPC2      pPC3
## C._aridus       0.59441  -0.40209   0.57109
## C._minor        0.65171  -0.32732   0.55692
```

---

[15]Simultaneously, across all the branches of our tree. OK, perhaps this is not that realistic most of the time; however, we already learned about a model, the EB model, in which the rate of evolution changes across all the branches of the phylogeny according to a constant function!

[16]The same EB model that we studied in chapter 4.

```
## C._imkeae          0.19958  -0.08978   0.56671
## C._mclachlani       0.62065   0.03746   0.86721
## C._macropholis      0.44875  -0.75942   0.09737
## C._cordylus        -0.07267   0.48294  -0.54394
```

The data consist of scores for the first three principal components from a phylogenetic PCA (Revell 2009). We'll just analyze the first of these, pPC1 in the Broeckhoven et al. (2016) data. Phylogenetic PC1 was positively correlated with spine lengths and negatively correlated with limb lengths (Broeckhoven et al. 2016).

```
cordylid.pc1<-setNames(cordylid.data$pPC1,rownames(cordylid.data))
```

Now, to fit our model, we just have to call the function rateshift and indicate the number of rate shifts that we hypothesize. We'll fit three models: a model without rate shifts,[17] a model with one rate shift and thus two different rates of evolution, and a model with two rate shifts. We need to keep in mind that Brownian motion has two parameters (σ^2 and x_0), and our rate shift model will have two additional parameters for each estimated shift: the *location* of the shift and the extra rate of evolution! For fun, we can also compare these three models to the EB model of chapter 4.

```
## fit single-rate model (no rate shift)
fit1<-rateshift(cordylid.tree,cordylid.pc1)

## Optimization progress:
## |.........|
## Done.

## fit two-rate model (one rate shift)
fit2<-rateshift(cordylid.tree,cordylid.pc1,nrates=2)

## Optimization progress:
## |.........|
## Done.

## fit three-rate model (two rate shifts)
fit3<-rateshift(cordylid.tree,cordylid.pc1,nrates=3)

## Optimization progress:
## |.........|
## Done.
```

This model is a bit difficult for R to optimize,[18] and this is especially true for models with more shifts.

---

[17] This is just a regular Brownian motion model—and it should have the same likelihood and parameter estimates as we would've obtained from fitContinuous in the *geiger* package.

[18] In fact, when you run rateshift on your own computer, you may have to try different starting values or computer seeds to get it to converge!

Let's now fit the EB model using `fitContinuous` and build a table comparing our four different fitted models.[19]

```
## fit EB model using geiger::fitContinuous
fitEB<-fitContinuous(cordylid.tree,cordylid.pc1,
    model="EB")
## compile our results into a list, sorted by
## the number of parameters estimated
fits<-list(fit1,fitEB,fit2,fit3)
## create a table summarizing model fits
data.frame(model=c("BM","EB","two-rate","three-rate"),
    logL=sapply(fits,logLik),
    k=sapply(fits,function(x) attr(logLik(x),"df")),
    AIC=sapply(fits,AIC),
    weight=unclass(aic.w(sapply(fits,AIC))))
```

```
##          model       logL k      AIC      weight
## 1           BM -25.95898 2 55.91795 0.05385102
## 2           EB -24.21673 3 54.43345 0.11312260
## 3    two-rate -21.91226 4 51.82451 0.41693974
## 4  three-rate -19.91430 6 51.82861 0.41608664
```

This shows that the best-supported model is the two-rate model; however, the weight of evidence is almost exactly equal between the two- and three-rate models.

Let's print out our three-rate model to see how it is parameterized.

```
fit3

## ML 3-rate model:
##       s^2(1) se(1) s^2(2)    se(2) s^2(3)    se(3) k
## value  11.61 7.279 0.0404 0.08544 0.5324 0.2293 6
##         logL
## value -19.91
##
## Shift point(s) between regimes (height above root):
##        1|2 se(1|2)    2|3 se(2|3)
## value 0.219       0 0.8349 0.01414
##
## Model fit using ML.
##
## Frequency of best fit: 0.5
##
## R thinks it has found the ML solution.
```

---

[19] We'll make the code to create our table a bit more compact by first compiling all our different fitted models into a list and then iterating over the elements of the list using `sapply`.

The order of the rates in the model object is from the root, forward in time toward the tips. We can see that the *highest* value of σ^2 in our fitted model is for the first part of the tree (from the root to 0.219,[20] then the rate declines to a much lower value for the middle part of the phylogeny, before increasing again slightly toward the tips of the tree; figure 5.6g,h).

To finish up this example, let's create a nice plot showing each fitted model both graphed onto the tree and as a line plot illustrating the change in rate through time (figure 5.6).

```
## compute the total height of our cordylid tree
h<-max(nodeHeights(cordylid.tree))
## split our plot window into eight panels
par(mfrow=c(4,2))
## panel a) single-rate model graphed on the tree
plot(fit1,mar=c(1.1,4.1,2.1,0.1),ftype="i",fsize=0.5,col="gray")
mtext("(a)",adj=0,line=0)
## panel b) line graph of single-rate model
par(mar=c(4.1,4.1,2.1,1.1))
plot(NA,xlim=c(0,h),
    ylim=c(0,12),xlab="time",
    ylab=expression(sigma^2),bty="n")
lines(c(0,h),
    rep(fit1$sig2,2),lwd=3,col="gray")
mtext("(b)",adj=0,line=0)
## panel c) compute EB model and graph it on the tree
## calculate sigma^2 through time under fitted model
s2<-fitEB$opt$sigsq*exp(fitEB$opt$a*
    seq(h/200,h-h/200,length.out=100))
s2.index<-round((s2-min(s2))/diff(range(s2))*
    100)+1
## use make.era.map to paint fitted EB model onto tree
tmp<-make.era.map(cordylid.tree,
    setNames(seq(0,h,length.out=101),
    s2.index))
## set colors for graphing
cols<-setNames(
    gray.colors(101,0.9,0),
    1:101)
## plot tree
plot(tmp,cols,mar=c(1.1,4.1,2.1,0.1),ftype="i",
    ylim=c(-0.1*Ntip(cordylid.tree),Ntip(cordylid.tree)),
    fsize=0.5)
## add color bar legend
add.color.bar(leg=0.5*h,cols=cols,prompt=FALSE,
    x=0,y=-0.05*Ntip(cordylid.tree),lims=round(range(s2),3),
    title=expression(sigma^2))
```

---

[20]Our tree in the example is scaled to be one unit long—but this is by no means a requirement of the method!

```
mtext("(c)",adj=0,line=0)
## panel d) line graph of EB model
par(mar=c(4.1,4.1,2.1,1.1))
plot(NA,xlim=c(0,h),
    ylim=c(0,12),xlab="time",
    ylab=expression(sigma^2),bty="n")
lines(seq(0,h,length.out=100),
    s2,lwd=3,col="gray")
mtext("(d)",adj=0,line=0)
## panel e) two-rate model projected on the tree
plot(fit2,mar=c(1.1,4.1,2.1,0.1),ftype="i",fsize=0.5,col=cols)
mtext("(e)",adj=0,line=0)
## panel f) line graph of two-rate model
par(mar=c(4.1,4.1,2.1,1.1))
plot(NA,xlim=c(0,h),
    ylim=c(0,12),xlab="time",
    ylab=expression(sigma^2),bty="n")
lines(c(0,fit2$shift,h),
    c(fit2$sig2,fit2$sig2[2]),
    type="s",lwd=3,col="gray")
mtext("(f)",adj=0,line=0)
## panel g) three-rate model projected on the tree
plot(fit3,mar=c(1.1,4.1,2.1,0.1),ftype="i",fsize=0.5,col=cols)
mtext("(g)",adj=0,line=0)
## panel h) line graph of three-rate model
par(mar=c(4.1,4.1,2.1,1.1))
plot(NA,xlim=c(0,h),
    ylim=c(0,12),xlab="time",
    ylab=expression(sigma^2),bty="n")
lines(c(0,fit3$shift,h),
    c(fit3$sig2,fit3$sig2[3]),
    type="s",lwd=3,col="gray")
mtext("(h)",adj=0,line=0)
```

This code chunk gets a little complicated,[21] but hopefully you can figure most of it out without our help.

In panels (a), (c), (e), and (g) of figure 5.6, we see the estimated rate (σ^2) mapped as a color gradient onto the branches of our cordylid lizard phylogeny. Then, in panels (b), (d), (f), and (h), we can see the reconstructed rate shown using a line graph.[22]

Overall, our results demonstrate that a one or two rate-shift model[23] fits better than either a constant-rate Brownian model or an EB model for the evolution of body armament in cordylid lizards. Neat!

---

[21]Especially panels (c) and (d) graphing the EB model!

[22]For the latter, we used the same vertical scale so that they would be easily comparable. It would've been nice to do the same for the former set of figure panels, but the difference in the range of estimated values of σ^2 was simply too large for that to be feasible while still making a useful graph!

[23]Thus with either two or three different values of σ^2, respectively.

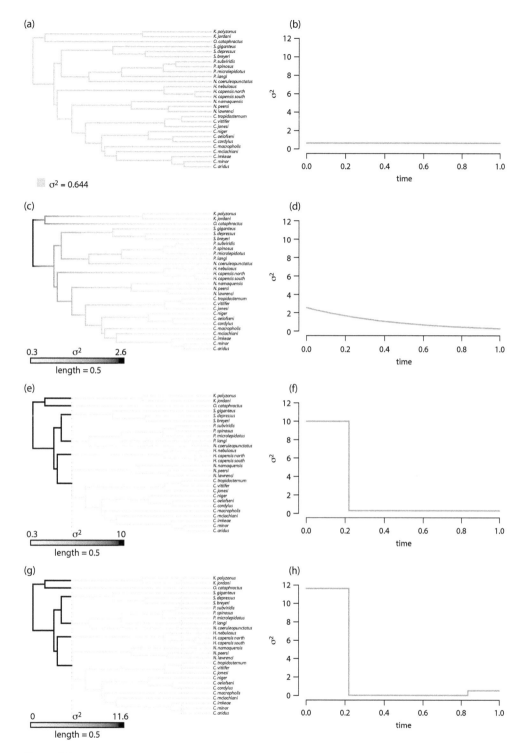

Figure 5.6
Fitted rate-shift models, compared to Brownian motion and EB. (a, b) The best-fit constant rate Brownian motion model. (c, d) The EB model. (e, f) Two-rate, rate-shift model. (g, h) Three-rate, rate-shift model.

5.4.2 Using rjMCMC to explore heterogeneity across branches and clades

In our previous example, we were able to allow our continuous character (the principal component dimension correlated with body armament in cordylid lizards) to dictate rate heterogeneity through time, absent any specific a priori hypothesis for rate changes in the tree. We fit this model using maximum likelihood and obtained a reasonable result showing that a rate-shift model fit better than some alternatives that we also considered.

More generally, it might be interesting to explore whether our continuous-trait data suggest that the rate or process of evolutionary change varies among edges or among clades in our tree. Furthermore, it could be interesting to analyze this model without having to specify even *how many* different regimes we have hypothesized for our trait (e.g., Revell 2021).

To fit this kind of model to our data, we're going to use a very different approach called *reversible-jump Markov chain Monte Carlo* or rjMCMC (Green 1995). rjMCMC is a Bayesian estimation procedure, so like regular Bayesian MCMC, rjMCMC samples values for the parameters of our evolutionary model in an attempt to approximate their joint posterior probability distribution.

rjMCMC is similar to regular MCMC, except that, in addition to merely sampling different parameter values for our evolutionary model from their posterior distribution, rjMCMC also[24] jumps between alternative model paramterization.

Going back to our scenario of a shifting evolutionary rate among branches or clades of the tree, this means that rjMCMC could be used to sample models (and model parameters) in which there is one rate shift, two, and so on. At the end of our rjMCMC analysis, we should have a posterior sample of both evolutionary models and parameter values under the models.

rjMCMC methods only began to appear in phylogenetic comparative analysis about a decade ago (Eastman et al. 2011; Pagel and Meade 2013); however, they have come to be an important new tool in studying how continuous traits evolve.

Although there are several different implementations of rjMCMC for phylogenetic comparative analysis, we're going to use the *bayou* package (Uyeda and Harmon 2014), which uses rjMCMC to fit a multi-regime OU model to phylogenetic comparative data. This model is equivalent to the one we analyzed using the *OUwie* R package earlier in the chapter. The important difference between *bayou* and *OUwie*, though, is that in *bayou*, we're going to let the data for our continuous trait define how our different evolutionary regimes are arrayed on the tree—rather than specifying them a priori using independent information, as we did for *OUwie*.

The *bayou* package is not presently on CRAN, so to follow along with this exercise, you'll need to install it directly from its GitHub development page. This sounds fancy, but it can be done very easily using the *devtools* package (Wickham et al. 2020).[25],[26]

To install *bayou*, we'll need to use the *devtools* function `install_github`. Keep in mind, just like the package installations we did in chapter 1, you only need to run this once,[27] not every time you will use the package in R.

---

[24] Reversibly, sensibly.

[25] Which *is* on CRAN, so can be installed in the usual way.

[26] We'll be installing at least one other package using *devtools* later in the book!

[27] Or, at most, once every time the package is updated by the package authors.

```
## load devtools
require(devtools)
## install bayou from GitHub
install_github("uyedaj/bayou")
library(bayou)
```

For our *bayou* analysis, we're going to explore body mass evolution in mammals. We'll use the same data set that we saw in an earlier chapter. The tree and data files[28] are mammalHR.phy and mammalHR.csv.

Let's load the tree and data and the extract body mass in kilograms as a new vector with names.

```
## read mammal tree from file
mammal.tree<-read.tree("mammalHR.phy")
print(mammal.tree,printlen=4)

##
## Phylogenetic tree with 49 tips and 48 internal nodes.
##
## Tip labels:
##    U._maritimus, U._arctos, U._americanus, N._narica, ...
##
## Rooted; includes branch lengths.
```

```
## read mammal data
mammal.data<-read.csv("mammalHR.csv",row.names=1)
## extra log body mass as a new vector
bodyMass<-setNames(log(mammal.data$bodyMass),
    rownames(mammal.data))
```

Great! This is the character trait that we're going to analyze using *bayou*.

bayou is a Bayesian method. To use it, we first need to set up prior probabilities for all our model parameters.[29]

To set our priors in *bayou*, we use a function called make.prior. When we run make.prior, it will generate the prior probability densities for our *bayou* rjMCMC run; however, it also creates a nice plot[30] so that we can visualize the different prior distributions that we have imposed (figure 5.7).

---

[28] Available from the book website, of course.

[29] All Bayesian methods use priors. Sometimes, Bayesian methods in R have sensible default priors, so this process may be hidden from us if we accept the default values; however, all good Bayesian method implementations should allow the user to adjust the priors if they want to. In general, we advocate a thoughtful approach to setting priors and do not shy away from informative priors, when appropriate. If you are unsure, try different priors and make sure your conclusions are robust!

[30] As long as we leave the argument plot.prior=TRUE: the default.

```
## turn of the box for our plot
par(bty="n")
## make our OU prior distribution
priorOU<-make.prior(mammal.tree,
    dists=list(dalpha="dhalfcauchy",
    dsig2="dhalfcauchy",
    dk="cdpois",dtheta="dnorm"),
    param=list(dalpha=list(scale=0.1),
    dsig2=list(scale=0.1),
    dk=list(lambda=10, kmax=50),
    dsb=list(bmax=1, prob=1),
    dtheta=list(mean=mean(bodyMass),
    sd=1.5*sd(bodyMass))),
    plot.prior=TRUE)
```

make.prior takes lots of different arguments, but they are fairly straightforward to figure out. For instance, the function argument dists is meant to be a list of the

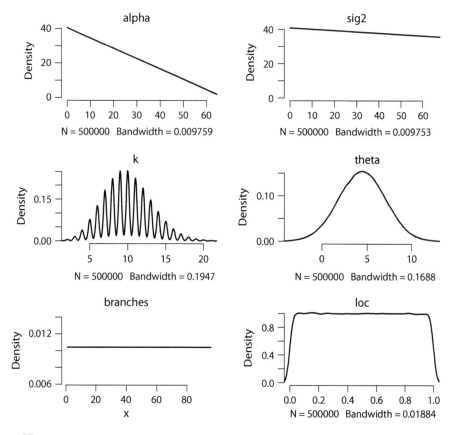

Figure 5.7
Prior densities that we'll use in our *bayou* Ornstein–Uhlenbeck model analysis.

different prior probability densities for the different parameters of our model. Setting `dalpha="dhalfcauchy"` assigns a half-Cauchy prior distribution to the OU parameter α.

Likewise, the argument `param` is a list of lists that can be used to set the parameters of each different prior distribution. So, for instance, setting `dalpha=list(scale=0.1)` tells `make.prior` to parameterize our half-Cauchy prior distribution for α using a scale parameter value of 0.1. Get it?

Most of the parameters of the OU model we're seeing here should be familiar based on what we did using *OUwie* earlier in the chapter, except for one: k. k is a parameter describing the number regime shifts on our tree. We have to assume a prior distribution for this parameter too—just like the others in our model. Here, we set the prior distribution on k to be a conditional Poisson (`dk="cdpois"`) with parameters `dk=list(lambda=10, kmax=50)`, which gives the λ parameter value of the Poisson distribution and the maximum number of regime shifts that we'll allow.

To run our rjMCMC, we're going to also need to select starting values for the various parameters in our model. A sensible way to do this is to pick values for each parameter at random from their prior distribution. *bayou* makes this easy for us with a function called `priorSim`.

```
## randomly select starting values for our MCMC
startpars<-priorSim(priorOU,mammal.tree,
    plot=FALSE)$pars[[1]]
startpars
```

```
## $alpha
## [1] 0.2351835
##
## $sig2
## [1] 2.289361
##
## $k
## [1] 17
##
## $theta
##   [1] -0.0873001   3.5688487   5.4020994   5.7992328
##   [5]  0.4357141   6.6196830   6.1414261   6.2003963
##   [9]  1.5701569   0.8426242   7.4996354   3.9164938
##  [13]  7.8399448   4.3847996   5.0318743   3.2433530
##  [17]  7.3885237   0.3994895
##
## $sb
##   [1] 83 51 69 18 24 58 30 56 88 17 43 82 74 60  6
##  [16] 22 47
##
## $loc
##   [1]  2.8733193   5.8846594   0.7674212   0.4347529
##   [5]  2.8611544   0.8690671   3.1981274  48.5493822
##   [9]  4.4009958   2.6168196   7.8686287   2.2968181
##  [13]  1.0703136   1.5477952   0.1680226   2.6539793
##  [17]  4.1756002
```

```
##
## $ntheta
## [1] 18
##
## $t2
##  [1]  2  3  4  5  6  7  8  9 10 11 12 13 14 15 16
## [16] 17 18
```

Looking at the starting values for the rjMCMC also tells us a lot about how our model is parameterized.

We can see that startpars includes single values for α and σ^2, a starting value for the number of regimes (k), and then a corresponding number of starting θ values—one for each regime.[31]

We can likewise compute the (log) prior probability of the starting values for our rjM-CMC using the priorOU function that we created using make.prior earlier in this section. Let's see.

```
priorOU(startpars)

## [1] -94.01578
```

Now we're just about ready to run our MCMC. We have one more step, which is to use the function bayou.makeMCMC to set up the MCMC run. This function takes as input arguments all of our data, our prior distributions (in the function priorOU), and some other run conditions for the MCMC.

```
mcmcOU<-bayou.makeMCMC(mammal.tree,bodyMass,
    prior=priorOU,plot.freq=NULL,file.dir=NULL,
    ticker.freq=1000000)

## Checking inputs for errors:..........
## seems fine...
##
## File directory specified as NULL. Results will be returned as
## output.
```

Only a couple of arguments in bayou.makeMCMC require additional explanation.

The first is file.dir, which can be used to specify a directory for writing files containing intermediate states of the MCMC. This could be useful if we have to run a very long MCMC and we might run out of memory on our system; otherwise, if we leave it as file.dir=NULL, the MCMC state will just be held in memory by R. ticker.freq gives the frequency with which we want to print intermediate states of the MCMC to the display buffer in R. We have just set this to a high value[32] so that the MCMC doesn't print at all! In your own analysis, it might

---

[31] Remember, just as in the previous example, the number of regimes will always be one more than the number of changes between regimes.

[32] Actually, we set it to a higher value than the number of generations that we intend to run!

be useful to monitor the progress of the MCMC, in which case you could set `ticker.freq` to a lower value.[33]

The `bayou.makeMCMC` checks our input and returns a special object type (an object of class `"bayouMCMCFn"`) that contains absolutely *everything* that we need to run our MCMC. All that we need to do is call the `run` function within our object on an integer indicating the number of generations that we want to use for our rjMCMC.

```
mammal.rjMCMC<-mcmcOU$run(100000)
```

Now let's take a closer look at our results.

At this point, the first thing we must to do is set a burn-in fraction. Burn-in is the set of generations of MCMC that we're going to assume were required to converge on the posterior distribution. Normally, it would be wise to evaluate convergence and decide on a appropriate burn-in.[34] In our case, we're going to arbitrarily set it to 30 percent of our sample. We'll do this using the *bayou* function `set.burnin`.[35]

```
mammal.rjMCMC<-set.burnin(mammal.rjMCMC,0.3)
```

bayou has a useful `summary` method for the object class of our MCMC run. To run it, you can type the following.[36]

```
summary(mammal.rjMCMC)
```

We can also pass the output of this `summary` method to another object and then print just part of the object.[37]

```
mammal.mcmc.result<-summary(mammal.rjMCMC)
```

Now let's just print the `statistics` part of the object.

```
mammal.mcmc.result$statistics
```

```
##                  Mean      SD Naive SE
## lnL          -71.1971  4.9397 0.059033
## prior        -57.8804 14.3830 0.171885
## alpha          0.0323  0.0312 0.000373
## sig2           0.1186  0.0435 0.000519
```

---

[33] In fact, during an interactive R session, that's what we'd probably recommend.

[34] There are multiple functions to evaluate convergence on the posterior distribution and effective sample size in the multifunctional R package *coda* (Plummer et al. 2006).

[35] Even though when we set burn-in using `set.burnin`, here we are overwriting our original object, we're not permanently deleting the first 30 percent of generations—we're just setting a burn-in fraction to be used by downstream functions. We can even change it later if we want.

[36] We won't do it ourselves—because the result is quite long and would fill several pages of the book!

[37] That we *will* do.

```
## k              10.9160   3.2577 0.038931
## ntheta         11.9160   3.2577 0.038931
## root.theta      4.7751   0.9106 0.010882
## all theta       4.4122   2.2221       NA
##              Time-series SE Effective Size
## lnL                 0.76585           41.6
## prior               1.12695          162.9
## alpha               0.00541           33.2
## sig2                0.00414          110.3
## k                   0.25897          158.2
## ntheta              0.25897          158.2
## root.theta          0.07992          129.8
## all theta                NA             NA
##              HPD95Lower HPD95Upper
## lnL            -7.82e+01   -60.7468
## prior          -8.63e+01   -29.9199
## alpha           7.43e-05     0.0878
## sig2            5.62e-02     0.1944
## k               5.00e+00    17.0000
## ntheta          6.00e+00    18.0000
## root.theta      2.87e+00     6.2647
## all theta            NA         NA
```

This table gives us a lot of information about our sample. For instance, we see that the mean number of shifts between regimes is around 12, with a standard deviation in the posterior sample,[38] of about 3.2. Likewise, the 95 percent high probability density (HPD) interval is 5 to 17.

bayou also helps us create a plot to summarize our posterior sample on the tree. To do that, we'll use the function plotSimmap.mcmc. This function runs on our MCMC output—not on our original tree.

We're going to call it twice: once to map the regimes that had higher than 0.25 posterior probability in our sample (figure 5.8a) and the second time to show the average value of the optimum parameter, θ, for each edge across the posterior sample (figure 5.8b).

```
## split our plotting area into two panels
par(mar=c(1.1,1.1,3.1,0.1),mfrow=c(1,2))
## plot regimes using different colors for all regime
## shifts with PP>0.25
plotSimmap.mcmc(mammal.rjMCMC,edge.type="regimes",
    lwd=2,pp.cutoff=0.25,cex=0.6)
mtext("(a)",adj=0,line=1)
## plot mean value of theta on each edge across posterior
## sample from rjMCMC
```

---

[38]This is a measure of the uncertainty around the estimate—somewhat akin to the standard error of frequentist statistical estimation.

```
plotSimmap.mcmc(mammal.rjMCMC,edge.type="theta",
    lwd=2,pp.cutoff=0.25,cex=0.6,legend_settings=
    list(x=0.2*max(nodeHeights(mammal.tree)),
    y=0.7*Ntip(mammal.tree)))
mtext("(b)",adj=0,line=1)
```

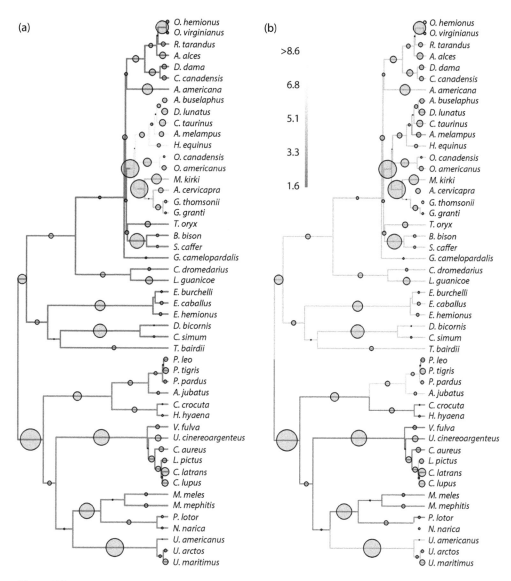

Figure 5.8
Phylogeny showing the results of a *bayou* rjMCMC analysis of the multi-regime OU model. Different-sized circles show the posterior probability of a regime shift on the corresponding edge. (a) All regimes with posterior probability of a regime shift >0.25 mapped using different colors. (b) Colors of each edge in the tree show the mean value of θ for that edge from the posterior sample.

In this plot (figure 5.8a), we can see that there are numerous places in the phylogeny with medium support for a regime shift in body size in mammals. In panel (b), we see that these shifts correspond sometimes to increases in body size, shown by lighter heat colors in the figure, and other times to smaller body sizes.

5.5 Practice problems

5.1 Repeat the *Anolis* analysis in this chapter with the *OUwie* package, but this time analyze PC1 instead of PC3. Describe and quantify how your results differ.

5.2 Create a new variable that classifies anoles into either the "trunk-ground" ecomorph or "other." Using this new binary variable, carry out an analysis of the evolution of PC1 to PC3 using `evol.vcv` of the *phytools* package. What do you find?

5.3 In chapter 3, we explored some data on microbial genome size. That data set also has a variable giving DNA GC content (`"GC_Content_Percent"`). Use that variable to divide the bacteria into two categories, `"low_GC"` and `"high_GC"` (for best results, you might choose a dividing point that eventually splits the taxa into two groups). Now, use some analyses from this chapter to see if these two regimes have distinct evolutionary models for genome size, mutation accumulation rate, or the bivariate model including both of these variables.

Modeling discrete character evolution on a phylogeny

6

6.1 Introduction

So far in this book, we've focused on analyzing the evolution of continuously valued traits, such as overall body size, limb length, or home range size.

Obviously, not all phenotypic traits of interest to evolutionary biologists are continuously distributed.

Many traits, such as the presence or absence of a feature, can only be recorded in a discontinuous or *discrete* fashion. Others, such as color or habitat type, could be continuous *in theory* but most often (or most conveniently) will be discretely coded by investigators.

These characters evolve, and just like the continuous characters we've studied in previous chapters, their evolution also occurs in the context of a phylogenetic tree. The analysis of discretely valued characteristics on a phylogeny will be the topic of the next two chapters. We'll focus on fitting models of discrete character evolution that span a range from simple to complex, and on using those models to describe how characters evolve.

In this chapter, we will:

1. Use simulation to introduce the Mk model, a model for the evolution of discrete characters on trees.
2. Show how to fit the Mk model to data and how to estimate model parameters.
3. Fit three different, commonly used discrete models, called the equal-rates, symmetric, and all-rates-different models, and show how to compare among models using likelihood and Akaike information criterion (AIC).
4. Describe how to implement a custom Mk model using a design matrix.
5. Apply all of these approaches to analyze the evolution of digit number in squamates.

6.2 The Mk model

The dominant model for the evolution of discrete characters on phylogenies is a model that has been named the Mk model (Lewis 2001). The Mk model is so denominated because it describes a continuous-time, discrete k-state **Markov** process. This model was in turn inspired directly by models of sequence evolution (e.g., those in Yang 2006).

The Mk process is one in which changes can occur between states at any time, and the rate of change[1] to other states depends only on the current state of the chain and not on any prior state or on how long the chain has been in its current state.

In other words, the process has no historical "memory." Instead, all that matters is the current state and the rate of change from that state to others.[2]

When evolution occurs by this process, the waiting times between changes will be exponentially distributed, with a shape parameter that depends only on the rate of change between states, typically denoted q (Yang 2006).

When the value of q is large, the rate is fast and, thus, waiting times between changes will tend to be short. Conversely, small values of q will lead to rarer changes and longer waiting times between them.[3]

The Mk model was originally defined by Lewis (2001) to include a single rate of transition between states; however, the term[4] has come to be used for all varieties of k-state Markov models, including variants of the model in which the rates of changes are allowed to differ between different states of the chain.[5]

In this more general case, we use a $k \times k$ matrix (normally denoted **Q**) to represent *all* the different rates of change between states (reviewed in Harmon 2019).

In this **Q** matrix, the element $Q_{i,j}$ gives the instantaneous rate of change between the states i and j. As such, you can think of the **Q** matrix as a "map" of the rates of change between character states, with each entry in row i and column j giving a rate of change from state (row) i to state (column) j.

6.2.1 Simulating character histories under the Mk model

We feel like one of the best ways to understand the properties of a mathematical model is to simulate data under the model.

In chapter 4, we simulated Brownian motion evolution for one or many lineages and on a phylogenetic tree to develop a better understanding of this important continuous-trait model. We can do much the same with the Mk model.

Multiple R functions can be used to simulate data under the Mk model. Here, we'll use the *phytools* function `sim.history`, which simulates the entire history of a character on the tree—not just the end state at the tips—and we'll use it to simulate data for a binary (0/1) character under three different evolutionary scenarios.[6]

Figure 6.1 shows the result.

---

[1] Or, equivalently, the probability of change in the next infinitesimal time interval.

[2] This is what makes the process *Markovian*, by definition.

[3] The mean waiting time is actually just $1/q$.

[4] Or, alternatively, the *extended* Mk model (e.g., Harmon 2019).

[5] Lewis (2001) also focused on estimating the branch lengths of the tree, while we will consider them fixed and estimate the rate of character evolution.

[6] For simulating lots of different discrete traits, or for conducting many replicate simulations, `phytools::sim.history` is very slow. Under those circumstances, we'd recommend a different simulator, such as `phytools::sim.Mk`, which is faster because it simulates only the states at the tips of the tree and not the full character history.

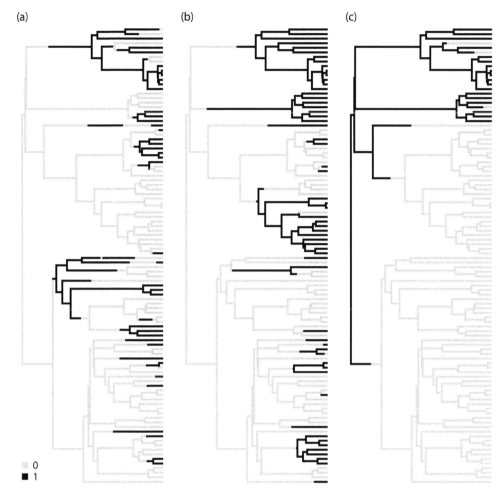

(a)   (b)   (c)

0
1

Figure 6.1
Simulated discrete character evolution with (a) an equal rate of transition between the two states, (b) a
higher rate of change from 0 (gray) to 1 (black) than the reverse, and (c) irreversible evolution from 1 to 0.

```
## load phytools
library(phytools)
## assign colors for the two states
colors<-setNames(c("gray","black"),0:1)
colors
```

```
##       0       1
## "gray"   "black"
```

```
## simulate a stochastic pure-birth tree with 100 taxa
tree<-pbtree(n=100,scale=1)
## divide plotting area into three panels
```

```
par(mfrow=c(1,3))
## Q matrix for simulation 1: equal backward & forward
## rates (ER)
Q1<-matrix(c(-1,1,1,-1),2,2,
    byrow=TRUE,dimnames=list(0:1,0:1))
Q1
```

```
##    0  1
## 0 -1  1
## 1  1 -1
```

```
## simulate ER character evolution and plot
plot(sim.history(tree,Q1,message=FALSE),
    colors,ftype="off",mar=c(1.1,1.1,1.1,0.1))
mtext("(a)",line=-1,adj=0)
legend(x="bottomleft",legend=c("0","1"),
    pt.cex=1.5,pch=15,col=colors,
    bty="n")
## Q matrix for simulation 2: different backward &
## forward transition rates (ARD)
Q2<-matrix(c(-1,1,0.25,-25),2,2,
    byrow=TRUE,dimnames=list(0:1,0:1))
Q2
```

```
##       0   1
## 0 -1.00   1
## 1  0.25 -25
```

```
## simulate ARD character evolution and plot
plot(sim.history(tree,Q2,
    direction="row_to_column",message=FALSE),
    colors,ftype="off",mar=c(1.1,1.1,1.1,0.1))
mtext("(b)",line=-1,adj=0)
## Q matrix for (effectively) irreversible trait
## evolution (changes from 1->0, but not the reverse)
Q3<-matrix(c(-1e-12,1e-12,1,-1),2,2,
    byrow=TRUE,dimnames=list(0:1,0:1))
Q3
```

```
##        0       1
## 0 -1e-12   1e-12
## 1  1e+00  -1e+00
```

```
## simulate irreversible character evolution and plot
plot(sim.history(tree,Q3,anc="1",
    direction="row_to_column",
    message=FALSE),colors,ftype="off",
    mar=c(1.1,1.1,1.1,0.1))
mtext("(c)",line=-1,adj=0)
```

To create figure 6.1, we first (a) simulated a rate of change from 0 to 1 and 1 to 0 that was equal to one another. Next, (b) we simulated a rate of change that was four times higher from 0 to 1 than the reverse. Finally, (c) we simulated the evolution of a discrete character that can only change from 1 to 0 but never the reverse.[7],[8]

In each case, you should be able to see the correspondence between the entries in the **Q** matrix and the rates of transition among character states. Likewise, different **Q** matrices lead to different patterns of character state evolution on a phylogenetic tree. For instance, in panel (a), both types of character transition occur, and 0 and 1 character states are more or less evenly distributed across the tips of the tree. By contrast, in panel (c), only transitions from 1 to 0 occur, resulting in entire clades fixed for the 0 state.

In the next section, we'll use maximum likelihood estimation and model selection to try and detect these differences in the process of trait evolution from data.

6.3 Fitting the M*k* model to data

Of course, not only can we simulate data under the M*k* (or extended M*k*) model, but we can also fit the model to our phylogeny and trait data.

The idea here is quite simple. For any *particular* transition matrix, **Q**, we compute its likelihood as the probability of a pattern of character states at the tips of the tree. If we proceed to find the value of **Q** that maximizes this probability, we'll have found the maximum likelihood estimate (MLE) of **Q**[9] (Yang 2006).

To learn how to fit models of discrete character evolution, we're going to proceed and use a phylogenetic tree and a data set from a study by Brandley et al. (2008).[10]

The phylogeny is a large tree of snakes and lizards, and the data consist of the number of digits (toes) in the hindfoot of these animals (Brandley et al. 2008).

As such, the two files we'll use are as follows: `squamate-data.csv` and `squamate.tre`. Both can be downloaded from the book webpage.[11]

Let's next load these two data files into R and then graph our phylogeny in a circular or "fan" style (figure 6.2).

```
library(geiger)
## read data matrix
sqData<-read.csv("squamate-data.csv",row.names=1)
## print dimensions of our data frame
dim(sqData)

## [1] 120   1
```

---

[7] Instead of simulating a rate of transition from 0 to 1 of zero, we simulated a rate so close to zero, the type of change will never occur. This is for computational reasons beyond the scope of this chapter.

[8] The *phytools* `sim.history` function allows us to specify **Q** in either row-column or column-row order using the argument `direction`—that is, such that the transition rate from *i* to *j* is in position $[i,j]$ or $[j,i]$, respectively.

[9] We talked a bit more about likelihood in chapter 4.

[10] We're actually using a reduced version of these data because the original tree and data are a bit large.

[11] http://www.phytools.org/Rbook/.

```
## read phylogenetic tree
sqTree<-read.nexus("squamate.tre")
print(sqTree,printlen=2)

##
## Phylogenetic tree with 258 tips and 257 internal nodes.
##
## Tip labels:
##    Abronia_graminea, Acontias_litoralis, ...
##
## Rooted; includes branch lengths.

## plot our tree
plotTree(sqTree,type="fan",lwd=1,fsize=0.3,ftype="i")
```

Seeing as they include different total numbers of taxa (120 in the data frame and 258 in the tree), it is clear that there must be some differences between the tips in our tree and the rows of our data matrix. Let's address this as we have in prior exercises.[12]

```
## check name matching
chk<-name.check(sqTree,sqData)
summary(chk)

## 139 taxa are present in the tree but not the data:
##      Abronia_graminea,
##      Acontias_litoralis,
##      Acontophiops_lineatus,
##      Acrochordus_granulatus,
##      Agamodon_anguliceps,
##      Agkistrodon_contortrix,
##      ....
## 1 taxon is present in the data but not the tree:
##      Trachyboa_boulengeri
##
## To see complete list of mis-matched taxa, print object.

## drop tips of tree that are missing from data matrix
sqTree.pruned<-drop.tip(sqTree,chk$tree_not_data)
## drop rows of matrix that are missing from tree
sqData.pruned<-sqData[!(rownames(sqData)%in%
    chk$data_not_tree),,drop=FALSE]
```

Now, let's pull out the character vector we're going to use as follows:

---

[12] The unusual syntax X[,drop=FALSE] is used to tell R not to drop the row names of our data frame, even if it only consists of a single column!

Figure 6.2
Phylogeny of 258 squamate reptiles from Brandley et al. (2008).

```
## extract discrete trait
toes<-setNames(as.factor(sqData.pruned[,"rear.toes"]),
    rownames(sqData.pruned))
head(toes)

##          Acontias_meleagris
##                           0
##          Acontias_percivali
##                           0
##    Alopoglossus_atriventris
##                           5
##          Alopoglossus_copii
```

```
##                              5
##              Ameiva_ameiva
##                              5
## Amphiglossus_igneocaudatus
##                              5
## Levels: 0 1 2 3 4 5
```

We're ready to fit our models now.

To do this, we'll use the function `fitDiscrete` in *geiger*.

Other R functions also fit an M*k* model, such as `fitMk` in the *phytools* package. Some differences[13] exist between the default model that is implemented in the two functions, but both are valid.

6.3.1 The equal-rates (ER) model

The first variant of the M*k* model that we're going to fit is called the "equal-rates" or ER model (figure 6.1a). This model is exactly as it sounds—it fits a single transition rate between all pairs of states for our discrete trait (Harmon 2019).

```
## fit ER model to squamate toe data using fitDiscrete
fitER<-fitDiscrete(sqTree.pruned,toes,model="ER")
print(fitER,digits=3)

## GEIGER-fitted comparative model of discrete data
##  fitted Q matrix:
##            0         1         2         3
##    0 -0.00517   0.00103   0.00103   0.00103
##    1  0.00103  -0.00517   0.00103   0.00103
##    2  0.00103   0.00103  -0.00517   0.00103
##    3  0.00103   0.00103   0.00103  -0.00517
##    4  0.00103   0.00103   0.00103   0.00103
##    5  0.00103   0.00103   0.00103   0.00103
##            4         5
##    0  0.00103   0.00103
##    1  0.00103   0.00103
##    2  0.00103   0.00103
##    3  0.00103   0.00103
##    4 -0.00517   0.00103
##    5  0.00103  -0.00517
##
##  model summary:
##  log-likelihood = -134.825059
##  AIC = 271.650118
```

---

[13] The main difference between `fitContinuous` and `fitMk` is the assumption that each function makes about the state at the root node of the tree. In most cases, this difference will be of little consequence, but occasionally, it can affect the inferred model to a larger extent. We recommend thinking carefully about this assumption rather than taking it for granted.

```
##  AICc = 271.684306
##  free parameters = 1
##
## Convergence diagnostics:
##  optimization iterations = 100
##  failed iterations = 0
##  number of iterations with same best fit = 100
##  frequency of best fit = 1.00
##
##  object summary:
##  'lik' -- likelihood function
##  'bnd' -- bounds for likelihood search
##  'res' -- optimization iteration summary
##  'opt' -- maximum likelihood parameter estimates
```

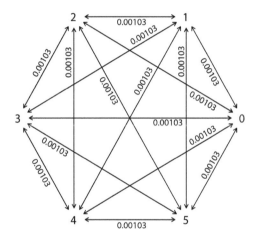

Figure 6.3
Fitted equal-rates (ER) model.

This shows the parameter estimate, log-likelihood AIC, and convergence diagnostics for our ER fit. For an ER model, there is just one model parameter, the transition rate, q. We estimate $q = 0.00103$.

How should we think about our estimated value of q? Remember, q is our instantaneous transition rate among states. As such, the expected number of transitions given a particular amount of time, t, can be calculated as the simple product of q and t (Yang 2006).[14]

In addition to printing out the results of our fitted model, we can also graph the model. To do this, we'll use another generic plotting method from the *phytools* package. The result can be seen in figure 6.3.

---

[14] In our particular case, since each character state has five other conditions that it can change to, all at the same rate q, the total number of expected changes on the tree is $5qT$ in which T is the sum of the edge lengths of the tree.

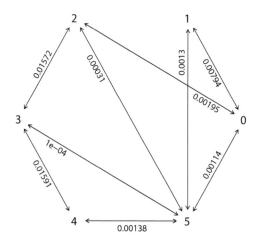

Figure 6.4
Fitted symmetric transition (SYM) model.

```
## plot fitted ER model
plot(fitER,mar=rep(0,4),signif=5)
```

6.3.2 The symmetric transition model (SYM)

In addition to the ER model, another popular discrete character model is a model called the "symmetric rates" model, SYM (Harmon 2019).

This model assumes that the rate of transition from each character state i to each state j is equal to the rate of change from state j to state i but that each *pair* of character states can have a different rate.

Let's go ahead and fit this model to our squamate toe data and then plot[15] our fitted model (figure 6.4).

```
## fit SYM model
fitSYM<-fitDiscrete(sqTree.pruned,toes,model="SYM")
print(fitSYM,digits=3)

## GEIGER-fitted comparative model of discrete data
##  fitted Q matrix:
##              0         1         2         3
##     0 -1.10e-02  7.94e-03  1.95e-03  1.11e-59
##     1  7.94e-03 -9.24e-03  8.48e-48  1.73e-76
##     2  1.95e-03  8.48e-48 -1.80e-02  1.57e-02
##     3  1.11e-59  1.73e-76  1.57e-02 -3.17e-02
##     4  1.18e-71  1.12e-81  2.46e-34  1.59e-02
```

---

[15]By setting the argument show.zeros to be FALSE, we will turn off all the arrows on our plot that represent rates that are very close to zero. Leaving only arrows between transitions that (our model thinks) occur gives us a better sense of the structure of the fitted model that is implied by our data.

```
##    5  1.14e-03  1.30e-03  3.14e-04  1.03e-04
##                4          5
##    0  1.18e-71  0.001139
##    1  1.12e-81  0.001299
##    2  2.46e-34  0.000314
##    3  1.59e-02  0.000103
##    4 -1.73e-02  0.001381
##    5  1.38e-03 -0.004236
##
##  model summary:
##  log-likelihood = -123.093649
##  AIC = 276.187298
##  AICc = 280.847492
##  free parameters = 15
##
## Convergence diagnostics:
##  optimization iterations = 100
##  failed iterations = 0
##  number of iterations with same best fit = 2
##  frequency of best fit = 0.02
##
##  object summary:
##  'lik' -- likelihood function
##  'bnd' -- bounds for likelihood search
##  'res' -- optimization iteration summary
##  'opt' -- maximum likelihood parameter estimates
```

```
## graph fitted SYM model
plot(fitSYM,show.zeros=FALSE,mar=rep(0,4),signif=5)
```

This time, you can see that our **Q** matrix has many more unique entries, all corresponding to transition rates. Notice also that this matrix is symmetric, with, for example, $Q_{1,2} = Q_{2,1}$, $Q_{2,3} = Q_{3,2}$, and so on. You can also see the likelihood and AIC score for this model and, if you want, compare these to the ER results.

6.3.3 All-rates-different model

Within the extended Mk framework, the most complicated model imaginable[16] is one in which every type of transition is allowed to occur with a different rate. This is called the "all-rates-different" model or ARD (Harmon 2019).

Just as we did for the ER and SYM model, let's try to fit this model to our squamate toe data.

```
## fit ARD model
fitARD<-fitDiscrete(sqTree.pruned,toes,model="ARD")
print(fitARD,digits=3)
```

---

[16]That is, assuming a constant process across all the branches and nodes of the phylogeny—we'll relax this assumption in chapter 7.

```
## GEIGER-fitted comparative model of discrete data
##   fitted Q matrix:
##               0           1           2           3
##     0  -5.05e-06   4.17e-06   3.69e-07   3.06e-07
##     1   1.35e-02  -1.35e-02   3.56e-06   7.91e-07
##     2   1.61e-02   1.04e-05  -3.07e-02   1.46e-02
##     3   2.21e-03   9.63e-04   8.91e-03  -5.87e-02
##     4   5.96e-06   6.64e-03   6.81e-03   1.57e-02
##     5   6.45e-08   1.46e-03   4.88e-05   4.79e-05
##               4           5
##     0   1.42e-07   6.51e-08
##     1   6.07e-07   4.83e-07
##     2   1.24e-05   7.10e-06
##     3   3.12e-02   1.54e-02
##     4  -6.85e-02   3.93e-02
##     5   3.50e-03  -5.05e-03
##
##   model summary:
##   log-likelihood = -109.937556
##   AIC = 279.875112
##   AICc = 301.011476
##   free parameters = 30
##
## Convergence diagnostics:
##   optimization iterations = 100
##   failed iterations = 0
##   number of iterations with same best fit = 2
##   frequency of best fit = 0.02
##
##   object summary:
##   'lik' -- likelihood function
##   'bnd' -- bounds for likelihood search
##   'res' -- optimization iteration summary
##   'opt' -- maximum likelihood parameter estimates
```

```
## plot fitted model
plot(fitARD,show.zeros=FALSE,mar=rep(0,4),signif=5)
```

For the ARD model, all of the entries in the $\mathbf{Q}$ matrix are unique, and the matrix is no longer symmetric (figure 6.5). Below we will compare the fit of all three of these models in various ways, but not before we expand our candidate model set a bit!

6.3.4 Custom transition models: An ordered evolution model

So far, we've already seen what are most likely the three most commonly used models for discrete character evolution in phylogenetic comparative analyses: ER, SYM, and ARD.

However, these three models do not by any means completely comprise the *range* of possible models we could imagine fitting to the evolution of a discrete character trait on the tree!

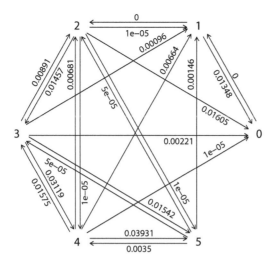

Figure 6.5
Fitted all-rates-different (ARD) model.

For instance, we might hypothesize that toes can be lost and not reacquired, or perhaps that toes can be lost and regained, but that only changes between adjacent states (e.g., from five to four, from two to three, and so on) can occur (Brandley et al. 2008).

fitDiscrete[17] gives us the flexibility to specify this kind of model. The way we need to do it is by first creating a *design matrix*.

Our design matrix will be the same size as **Q**, but we'll populate it with integers.

Each position in the matrix with a nonzero positive integer indicates a type of change that can occur, and any cells in the matrix with the same integer will have the same rate of change. We put zeros on the diagonal of our matrix, as well as in cells of the matrix that correspond to changes that are not permitted to occur.[18]

Let's fit two different models now. The first of these is an *ordered* model in which changes are permitted between all pairs of adjacent states and in which we will allow these changes to occur at different rates.

The second of these will be a *directional* model in which we suppose that digits can be lost but cannot be reacquired. We further suppose that every different type of digit loss (i.e., from five to four digits, from four to three digits, etc.) is allowed to proceed at a different rate.

We can start by building our design matrices. Remember, the integers in this matrix just correspond to the different parameters that we want R to estimate from our data—and their relative values thus indicate nothing at all about the values of those parameters.

```
## create design matrix for bi-directional
## ordered model
```

---

[17] And, likewise, phytools::fitMk.

[18] Under our model. Remember, this is just a *hypothesis* about how evolution transpired.

Discrete character evolution

```
ordered.model<-matrix(c(
    0,1,0,0,0,0,
    2,0,3,0,0,0,
    0,4,0,5,0,0,
    0,0,6,0,7,0,
    0,0,0,8,0,9,
    0,0,0,0,10,0),6,6,byrow=TRUE,
    dimnames=list(0:5,0:5))
ordered.model
```

```
##   0 1 2 3  4 5
## 0 0 1 0 0  0 0
## 1 2 0 3 0  0 0
## 2 0 4 0 5  0 0
## 3 0 0 6 0  7 0
## 4 0 0 0 8  0 9
## 5 0 0 0 0 10 0
```

```
## create design matrix for directional ordered
## model
directional.model<-matrix(c(
    0,0,0,0,0,0,
    1,0,0,0,0,0,
    0,2,0,0,0,0,
    0,0,3,0,0,0,
    0,0,0,4,0,0,
    0,0,0,0,5,0),6,6,byrow=TRUE,
    dimnames=list(0:5,0:5))
directional.model
```

```
##   0 1 2 3 4 5
## 0 0 0 0 0 0 0
## 1 1 0 0 0 0 0
## 2 0 2 0 0 0 0
## 3 0 0 3 0 0 0
## 4 0 0 0 4 0 0
## 5 0 0 0 0 5 0
```

You can count the number of parameters in a design matrix by counting the number of unique positive integers. So, we can already see that fitting our ordered model will result in a total of ten estimated parameters, whereas fitting our directional model will result in the estimation of only five parameters.

We're now ready to fit these two models to our data. Let's do that and then plot them.[19]

---

[19] We're setting the fitDiscrete argument surpressWarnings=TRUE, but this is just because one of the warnings printed by fitDiscrete is just to tell us that some of our parameter estimates could be at their bounds. Since we are forcing some of the transition rates in our ordered models to be zero, and since zero always corresponds to the lower bound for q, we're not concerned about this particular warning here. On the other hand, we would strongly *recommend against* setting supressWarnings to be TRUE under other circumstances as it could cause you to miss an important warning!

```
## fit bi-directional ordered model
fitOrdered<-fitDiscrete(sqTree.pruned,toes,
    model=ordered.model,surpressWarnings=TRUE)
print(fitOrdered,digits=3)
```

```
## GEIGER-fitted comparative model of discrete data
##   fitted Q matrix:
##               0          1          2        3
##     0 -5.78e-11  5.78e-11  0.00e+00   0.0000
##     1  2.23e-02 -2.23e-02  1.18e-12   0.0000
##     2  0.00e+00  6.63e-02 -2.72e+00   2.6505
##     3  0.00e+00  0.00e+00  3.38e+00  -3.3754
##     4  0.00e+00  0.00e+00  0.00e+00   0.0264
##     5  0.00e+00  0.00e+00  0.00e+00   0.0000
##               4          5
##     0  0.00e+00  0.00000
##     1  0.00e+00  0.00000
##     2  0.00e+00  0.00000
##     3  3.58e-20  0.00000
##     4 -6.63e-02  0.03991
##     5  5.82e-03 -0.00582
##
##   model summary:
##   log-likelihood = -114.971488
##   AIC = 249.942977
##   AICc = 251.980014
##   free parameters = 10
##
## Convergence diagnostics:
##   optimization iterations = 100
##   failed iterations = 0
##   number of iterations with same best fit = 4
##   frequency of best fit = 0.04
##
##   object summary:
##   'lik' -- likelihood function
##   'bnd' -- bounds for likelihood search
##   'res' -- optimization iteration summary
##   'opt' -- maximum likelihood parameter estimates
```

```
## fit directional (loss only) ordered model
fitDirectional<-fitDiscrete(sqTree.pruned,toes,
    model=directional.model,surpressWarnings=TRUE)
print(fitDirectional,digits=3)
```

```
## GEIGER-fitted comparative model of discrete data
##   fitted Q matrix:
##               0          1        2        3        4
```

Discrete character evolution

```
##      0 0.0000   0.0000   0.000   0.0000   0.00000
##      1 0.0242  -0.0242   0.000   0.0000   0.00000
##      2 0.0000   0.1031  -0.103   0.0000   0.00000
##      3 0.0000   0.0000   0.109  -0.1092   0.00000
##      4 0.0000   0.0000   0.000   0.0833  -0.08329
##      5 0.0000   0.0000   0.000   0.0000   0.00437
##                5
##      0  0.00000
##      1  0.00000
##      2  0.00000
##      3  0.00000
##      4  0.00000
##      5 -0.00437
##
##  model summary:
##  log-likelihood = -118.387049
##  AIC = 246.774098
##  AICc = 247.305071
##  free parameters = 5
##
## Convergence diagnostics:
##  optimization iterations = 100
##  failed iterations = 0
##  number of iterations with same best fit = 13
##  frequency of best fit = 0.13
##
##  object summary:
##  'lik' -- likelihood function
##  'bnd' -- bounds for likelihood search
##  'res' -- optimization iteration summary
##  'opt' -- maximum likelihood parameter estimates
```

We can also plot these two fitted models, just as we did for our ER, SYM, and ARD models earlier. Why don't we combine them into a single, two-panel figure using par?

```
## split plot area into two panels
par(mfrow=c(1,2))
## plot ordered and directional models
plot(fitOrdered,show.zeros=FALSE,signif=5,
    mar=c(0.1,1.1,0.1,0.1))
mtext("(a)",line=-2,adj=0,cex=1.5)
plot(fitDirectional,show.zeros=FALSE,signif=5,
    mar=c(0.1,1.1,0.1,0.1))
mtext("(b)",line=-2,adj=0,cex=1.5)
```

Comparing these two models, you can see (figure 6.6) that the ordered model allows toe number to increase or decrease by one step at a time, while the directional model does

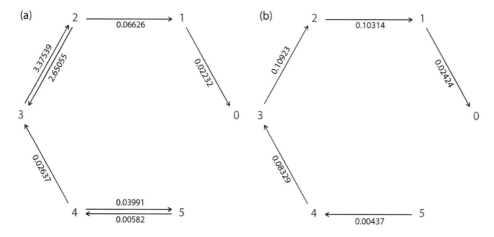

Figure 6.6
(a) Fitted ordered model. (b) Fitted directional model.

not—just as designed. You can also see that the parameter estimates are quite different between the two ML model fits.

The number of different models we can fit to discrete character data is virtually limited only by the bounds of our imagination; however, as a general rule, we typically recommend fitting models that are relatively simple, biologically justifiable, or (ideally) both.

6.4 Comparing alternative discrete character models

We've now seen how to fit different models to our data vector using likelihood.

With likelihood, it's also relatively straightforward to *compare* alternative models and thus identify the model that is best supported by our data.

One simple way to do this is using the likelihood ratio test. According to the theory of likelihoods, two times the difference in log-likelihoods of *nested* models[20] should be distributed as a χ^2 with degrees of freedom equal to the difference in the number of parameters between the two fitted models (Wilks 1938).

To run a likelihood ratio test in R we will use the package *lmtest* (Zeileis and Hothorn 2002). Let's load[21] *lmtest* and then run our test.

We can start by comparing the ER, SYM, and ARD models as follows.

```
library(lmtest)
## likelihood-ratio test comparing ER & SYM
lrtest(fitER,fitSYM)
```

---

[20]Nested models consist of a pair of models in which model A has model B as a special case. That is, we can write down model B as a particular example of model A. Among the models we've seen in this chapter, model ARD has model ER as a special case—an ARD model in which all rates are equal—and thus these two models can be compared using a likelihood ratio test. By contrast, our ordered model *does not* have the ER model as a special case, and as such, these models *are not* nested and thus cannot be compared using a likelihood ratio.

[21]If *lmtest* has not been installed, you should obviously first install it from CRAN.

```
## Likelihood ratio test
##
## Model 1: fitER
## Model 2: fitSYM
##   #Df  LogLik Df  Chisq Pr(>Chisq)
## 1    1 -134.82
## 2   15 -123.09 14 23.463    0.05314 .
## ---
## Signif. codes:
## 0 '***' 0.001 '**' 0.01 '*' 0.05 '.' 0.1 ' ' 1
```

```
## likelihood-ratio test comparing ER & ARD
lrtest(fitER,fitARD)
```

```
## Likelihood ratio test
##
## Model 1: fitER
## Model 2: fitARD
##   #Df  LogLik Df  Chisq Pr(>Chisq)
## 1    1 -134.82
## 2   30 -109.94 29 49.775   0.009549 **
## ---
## Signif. codes:
## 0 '***' 0.001 '**' 0.01 '*' 0.05 '.' 0.1 ' ' 1
```

```
## likelihood-ratio test comparing SYM & ARD
lrtest(fitSYM,fitARD)
```

```
## Likelihood ratio test
##
## Model 1: fitSYM
## Model 2: fitARD
##   #Df  LogLik Df  Chisq Pr(>Chisq)
## 1   15 -123.09
## 2   30 -109.94 15 26.312    0.03487 *
## ---
## Signif. codes:
## 0 '***' 0.001 '**' 0.01 '*' 0.05 '.' 0.1 ' ' 1
```

In this set of hypothesis tests, we first compared the ER and SYM models. Here, since $P > 0.05$, we *failed* to reject the simpler ER model.

Next, we compared the ER and ARD models. In this second case, since $P < 0.05$, we rejected the simpler ER model in favor of the ARD model.

Finally, we compared the SYM and ARD models, also rejecting the simpler SYM model.

In addition to this set of comparisons, we can also compare our reversible ordered and our directional models, and we can compare either of these models to the ARD model.[22] Let's do just that.

---

[22] Because both are also special cases of ARD. On the other hand, we *could not* compare our ordered models and the ER model, even though the latter has greater complexity, because the ER model does not have either of our ordered models as a special case.

```
## compare directional and ordered
lrtest(fitDirectional,fitOrdered)

## Likelihood ratio test
##
## Model 1: fitDirectional
## Model 2: fitOrdered
##    #Df  LogLik Df  Chisq Pr(>Chisq)
## 1    5 -118.39
## 2   10 -114.97  5 6.8311     0.2335

## compare direction and ARD
lrtest(fitDirectional,fitARD)

## Likelihood ratio test
##
## Model 1: fitDirectional
## Model 2: fitARD
##    #Df  LogLik Df  Chisq Pr(>Chisq)
## 1    5 -118.39
## 2   30 -109.94 25 16.899     0.8854

## compare ordered and ARD
lrtest(fitOrdered,fitARD)

## Likelihood ratio test
##
## Model 1: fitOrdered
## Model 2: fitARD
##    #Df  LogLik Df  Chisq Pr(>Chisq)
## 1   10 -114.97
## 2   30 -109.94 20 10.068     0.9669
```

Here in all three comparisons, we *failed* to reject the simpler model (the directional model in the former two comparisons and the reversible ordered model in the lattermost).

Via this set of comparisons, we're starting to build a picture that the ordered models may be better supported by the data than the ER, SYM, and ARD models.

This makes some degree of both biological and mathematical sense.

First, it's logical to imagine based on developmental biology that a model in which evolution tends to involve successive losses (or gains) of digits may be more consistent with our data than a model in which any and all types of changes can occur—particularly if that model also supposes that these changes occur at the same rate (Brandley et al. 2008)!

Second, these two models generally involve the estimation of fewer parameters (five or ten) than the considerably more complex SYM (fifteen) and ARD (thirty) models.

Nonetheless, it would be convenient if we could compare all models to each other. In fact, we can do exactly that using the Akaike information criterion, AIC. Let's try.

Here, we'll sort the models in order of complexity, but we don't have to do that—we could have listed our models in any order.

```
## accumulate AIC scores of all five models into
## a vector
aic<-setNames(c(AIC(fitER),AIC(fitDirectional),
    AIC(fitOrdered),AIC(fitSYM),AIC(fitARD)),
    c("ER","Directional","Ordered","SYM","ARD"))
aic
```

```
##          ER Directional      Ordered          SYM
##    271.6501     246.7741     249.9430     276.1873
##         ARD
##    279.8751
```

This result tells us that the "directional" model is the model best supported by the data because it has the lowest AIC score.

AIC already takes the number of parameters estimated from the data into account, so there is no additional correction necessary to reach this conclusion. Nonetheless (as a rough rule of thumb), many investigators consider AIC scores within around two units of each other to indicate similar or ambiguous support for the models under comparison (Burnham and Anderson 2003). Since our AIC score for the directional model is about three units better than the next best-supported model, we can feel reasonably confident that the directional model is the best of this set.

From AIC scores, we can also compute Akaike weights. Akaike weights show the weight of evidence in support of each model in our data. Let's compute these using a function called aic.w in the *phytools* package.

```
aic.w(aic)
```

```
##          ER Directional      Ordered          SYM
##  0.00000329  0.82982929  0.17016703  0.00000034
##         ARD
##  0.00000005
```

Finally, let's combine all of these different comparisons into a single table.

```
round(data.frame(
    k=c(fitER$opt$k,fitDirectional$opt$k,
    fitOrdered$opt$k,fitSYM$opt$k,fitARD$opt$k),
    logL=c(logLik(fitER),logLik(fitDirectional),
    logLik(fitOrdered),logLik(fitSYM),logLik(fitARD)),
    AIC=aic,Akaike.w=as.vector(aic.w(aic))),3)
```

```
##               k      logL      AIC Akaike.w
## ER            1  -134.825  271.650     0.00
## Directional   5  -118.387  246.774     0.83
## Ordered      10  -114.971  249.943     0.17
## SYM          15  -123.094  276.187     0.00
## ARD          30  -109.938  279.875     0.00
```

The weight for the directional model is much higher than the rest. This affirms our earlier conclusion that a directional model (in which digits tend to be lost but not reacquired) is best supported in the set, but an alternative model in which digits are both lost and regained, albeit in an ordered fashion, also garners some support. The other models (ER, SYM, and ARD) are relatively poorly supported by the data.

In summary, the Mk model and its relatives provide a flexible and powerful way to evaluate the evolution of a discrete character on a phylogenetic tree. In the next chapters, we will add even more to our discrete character toolbox.

6.5 Practice problems

6.1 Using the same squamate data set and tree as in the chapter, fit both an ordered and a directional model, but in which you constrain the ordered model to have just one rate of digit loss and a separate rate of digit gain ($k = 2$) and in which you constrain the directional model to a single rate of digit loss ($k = 1$). How can you compare these two models to the multi-rate ordered and directional models that we fit to the same data? What does this comparison reveal?

6.2 Now fit a model where digits can only be gained and never lost. Compare this model to the other candidates. What do you conclude?

6.3 Using a simulation study, determine the type I error rate of fitting an ARD model to a two-state character evolved under an ER model on a phylogenetic tree with 100 species.

Other models of discrete character evolution

7

7.1 Introduction

In chapter 6, we learned about using the extended Mk model to study the evolution of a single discrete character trait on a phylogenetic tree.

Although we discussed a range of different flavors of this model (in which, for instance, backward and forward rates of change between character states could assume different values, or in which some types of character change are not permitted), we haven't covered a number of *other* important scenarios for the evolution of discrete character traits. For instance, what if your data consist of values for more than one character? What if some species are *polymorphic* (that is, exhibit more than one character condition) for your discrete trait? What if your character evolves more quickly in one part of the tree than another?

In this chapter, we'll:

1. Learn how to fit a model for the correlated evolution of two binary traits on the tree.
2. See how to test a hypothesis for a change in the rate or evolutionary process for a discrete trait in different parts of the phylogeny.
3. Learn about a hidden-rate model of discrete trait evolution in which our discrete character evolution can speed up or slow down without specifying an a priori hypothesis of rate heterogeneity.
4. Examine how to fit a model to the evolution of a polymorphic character trait.
5. Finally, discuss a philosophically different model, called the threshold model, and see how it can be used to measure the evolutionary correlation between discrete and continuous characters on a phylogenetic tree.

7.2 Correlated binary traits

The first method we'll discuss in this chapter is one that was first developed by Mark Pagel (1994) to test for an evolutionary relationship between two binary characters. This approach, often called the Pagel94 model, is really quite simple at its core.

According to the method, we'll fit two different models to our data.

The first of these is called the *independent* model. Under this model, sensibly, our two different binary characters evolve independently from one another on the tree. This is exactly equivalent to fitting two, separate M*k* models (from chapter 6) to our two binary traits. Since the characters are being treated as if they evolve independently, the total likelihood of the fitted model is just the product of the likelihoods for each trait.[1]

The second model, by contrast, is usually referred to as the *dependent* model. Under this model, the rate of evolution for character one is allowed to *depend on*[2] the state for character two and/or vice versa (Pagel 1994).

This latter model is also often characterized as a model for *correlated*[3] binary trait evolution. How come?

To understand why, let's imagine the circumstance of two binary (0/1) character traits, *A* and *B*, evolving on the tree.

If it's true that when character *A* changes state $0 \rightarrow 1$, character *B* tends to follow suit[4] (and vice versa), then the rates of change for *B* will depend on *A*—and characters *A* and *B* could be said to be evolving in a correlated manner because certain combinations of states for the two characters (0 + 0 and 1 + 1, in this case) will tend to accumulate disproportionately via the evolutionary process compared to other combinations (i.e., 0 + 1 or 1 + 0).

On the other hand, sometimes Pagel's (1994) model can fit our model much better than an independent evolution model *even if* our two characters cannot genuinely be said to be evolving in a correlated fashion.

This might be the case, for instance, if the rate of evolution for character *A* depends on the *state* for *B*, but with no particular tendency to evolve a particular combination of character values for the two traits.

We'll learn more about this a bit further on in the chapter.

7.2.1 An empirical example of Pagel's (1994) model: Paternal care evolution

For an empirical example of Pagel's (1994) method, we'll use a data set originally published by Benun Sutton and Wilson (2019).

In this study, the authors proposed that male parental care in bony fish might be related to the behavioral trait of pair spawning.[5] This is presumably because pair spawning should provide males with greater confidence in the paternity of their offspring (Benun Sutton and Wilson 2019).

Benun Sutton and Wilson (2019) tested this hypothesis using Pagel's (1994) method, so let's go ahead and do the same.[6]

To follow along, you should first download the tree and phenotypic datafiles (`bonyfish.tre` and `bonyfish.csv`) from the book website[7] and then read them into R.

---

[1] Thus, the total log-likelihood is just the *sum* of log-likelihoods.

[2] That is, vary as a function of.

[3] Rather than, for instance, *state-dependent* binary trait evolution.

[4] That is, also change from 0 to 1.

[5] As opposed to *group* spawning.

[6] For computational expediency, we're using a reduced version of their data set; however, readers who wish to reanalyze the original data set can obtain it from https://datadryad.org/.

[7] http://www.phytools.org/Rbook/.

```
## load the phytools package
library(phytools)
## read the Benun Sutton & Wilson phylogeny
bonyfish.tree<-read.tree(file="bonyfish.tre")
print(bonyfish.tree,printlen=3)
```

```
##
## Phylogenetic tree with 90 tips and 89 internal nodes.
##
## Tip labels:
##   Xenomystus_nigri, Chirocentrus_dorab, Talismania_bifurcata, ...
##
## Rooted; includes branch lengths.
```

```
## read the phenotypic trait data
bonyfish.data<-read.csv(file="bonyfish.csv",row.names=1,
    stringsAsFactors=TRUE)
head(bonyfish.data)
```

```
##                         spawning_mode
## Xenomystus_nigri                 pair
## Chirocentrus_dorab              group
## Talismania_bifurcata            group
## Alepocephalus_tenebrosus        group
## Misgurnus_bipartitus             pair
## Opsariichthys_bidens             pair
##                         paternal_care
## Xenomystus_nigri                 male
## Chirocentrus_dorab               none
## Talismania_bifurcata             none
## Alepocephalus_tenebrosus         none
## Misgurnus_bipartitus             none
## Opsariichthys_bidens             none
```

As we've done in previous chapters, here we set the read.csv argument stringsAsFactors = TRUE to ensure that our discrete character data are read into R as a factor instead of as a simple character string.

Next, let's plot our tree and discrete character data together.

For this we'll use a plotting function from *phytools* that we haven't seen yet, called plotTree.datamatrix, that can be very handy for plotting two or more discrete characters. The result is shown in figure 7.1.

```
## plot the tree with adjacent data matrix
object<-plotTree.datamatrix(bonyfish.tree,bonyfish.data,
    fsize=0.5,yexp=1,header=FALSE,xexp=1.45,
    palettes=c("YlOrRd","PuBuGn"))
## add a legend for trait 1
```

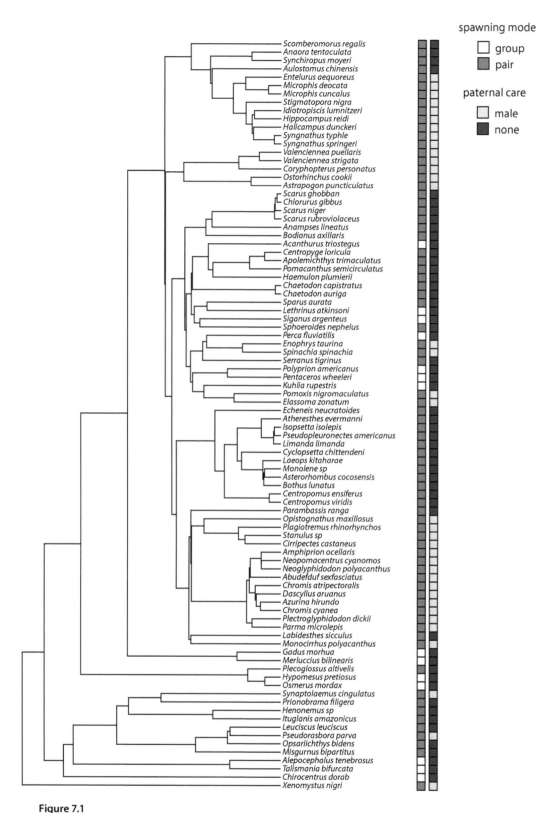

Figure 7.1

A phylogeny for ninety species of bony fishes with the states for two different discrete characters: spawning mode (pair spawning vs. group spawning, in column 1) and paternal care (present or none, in column 2). Phylogeny and data from Benun Sutton and Wilson (2019).

```
leg<-legend(x="topright",names(object$colors$spawning_mode),
    cex=0.7,pch=22,pt.bg=object$colors$spawning_mode,
    pt.cex=1.5,bty="n",title="spawning mode")
## add a second legend for trait 2
leg<-legend(x=leg$rect$left+4.7,y=leg$rect$top-leg$rect$h,
    names(object$colors$paternal_care),cex=0.7,
    pch=22,pt.bg=object$colors$paternal_care,pt.cex=1.5,
    bty="n",title="paternal care")
```

In this code chunk, the call to plotTree.datamatrix graphs the tree and adjacent character matrix. This function *also* invisibly[8] returns an object to the user that contains the color palette used to represent the different states of each trait.

Then, the two subsequent legend calls add the plot legends for each character to the right of the tree (figure 7.1). The base R function legend returns[9] a list containing the coordinates of the plotted legend. We use these coordinates to ensure that our second legend is positioned directly below the first.

Based on this plot (figure 7.1), it's a little hard to tell whether the two different binary traits are evolving dependently or independently.

Let's fit the Pagel (1994) model to find out.

To do that, we'll use the *phytools* function fitPagel.

fitPagel takes the tree and two factor vectors as input, so we'll need to extract these one by one from our bonyfish.data data frame first. Just as we've seen in prior chapters, a convenient way to do this is using the handy R function setNames.

```
spawning_mode<-setNames(bonyfish.data[,1],
    rownames(bonyfish.data))
paternal_care<-setNames(bonyfish.data[,2],
    rownames(bonyfish.data))
```

Then we proceed to fit our model and print the result. fitPagel fits both the dependent *and* the independent models and then compares the likelihoods of the two models.

```
parentalCare.fit<-fitPagel(bonyfish.tree,paternal_care,
    spawning_mode)
print(parentalCare.fit)

##
## Pagel's binary character correlation test:
##
## Assumes "ARD" substitution model for both characters
##
## Independent model rate matrix:
```

---

[8] That's why if we hadn't passed it to the variable object, we'd never see it.

[9] Also invisibly.

```
##              male|group male|pair none|group
## male|group    -0.00303    0.00303    0.00000
## male|pair      0.00194   -0.00194    0.00000
## none|group     0.00223    0.00000   -0.00526
## none|pair      0.00000    0.00223    0.00194
##               none|pair
## male|group     0.00000
## male|pair      0.00000
## none|group     0.00303
## none|pair     -0.00416
##
## Dependent (x & y) model rate matrix:
##              male|group male|pair none|group
## male|group   -0.28748    0.10582    0.18166
## male|pair     0.00000    0.00000    0.00000
## none|group    0.00000    0.00000   -0.00402
## none|pair     0.00000    0.00311    0.00287
##               none|pair
## male|group     0.00000
## male|pair      0.00000
## none|group     0.00402
## none|pair     -0.00598
##
## Model fit:
##              log-likelihood       AIC
## independent       -62.00739 132.0148
## dependent         -55.35818 126.7164
##
## Hypothesis test result:
##    likelihood-ratio:  13.298
##    p-value:  0.0099061
##
## Model fitting method used was fitMk
```

This result shows us that the *dependent* model (that is, the model in which spawning mode can affect paternal care evolution and vice versa) fits significantly better than our independent model.

In our opinion, however, a strong inference of correlated evolution would only be supported if the fitted parameters of the evolutionary process implied that certain character combinations (e.g., pair spawning fish with paternal care, in this example) should accumulate disproportionately over time when compared to other character combinations.

To get a better sense of this, we can also *graph* our fitted models—just as we did with Mk models in chapter 6.

```
plot(parentalCare.fit,signif=4,cex.main=1,
    cex.sub=0.8,cex.traits=0.7,cex.rates=0.7,
    lwd=1)
```

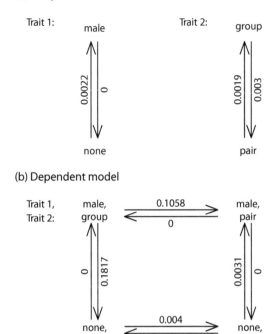

(a) Independent model

(b) Dependent model

Figure 7.2
Fitted Pagel (1994) binary character evolution model for spawning mode and paternal care evolution in a phylogeny of ninety bony fish species. Data and analysis based on Benun Sutton and Wilson (2019).

Neat.

This fitted model (figure 7.2) indeed seems to suggest that the character combinations of pair spawning *with* paternal care and group spawning *without* paternal care should tend to accumulate over time under the process.

We can see this because the rates linked to the arrows that point *toward* our character combinations *male + pair* and *none + group* are, on average, *larger* than the arrows pointing toward the other state combinations. That makes sense.

7.2.2 What about unique evolutionary events?

A few years ago, Maddison and FitzJohn (2015) identified some interesting but underappreciated properties of the Pagel (1994) method. Most significantly, they pointed out that unique or *singular* evolutionary events could lead to significant model fit of the dependent model compared to the independent model.

To understand what we mean, let's imagine the following data pattern (figure 7.3) for three different binary traits on the tree.

```
## set seed to make the example reproducible
set.seed(6)
## simulate a 26 taxon tree with labels
```

```
tree<-pbtree(n=26,tip.label=LETTERS)
## generate three different data patterns on this
## tree (as shown in the figure)
x<-as.factor(setNames(c("a","a",c(rep("b",12),
    rep("a",12))),LETTERS))
y<-as.factor(setNames(c("c","c",c(rep("d",12),
    rep("c",12))),LETTERS))
z<-as.factor(setNames(c("e","e",rep(c("f","e"),6),
    rep("e",12)),LETTERS))
## graph the tree and data using plotTree.datamatrix
object<-plotTree.datamatrix(tree,data.frame(x,y,z),
    fsize=1,space=0.2,xexp=1.4,
    palettes=c("YlOrBr","Greens","Purples"))
## add a point to the edge of the tree where
## coincident changes in traits x & y may have
## occurred
pp<-get("last_plot.phylo",envir=.PlotPhyloEnv)
points(mean(pp$xx[c(29,31)]),pp$yy[31],
    pch=21,cex=1.5,bg="lightblue")
## add a legend
add.simmap.legend(colors=setNames(unlist(object$colors),
```

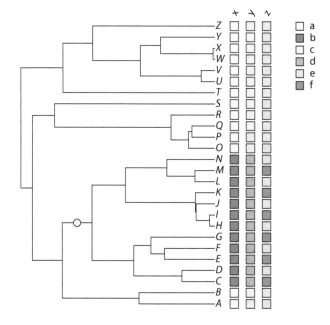

Figure 7.3
Hypothetical tree and data set showing two different types of singular evolutionary events: one in which singular changes in two different characters (*x* and *y*) coincide on the same branch of the tree (indicated by the circle) and a second in which a single change in state for one character (*x*) coincides with a single change in rate for a second (*z*).

```
      letters[1:6]),prompt=FALSE,x=1.05*object$end.x,
      y=Ntip(tree))
```

First, let's test for nonindependent evolution between the character traits x and y.

```
fit.xy<-fitPagel(tree,x,y)
print(fit.xy,digits=4)

##
## Pagel's binary character correlation test:
##
## Assumes "ARD" substitution model for both characters
##
## Independent model rate matrix:
##            a|c      a|d      b|c      b|d
## a|c -0.1718   0.0859   0.0859 0.0000
## a|d  0.0000  -0.0859   0.0000 0.0859
## b|c  0.0000   0.0000  -0.0859 0.0859
## b|d  0.0000   0.0000   0.0000 0.0000
##
## Dependent (x & y) model rate matrix:
##          a|c      a|d      b|c      b|d
## a|c 0.000   0.0000   0.0000 0.0000
## a|d 2.374  -3.0397   0.0000 0.6657
## b|c 2.374   0.0000  -3.0397 0.6657
## b|d 0.000   0.0000   0.0000 0.0000
##
## Model fit:
##                 log-likelihood      AIC
## independent          -10.8808 29.7615
## dependent             -4.9194 25.8387
##
## Hypothesis test result:
##    likelihood-ratio:  11.92
##    p-value:  0.01793
##
## Model fitting method used was fitMk
```

This shows that the nonindependent model fits our data for x and y better than the independent model—and not by small margin! In fact, the difference in log-likelihood is nearly 6 and the P-value of our significance test is quite a bit below our typical α threshold of 0.05.

Next, let's test for nonindependent evolution between the character traits x and z.

```
fit.xz<-fitPagel(tree,x,z)
print(fit.xz,digits=4)
```

```
##
## Pagel's binary character correlation test:
##
## Assumes "ARD" substitution model for both characters
##
## Independent model rate matrix:
##           a|e       a|f      b|e        b|f
## a|e -9.3477    9.2618   0.0859    0.0000
## a|f 30.2333  -30.3192   0.0000    0.0859
## b|e  0.0000    0.0000  -9.2618    9.2618
## b|f  0.0000    0.0000  30.2333  -30.2333
##
## Dependent (x & y) model rate matrix:
##           a|e       a|f       b|e        b|f
## a|e -0.0540    0.0000    0.0540     0.0000
## a|f  2.7305   -3.2702    0.0000     0.5396
## b|e  0.0000    0.0000 -176.6114   176.6114
## b|f  0.0000    0.0000  176.6114  -176.6114
##
## Model fit:
##              log-likelihood      AIC
## independent         -19.4125  46.8250
## dependent           -13.6476  43.2952
##
## Hypothesis test result:
##    likelihood-ratio:   11.53
##    p-value:   0.02121
##
## Model fitting method used was fitMk
```

Once again, our nonindependent model is *highly* significant.
When we plot this fitted model, the result makes perfect sense.

```
plot(fit.xz,signif=2,cex.main=1,
    cex.sub=0.8,cex.traits=0.7,cex.rates=0.7,
    lwd=1)
```

What we should see in figure 7.4 is a very high rate of transition for character z when trait x is in state b, but not when it's in state a.

So what's the problem?

The reason that Maddison and FitzJohn's (2015) article has caused such consternation is because most investigators never thought that the method would find singular evolutionary events to be significant. In fact, we ought not to have been surprised.

If we think about it objectively, the *probability*[10] of the pattern shown for characters x and y or, likewise, for characters x and z is extremely low if the characters are independent.

---

[10]Which is, of course, the quantity that we compute when evaluating the likelihood of a model.

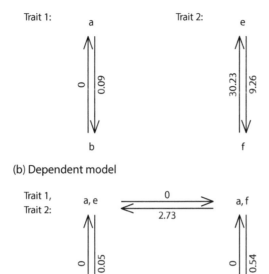

(a) Independent model

(b) Dependent model

Figure 7.4
Fitted Pagel (1994) model for the tree and data for characters *x* and *z* in figure 7.3.

One way to understand this is to consider picking two different characters that were *genuinely* evolving independently—out of all possible characters for our set of taxa.

The probability that these two randomly chosen characters, having each changed in state only once, happened to have changed along the same branch of the tree is equal to the length of that branch over the sum of all the branches of the tree! Normally (except for very small trees or very long branches), this is likely to be a relatively small number.

The reason that this is problematic in practice, however, is that biologists almost *never* choose characters for analysis with Pagel's (1994) method absent some foreknowledge about how their characters might be codistributed across the tips and clades of the tree.

For instance, we might *start* with a field observation that forest birds tend to have red plumage. Beginning with this observation, we then proceed to test for an evolutionary association between habitat and plumage color!

What to do, then? Well, rather than recommend that Pagel's (1994) method not be used, we suggest that it be used *with caution* in combination with graphing the data (as in figures 7.1 and 7.3) and the fitted models (as in figures 7.2 and 7.4).

These two things combined together—statistical model fitting and graphing our data—will give us a much better picture than either alone of what's going on in our evolutionary system of interest.

7.3 Modeling heterogeneity in the evolutionary rate for a discrete trait

In the previous section, we learned about a method to test for state-dependent evolution of a binary character: Pagel's (1994) model.

As we discussed, this approach is often used to investigate the evolutionary correlation between two discrete characters because, under some circumstances, state-dependent binary trait evolution can tend to result in a disproportionate accumulation of certain combinations of character states for the two traits compared to what we'd expect if the traits evolved independently (Maddison et al. 2007).

On the other hand, if we merely hypothesize that the evolutionary rate[11] varies from one part of the tree to another, it's also possible for us to just fit this model directly!

In this section and the one that follows, we'll learn about two different approaches to do exactly that.

The first of these, and the one we'll discuss in this section, is a model in which we propose different *regimes* for our discrete character's evolution and paint these regimes onto the phylogeny (Revell et al. 2021). As such, the model is philosophically very similar to the multi-rate Brownian motion and multi-optimum Ornstein–Uhlenbeck models we studied in chapter 5.

7.3.1 Multi-regime M*k* model: An empirical example with lizard throat fans

To understand this model, we'll apply it to a data set for dominant color of the *Anolis* lizard dewlap: an extensible throat fan used by anoles and other lizards to display to mates and competitors. We've already seen examples that use data from this diverse tropical lizard genus in chapters 1 and 5.

The tree file we'll use (anolis.mapped.nex) is the same as in chapter 5 (Mahler et al. 2010).

The data file is dewlap.colors.csv (from Ingram et al. 2016). As always, these files can be obtained from the book website. To start, we'll just read our tree and our data from file.

```
## read tree from file
anolis.tree<-read.simmap(file="anolis.mapped.nex",
    version=1.5,format="nexus")
anolis.tree

##
## Phylogenetic tree with 82 tips and 81 internal nodes.
##
## Tip labels:
##  ahli, allogus, rubribarbus, imias, sagrei, bremeri, ...
##
## The tree includes a mapped, 6-state discrete character
## with states:
##  CG, GB, TC, TG, Tr, Tw
##
## Rooted; includes branch lengths.

## read data from file
dewlap.data<-read.csv("dewlap.colors.csv",
    row.names=1,stringsAsFactors=TRUE)
head(dewlap.data)
```

---

[11]Or the process of evolution.

```
##                     island  pattern numcols domcol
## grahami            Jamaica gradient       2    red
## conspersus    SmallIsland    solid       2  black
## garmani            Jamaica    solid       2    red
## opalinus           Jamaica    basal       3    red
## valencienni        Jamaica    solid       3   pink
## lineatopus         Jamaica gradient       3    red
```

In the preceding code chunk, we used the *phytools* function read.simmap to read our input file, because the tree (as we'll see below) has an encoded mapped regime.

This phylogeny and our data for dewlap color come from different studies, so let's use name.check in the *geiger* package to verify that they match.

```
## load geiger package
library(geiger)
## run name.check
chk<-name.check(anolis.tree,dewlap.data)
summary(chk)
```

```
## 7 taxa are present in the tree but not the data:
##     bremeri,
##     isolepis,
##     longitibialis,
##     oporinus,
##     paternus,
##     rubribarbus,
##     ....
## 109 taxa are present in the data but not the tree:
##     acutus,
##     aequatorialis,
##     altae,
##     annectens,
##     anoriensis,
##     aquaticus,
##     ....
##
## To see complete list of mis-matched taxa, print object.
```

There are a number of differences between our data and tree, so we can proceed to prune our tree as well as subsample our data.

Here, since our phylogenetic tree is an object of class "simmap",[12] we'll use the *phytools* function drop.tip.simmap to make sure that we don't lose our regime mappings!

```
anolis.pruned<-drop.tip.simmap(anolis.tree,
    chk$tree_not_data)
```

---

[12] As well as "phylo". R objects are allowed to have more than one class!

We can likewise subsample our data to include only the taxa now represented in our pruned tree and then run name.check again.

```
dewlap.pruned<-dewlap.data[anolis.pruned$tip.label,]
name.check(anolis.pruned,dewlap.pruned)

## [1] "OK"
```

Now let's plot our tree with the hypothesized regimes as well as the discrete character we intend to model.

```
## set our colors for the discrete trait
dewlap.colors<-setNames(levels(dewlap.pruned[,4]),
    levels(dewlap.pruned[,4]))
## plot our tree with dewlap colors at the tips
plotTree.datamatrix(anolis.pruned,
    X=dewlap.pruned[,4,drop=FALSE],
    colors=list(dewlap.colors),
    yexp=1,header=FALSE,xexp=1.2,offset=0.5)
## set the colors for our regimes
cols<-setNames(rainbow(n=6),
    c("CG","GB","TC","TG","Tr","Tw"))
## graph our tree with mapped regimes on top of
## the previous tree
plot(anolis.pruned,colors=cols,ftype="off",
    outline=TRUE,add=TRUE,
    xlim=get("last_plot.phylo",envir=.PlotPhyloEnv)$x.lim,
    ylim=get("last_plot.phylo",envir=.PlotPhyloEnv)$y.lim)
## add legends for both the tip states of dewlap
## color and the mapped regimes
leg<-legend(x="topright",legend=names(dewlap.colors),
    pch=22,pt.bg=dewlap.colors,pt.cex=1.5,
    title="Dewlap color",cex=0.7,bty="n")
leg<-legend(x=leg$rect$left,y=leg$rect$top-leg$rect$h,
    legend=names(cols),pch=22,pt.bg=cols,pt.cex=1.5,
    title="Ecomorph",cex=0.7,bty="n")
```

Note that in this code chunk, we first used plotTree.datamatrix, a function that we've now seen a couple of times already, to graph the tree with our discrete character adjacent.

Next, we used plot.simmap,[13] with add=TRUE, to graph our tree with mapped regimes *on top* of our plotTree.datamatrix phylogeny.[14]

Finally, we added two different color legends for (first) our discrete tip states of dewlap color and then (second) for our mapped regimes (figure 7.5).

To fit our model, we now need to go ahead and first pull out the discrete character, dominant dewlap color, into a new vector, just as we've done in prior sections of this chapter and book.

---

[13]Remember, this is the S3 plot method for our object of class "simmap".

[14]The function call get("last_plot.phylo",envir=.PlotPhyloEnv) pulls the parameters of the most recently plotted tree. We'll learn more about this in chapter 13!

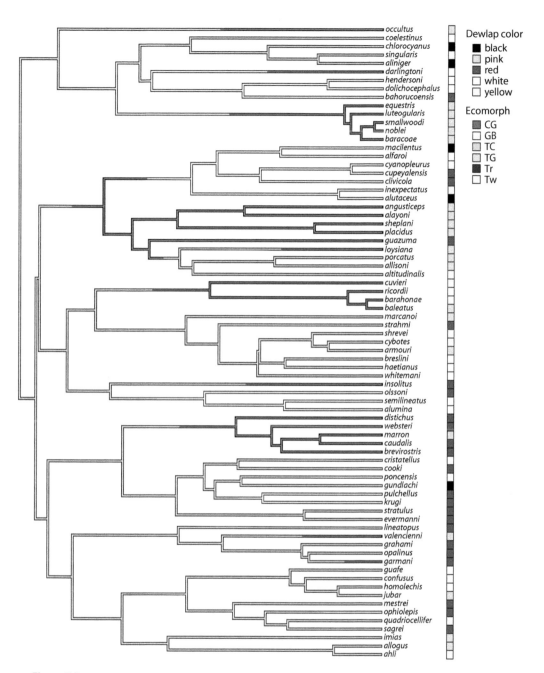

Figure 7.5
Tree of anole lizards with ecomorph regime and dominant dewlap color.

```
domcol<-setNames(dewlap.data$domcol,
    rownames(dewlap.data))
```

Finally, we're ready to fit our multi-regime model.

Since our tree is on the small side (only seventy-five taxa, after pruning), we'll just fit the *simplest* multi-regime model imaginable: one in which we assume a single constant rate of transition among *states* (ER) for each of the regimes mapped onto the tree. This model should have a number of parameters equal to the number of mapped regimes (six in our example) and has a special case, the standard ER model from chapter 6.

```
## fit the model using fitmultiMk
fit.multi<-fitmultiMk(anolis.pruned,domcol)
## print the results
print(fit.multi,digits=2)
```

```
## Object of class "fitmultiMk".
##
## Fitted value of Q[CG]:
##          black  pink   red white yellow
## black    -3.52  0.88  0.88  0.88   0.88
## pink      0.88 -3.52  0.88  0.88   0.88
## red       0.88  0.88 -3.52  0.88   0.88
## white     0.88  0.88  0.88 -3.52   0.88
## yellow    0.88  0.88  0.88  0.88  -3.52
##
## Fitted value of Q[GB]:
##          black  pink   red white yellow
## black    -3.23  0.81  0.81  0.81   0.81
## pink      0.81 -3.23  0.81  0.81   0.81
## red       0.81  0.81 -3.23  0.81   0.81
## white     0.81  0.81  0.81 -3.23   0.81
## yellow    0.81  0.81  0.81  0.81  -3.23
##
## Fitted value of Q[TC]:
##          black  pink   red white yellow
## black    -0.95  0.24  0.24  0.24   0.24
## pink      0.24 -0.95  0.24  0.24   0.24
## red       0.24  0.24 -0.95  0.24   0.24
## white     0.24  0.24  0.24 -0.95   0.24
## yellow    0.24  0.24  0.24  0.24  -0.95
##
## Fitted value of Q[TG]:
##          black  pink   red white yellow
## black    -3.91  0.98  0.98  0.98   0.98
## pink      0.98 -3.91  0.98  0.98   0.98
## red       0.98  0.98 -3.91  0.98   0.98
## white     0.98  0.98  0.98 -3.91   0.98
## yellow    0.98  0.98  0.98  0.98  -3.91
##
## Fitted value of Q[Tr]:
##          black  pink   red white yellow
## black    -0.60  0.15  0.15  0.15   0.15
```

```
## pink      0.15 -0.60  0.15  0.15     0.15
## red       0.15  0.15 -0.60  0.15     0.15
## white     0.15  0.15  0.15 -0.60     0.15
## yellow    0.15  0.15  0.15  0.15    -0.60
##
## Fitted value of Q[Tw]:
##          black  pink   red white yellow
## black    -0.48  0.12  0.12  0.12     0.12
## pink      0.12 -0.48  0.12  0.12     0.12
## red       0.12  0.12 -0.48  0.12     0.12
## white     0.12  0.12  0.12 -0.48     0.12
## yellow    0.12  0.12  0.12  0.12    -0.48
##
## Fitted (or set) value of pi:
##  black   pink    red  white yellow
##    0.2    0.2    0.2    0.2    0.2
##
## Log-likelihood: -107.49
##
## Optimization method used was "nlminb"
```

We can see that the maximum likelihood (ML) estimated rates for each regime are *different*—but the important question is whether or not this model explains our data significantly better than a simpler model, such as a single-regime model.

The only way to find out is by fitting that model too and then comparing the two different results.

For this analysis, we'll use a likelihood ratio test from the package *lmtest* (Zeileis and Hothorn 2002), which means that this package needs to be installed and loaded to follow along.

```
library(lmtest)
```

Now let's fit a single-rate ER model using fitMk from *phytools*.

```
fit.single<-fitMk(anolis.pruned,domcol,model="ER")
print(fit.single,digits=2)

## Object of class "fitMk".
##
## Fitted (or set) value of Q:
##          black  pink   red white yellow
## black    -1.83  0.46  0.46  0.46     0.46
## pink      0.46 -1.83  0.46  0.46     0.46
## red       0.46  0.46 -1.83  0.46     0.46
## white     0.46  0.46  0.46 -1.83     0.46
## yellow    0.46  0.46  0.46  0.46    -1.83
##
```

```
## Fitted (or set) value of pi:
##  black    pink     red   white yellow
##    0.2     0.2     0.2     0.2    0.2
## due to treating the root prior as (a) flat.
##
## Log-likelihood: -111.41
##
## Optimization method used was "nlminb"
```

```
lrtest(fit.single,fit.multi)
```

```
## Likelihood ratio test
##
## Model 1: fit.single
## Model 2: fit.multi
##    #Df  LogLik Df  Chisq Pr(>Chisq)
## 1    1 -111.41
## 2    6 -107.49  5 7.8341     0.1656
```

The function gives us a warning message;[15] however, this is not a problem.

The result, though, clearly indicates that our multi-regime model is *not* supported by the data.

How can we reconcile this with the fact that the numerical values of the maximum likelihood estimations (MLEs) for our different regimes were so different—by up to nearly a factor of 9—from one another?

To understand this, we'll compute[16] estimates of the standard errors of our rates by using the curvature matrix of our likelihood surface, called the Hessian matrix.

To obtain this matrix, we'll use the function hessian in *numDeriv* (Gilbert and Varadhan 2019).

```
## compute the Hessian matrix
H<-numDeriv::hessian(fit.multi$lik,fit.multi$rates)
## take its negative inverse
v<-diag(solve(-H))
## extra estimated standard errors
se<-sqrt(v)
se
```

```
## [1] 0.6615845 0.6150499 0.1468424 1.1705230
## [5] 0.1549024 0.1242525
```

---

[15]The warning says original model was of class "fitMk", updated model is of class "fitmultiMk", which makes sense because we fit our two models using different functions. It's OK. We can still compare their likelihoods.

[16]Approximate. The negative inverse Hessian (curvature) matrix of a likelihood surface gives the variance–covariance matrix of the MLEs; however, this is an asymptotic property of likelihood surfaces as the amount of data we have about our parameters increases. In our case, we don't have much data, so we're probably not that close to the asymptote!

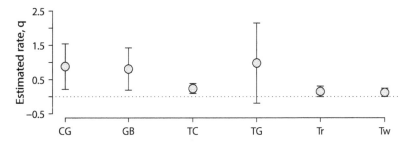

Figure 7.6
Estimates and standard errors of the rate of dewlap evolution for each ecomorph category in the tree of figure 7.5.

Finally, let's use the base R function `stripchart` to graph our estimated rates for each regime, along with approximate confidence intervals (based on the standard errors we just computed) for each estimate (figure 7.6).

```
## remove the box that would otherwise be graphed
## around our chart
par(bty="n")
## create a simple strip chart
stripchart(fit.multi$rates~fit.multi$regimes,vertical=TRUE,
    bty="n",ylim=c(-0.5,2.5),pch=21,cex=1.2,bg="gray",
    ylab="Estimated rate, q",cex.lab=0.8,cex.axis=0.7)
abline(h=0,lty="dotted")
## add confidence intervals
for(i in 1:length(se)){
    lines(x=rep(i,3),y=c(fit.multi$rates[i]-se[i],
        fit.multi$rates[i],fit.multi$rates[i]+se[i]))
    points(i,fit.multi$rates[i],pch=21,cex=1.2,bg="gray")
    lines(c(i-0.05,i+0.05),rep(fit.multi$rates[i]-se[i],2))
    lines(c(i-0.05,i+0.05),rep(fit.multi$rates[i]+se[i],2))
}
```

Our standard errors for each parameter estimate are broad and overlapping, which helps explain why we were not able to reject a constant-rate model!

7.4 Modeling rate variation using the hidden-rates model

Just as for discrete characters, sometimes we lack a specific prior hypothesis about how the rate of character evolution varies across the branches and nodes of our phylogeny.

To that end, Beaulieu et al. (2013) proposed a model that they referred to as the hidden-rates model.

The premise of this model is pretty straightforward. Imagine, in the simplest case, a model with two observable states for the character: 0 and 1.

Normally, we would model this trait's evolution as a continuous-time Markov chain with at most two transition rates: a rate of change in the trait from $0 \rightarrow 1$ and a second rate of transition from $1 \rightarrow 0$.

Now consider that instead of one rate of evolution for each state, *sometimes* when our trait is in state 1, it evolves *rapidly* (to state 0), and sometimes it evolves *slowly* or not at all.

Evolution by this process can create considerably more heterogeneity in the distribution of our character trait across the tips in the phylogeny.

Some clades will switch back and forth frequently between states, while other clades change little.

To get a sense of this, let's see what evolution looks like under a constant-rate Markov process, in which the trait changes back and forth between two states with constant probability. We can then compare this to evolution via a process in which some lineages evolve to a third, *hidden* state, from which changes are more difficult, that is, under this simple, *hidden-rate* process that we have just described.

To do this, we'll simulate using the function sim.history, just as we did in chapter 6. For reference, we can start by simulating a binary character *without* a hidden rate.

```
## set seed to make the code reproducible
set.seed(7)
## create a transition matrix between states under
## a simple, Mk model
Q.mk<-matrix(c(-1,1,1,-1),2,2,dimnames=list(0:1,0:1))
## simulate a character history under this constant
## rate model
mk.tree<-sim.history(tree<-pbtree(n=100,scale=2),Q.mk,
    anc="0")
```

```
## Done simulation(s).
```

Next, we can create a transition matrix between character states, **Q**, that contains three states—but then we'll imagine the scenario of only two *observed* states by merging the conditions of our second and third simulated character values.

```
## create a hidden-rate transition matrix -- this matrix
## has two different values for character 1: 1 and 1*
Q.hrm<-matrix(c(-1,1,0,1,-1.5,0.5,0,0.1,-0.1),3,3,
    byrow=TRUE,dimnames=list(c(0:1,"1*"),
    c(0:1,"1*")))
Q.hrm
```

```
##      0    1   1*
## 0   -1  1.0  0.0
## 1    1 -1.5  0.5
## 1*   0  0.1 -0.1
```

```
## simulate under the hidden rate model
hrm.tree<-sim.history(tree,Q.hrm,anc="0",message=FALSE)
```

Now we'll visualize our three different character histories. The first character history is the one that we simulated under a standard, *Mk* model of trait evolution.

The second is one in which there are two different values for character state 1: 1 and 1*.

Finally, we'll use a *phytools* function called `mergeMappedStates` to combine the two hidden conditions (1 and 1*) for the second of our two traits.

```
## subdivide plot area
par(mfrow=c(1,3))
## set colors for graphing
cols<-setNames(c("lightgray","black"),0:1)
## plot simple Mk model
plot(mk.tree,colors=cols,ftype="off",mar=c(1.1,2.1,3.1,0.1))
legend("bottomleft",names(cols),pch=15,col=cols,pt.cex=2,
    bty="n")
mtext("(a)",line=0,adj=0)
## set colors for hidden rate model
cols<-setNames(c("lightgray","black","slategray"),c(0:1,"1*"))
## plot HRM, but with "hidden" state shown
plot(hrm.tree,colors=cols,ftype="off",mar=c(1.1,2.1,3.1,0.1))
legend("bottomleft",names(cols),pch=15,col=cols,pt.cex=2,
    bty="n")
mtext("(b)",line=0,adj=0)
## plot HRM but with two 1 states (1 and 1*) merged
cols<-setNames(c("lightgray","black"),0:1)
plot(tree<-mergeMappedStates(hrm.tree,c("1","1*"),"1"),color=cols,
    ftype="off",mar=c(1.1,2.1,3.1,0.1))
legend("bottomleft",c("0","1/1*"),pch=15,col=cols,pt.cex=2,
    bty="n")
mtext("(c)",line=0,adj=0)
```

The key difference that you should see between panels (a) and (c) of figure 7.7 is that in panel (c), sometimes the character changes frequently between states—but elsewhere in the tree, a lineage evolves to character state 1 and then gets stuck there. Panel (b) shows us that this occurs whenever the hidden state, 1*, is reached, just as per our simulation design.

7.4.1 Fitting the hidden-rates model using `fitHRM`

To get a better appreciation of the hidden-rates model, we thought it would be interesting to revisit the squamate toe number evolution analysis from chapter 6.

Remember, last chapter, we found that in the second best-supported model, it was possible to evolve toes in a toeless (and thus, presumably, limbless) lizard.

Although this may be plausible in a lineage of skinks that lost their limbs very recently, in other major squamate groups (such as snakes), it would seem to be quite untenable.

This hypothesis[17] is a hidden-rates model.

---

[17] In which two different lineages—both with the same observed states—evolve to other states with different rates.

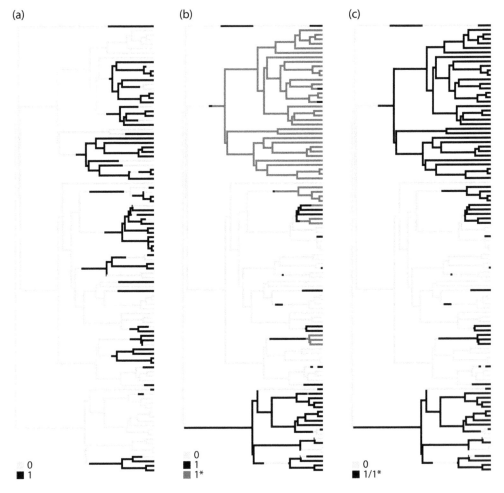

(a)   (b)   (c)

0   0   0
■ 1   ■ 1   ■ 1/1*
           ■ 1*

Figure 7.7
Simulation under (a) a constant-rate M*k* model and (b) a hidden-rates model, but in which the hidden state is shown (slate gray branches). (c) The same as (b), but in which the hidden state is no longer shown.

Mostly for computational reasons,[18] we decided to fit our model to the loss (and possible re-evolution) of *limbs* rather than number of hindfoot toes. We'll do so using the same data as in chapter 5 by assuming that any species *without* hindfoot toes is limbless,[19] while any species *with* toes must be limbed.

For this analysis, we'll use the same `squamate-data.csv` and `squamate.tre` datafiles as we did in chapter 6 and then clean up our data.

```
## read data from file
sqData<-read.csv("squamate-data.csv",row.names=1)
## read tree from file
```

---

[18]The hidden-rates model with all six digit states included ended up with a lot of parameters.

[19]There are some species of squamates with *only* forefeet, but we'll have to ignore that nuance here.

```
sqTree<-read.nexus("squamate.tre")
## match tree and data
chk<-name.check(sqTree,sqData)
sqTree.pruned<-drop.tip(sqTree,chk$tree_not_data)
sqData.pruned<-sqData[sqTree.pruned$tip.label,,drop=FALSE]
## extra number of hindfoot toes
toes<-setNames(sqData.pruned[,"rear.toes"],
    rownames(sqData.pruned))
```

Now that we have our data for hindfoot toe number, we can proceed to convert it to the binary condition `"limbed"` (for `toes>0`) and `"limbless"` (otherwise).

```
## create vector of NAs
limbs<-setNames(rep(NA,length(toes)),names(toes))
## set all values of toes > 0 to "limbed"
limbs[toes>0]<-"limbed"
## do the converse for toes == 0
limbs[toes==0]<-"limbless"
## convert to factor
limbs<-as.factor(limbs)
```

Now, we can just go ahead and fit our hidden-rate models.

To do this, we'll use the *phytools* function `fitHRM` because it allows a particular class of ordered hidden-rates model that corresponds well with our biological hypothesis.

The hidden-rates discrete character evolution model is also implemented in the powerful R package `corHMM` (Beaulieu et al. 2020), which we'll use later in this section as well as in chapter 8.

In total, we'll fit four models and compare them.

The first is a model in which we imagine that each state for our discrete character (*limbed* and *limbless*) has one hidden state and that no changes are permitted between the hidden states or from a hidden state to the other observed state, just as we described above.

We specify this model using the options `umbral=TRUE` (which tells the function that changes are *not allowed* between hidden states) and `ncat=2` (two rate categories per state).[20]

Let's see:

```
limb.HRM1<-fitHRM(sqTree.pruned,limbs,ncat=2,model="ARD",
    umbral=TRUE,pi="fitzjohn",niter=5,opt.method="nlminb")

##
## This is the design matrix of the fitted model.
## Does it make sense?
##
##              limbed limbed* limbless limbless*
## limbed            0       1        2         0
## limbed*           3       0        0         0
```

---

[20] As we'll see in a second, this could also have been given as `ncat=c(2,2)`.

```
## limbless        4        0        0        5
## limbless*       0        0        6        0
##
## log-likelihood from current iteration: -43.907
##  --- Best log-likelihood so far: -43.907 ---
## log-likelihood from current iteration: -43.907
##  --- Best log-likelihood so far: -43.907 ---
## log-likelihood from current iteration: -45.3539
##  --- Best log-likelihood so far: -43.907 ---
## log-likelihood from current iteration: -43.907
##  --- Best log-likelihood so far: -43.907 ---
## log-likelihood from current iteration: -43.907
##  --- Best log-likelihood so far: -43.907 ---
```

print(limb.HRM1,digits=4)

```
## Object of class "fitHRM".
##
## Observed states: [ limbed, limbless ]
## Number of rate categories per state: [ 2, 2 ]
##
## Fitted (or set) value of Q:
##            limbed limbed* limbless limbless*
## limbed    -0.0168  0.0053   0.0115    0.0000
## limbed*    0.0000  0.0000   0.0000    0.0000
## limbless   0.0995  0.0000  -0.1329    0.0334
## limbless*  0.0000  0.0000   0.0058   -0.0058
##
## Fitted (or set) value of pi:
##    limbed   limbed*  limbless limbless*
##    0.5175    0.0000    0.3968    0.0857
## due to treating the root prior as (a) nuisance.
##
## Log-likelihood: -43.907
##
## Optimization method used was "nlminb"
```

The argument $niter$ tells the function how many optimization iterations to run. The argument opt.method tells R which optimization routine to use to try to find the maximum likelihood solution. Depending on the size of the data set and difficulty of the problem, users may need to adjust these different function arguments to ensure convergence to the correct solution.

As an alternative to this first hidden-state model in which both *limbed* and *limbless* have hidden states, perhaps we should consider the possibility that only the *limbless* state has a hidden rate.

To fit this model, we leave everything the same but adjust ncat to be c(1,2): two rates for *limbless* but only one for *limbed*.[21] Let's try it.

---

[21] The order of ncat should be the same as the order of levels(limbs), which, in our case, is "limbed" "limbless".

```
limb.HRM2<-fitHRM(sqTree.pruned,limbs,ncat=c(1,2),model="ARD",
    umbral=TRUE,pi="fitzjohn",niter=5,opt.method="nlminb")
```

```
##
## This is the design matrix of the fitted model.
## Does it make sense?
##
##           limbed limbless limbless*
## limbed         0        1         0
## limbless       2        0         3
## limbless*      0        4         0
##
## log-likelihood from current iteration: -45.4463
##  --- Best log-likelihood so far: -45.4463 ---
## log-likelihood from current iteration: -45.4463
##  --- Best log-likelihood so far: -45.4463 ---
## log-likelihood from current iteration: -45.4463
##  --- Best log-likelihood so far: -45.4463 ---
## log-likelihood from current iteration: -45.4463
##  --- Best log-likelihood so far: -45.4463 ---
## log-likelihood from current iteration: -45.4463
##  --- Best log-likelihood so far: -45.4463 ---
```

```
print(limb.HRM2,digits=4)
```

```
## Object of class "fitHRM".
##
## Observed states: [ limbed, limbless ]
## Number of rate categories per state: [ 1, 2 ]
##
## Fitted (or set) value of Q:
##           limbed limbless limbless*
## limbed   -0.0041   0.0041    0.0000
## limbless  0.0823  -0.1329    0.0505
## limbless* 0.0000   0.0068   -0.0068
##
## Fitted (or set) value of pi:
##    limbed  limbless limbless*
##    0.5749    0.3461    0.0790
## due to treating the root prior as (a) nuisance.
##
## Log-likelihood: -45.4463
##
## Optimization method used was "nlminb"
```

Last, let's fit the converse of this[22] as well as the standard Mk ARD model.

---

[22]One rate for *limbless* and two rates for *limbed*, which we do by setting ncat=c(2,1).

The lattermost model we could have just as easily fit with fitDiscrete from *geiger* (as we learned in chapter 6) or phytools::fitMk, but we can likewise do with our fitHRM function by just setting ncat=1.

This also helps demonstrate that the standard M*k* model is really just a special case of the hidden-rates model. For this model, we'll also adjust niter to be 1 because it is much easier to fit than our previous, *genuinely* hidden-rate models!

```
## fit a HRM with a hidden state for limbed but not limbless
limb.HRM3<-fitHRM(sqTree.pruned,limbs,ncat=c(2,1),model="ARD",
    umbral=TRUE,pi="fitzjohn",niter=5,opt.method="nlminb")
```

```
##
## This is the design matrix of the fitted model.
## Does it make sense?
##
##           limbed limbed* limbless
## limbed         0       1        2
## limbed*        3       0        0
## limbless       4       0        0
##
## log-likelihood from current iteration: -45.7083
##  --- Best log-likelihood so far: -45.7083 ---
## log-likelihood from current iteration: -45.7083
##  --- Best log-likelihood so far: -45.7083 ---
## log-likelihood from current iteration: -46.1112
##  --- Best log-likelihood so far: -45.7083 ---
## log-likelihood from current iteration: -44.5816
##  --- Best log-likelihood so far: -44.5816 ---
## log-likelihood from current iteration: -45.7083
##  --- Best log-likelihood so far: -44.5816 ---
```

```
print(limb.HRM3,digits=4)
```

```
## Object of class "fitHRM".
##
## Observed states: [ limbed, limbless ]
## Number of rate categories per state: [ 2, 1 ]
##
## Fitted (or set) value of Q:
##            limbed limbed* limbless
## limbed   -0.0087  0.0045   0.0041
## limbed*   0.0000  0.0000   0.0000
## limbless  0.0049  0.0000  -0.0049
##
## Fitted (or set) value of pi:
##    limbed limbed* limbless
##    0.8284  0.0000   0.1716
## due to treating the root prior as (a) nuisance.
```

```
##
## Log-likelihood: -44.5816
##
## Optimization method used was "nlminb"
```

```
## fit a standard ARD model using fitHRM
limb.Mk<-fitHRM(sqTree.pruned,limbs,ncat=1,model="ARD",
    umbral=TRUE,pi="fitzjohn",niter=1,opt.method="nlminb")
```

```
##
## This is the design matrix of the fitted model.
## Does it make sense?
##
##          limbed limbless
## limbed        0        1
## limbless      2        0
##
## log-likelihood from current iteration: -45.9965
##  --- Best log-likelihood so far: -45.9965 ---
```

```
print(limb.Mk,digits=4)
```

```
## Object of class "fitHRM".
##
## Observed states: [ limbed, limbless ]
## Number of rate categories per state: [ 1, 1 ]
##
## Fitted (or set) value of Q:
##           limbed limbless
## limbed   -0.0021   0.0021
## limbless  0.0046  -0.0046
##
## Fitted (or set) value of pi:
##    limbed limbless
##    0.8539   0.1461
## due to treating the root prior as (a) nuisance.
##
## Log-likelihood: -45.9965
##
## Optimization method used was "nlminb"
```

When we're done, we can plot our four models to ensure that they correspond with the four different biological scenarios that we'd imagined. The object class created by fitHRM has its own custom plot method, so this should be easy. The only plotting function arguments we'll adjust are mar (which sets the margin around our plot) and spacer (which sets the amount of space left between our arrowheads and the character names; see figure 7.8).[23]

---

[23] Often the default value of spacer should work fine, but here our trait value names are quite long.

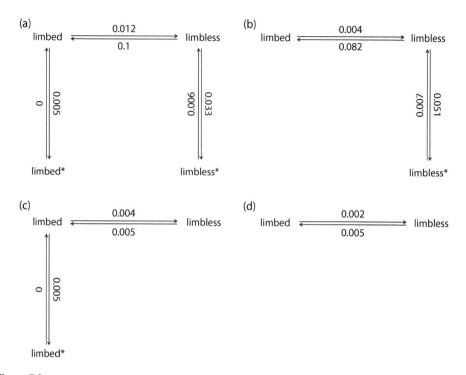

Figure 7.8
Fitted hidden-rate models for data for limbs versus limblessness in 119 species of squamate reptiles. (a) Hidden-rate model in which each of the two binary states had a single hidden rate. (b) Hidden-rate model with a hidden rate for *limbless* but not *limbs*. (c) The converse of (b). (d) Fitted M*k* ARD model.

```
## subdivide our plotting area
par(mfrow=c(2,2))
## plot each of our four different fitted models
plot(limb.HRM1,spacer=0.25,mar=c(0.1,1.1,2.1,0.1))
mtext("(a)",line=0,adj=0)
plot(limb.HRM2,spacer=0.25,mar=c(0.1,1.1,2.1,0.1))
mtext("(b)",line=0,adj=0)
plot(limb.HRM3,spacer=0.25,mar=c(0.1,1.1,2.1,0.1))
mtext("(c)",line=0,adj=0)
plot(limb.Mk,spacer=0.25,mar=c(0.1,1.1,2.1,0.1))
mtext("(d)",line=0,adj=0)
```

Our first model, the four-state hidden-rate model, has all the other models as a special case. As such, it should *always*[24] have a higher likelihood than the other three models.

Likewise, the M*k* model is itself a special case of all of the three previous models and, as such, should always have the *lowest* likelihood of the four.

---

[24]Barring a failure in optimization, which certainly can and does occur!

Let's go ahead and tabulate log-likelihoods and Akaike information criterion (AIC) values for each of the four models to verify that this is indeed the case. We'll put these together into a single data frame as follows:

```
data.frame(model=c("4-state HRM","limbless hidden",
    "limbed hidden","Mk model"),
    logL=sapply(list(limb.HRM1,limb.HRM2,
    limb.HRM3,limb.Mk),logLik),
    k=sapply(list(limb.HRM1,limb.HRM2,
    limb.HRM3,limb.Mk),function(x) length(x$rates)),
    AIC=sapply(list(limb.HRM1,limb.HRM2,
    limb.HRM3,limb.Mk),AIC))
```

```
##                  model      logL k      AIC
## 1         4-state HRM -43.90703 6 99.81407
## 2     limbless hidden -45.44625 4 98.89251
## 3       limbed hidden -44.58160 4 97.16319
## 4            Mk model -45.99652 2 95.99304
```

Here we see that although the four-state model has the highest likelihood, just as we expected, the AIC values indicate that model that is best supported by the data[25] is actually the simplest, M*k* model.[26]

7.4.2 Fitting the hidden-rates model using `corHMM`

Although we've been using *phytools* version of the hidden-rates model of Beaulieu et al. (2013), the most important package for hidden-rate analyses is the library by Beaulieu et al. (2020) called *corHMM*.

In the rest of this section, we'll see exactly how this package works.

We'll also be using *corHMM* in chapter 8, so if you don't have it installed already, we suggest that you install *corHMM* from CRAN and then load it in the normal way.

```
library(corHMM)
```

For this part of the section, we'll use data from Williams et al. (2014), in which the authors investigate the evolution of tri- and bicellular pollen grains.

Their hidden-rates model analysis included data for over 2,500 species! We're going to use a much *smaller* data set containing a phylogeny and character data for only 511 taxa subsampled from their study.

The tree and data are in the files `pollen-tree.phy` and `pollen-data.csv`, respectively, both of which can be found on the book website.

As usual, we'll start by reading in the tree and data from file, as well as checking that they match:

---

[25] Taking parameterization into account.

[26] Although there's not that much difference among our different models!

```
## read tree
pollen.tree<-read.tree(file="pollen-tree.phy")
print(pollen.tree,printlen=3)

##
## Phylogenetic tree with 511 tips and 510 internal nodes.
##
## Tip labels:
##    Nuphar_lutea, Nuphar_pumila, Nuphar_advena, ...
##
## Rooted; includes branch lengths.
```

```
## read data
pollen.data<-read.csv(file="pollen-data.csv",row.names=1)
head(pollen.data)

##                         V1
## Nuphar_lutea             3
## Nuphar_pumila            2
## Nuphar_advena            2
## Austrobaileya_scandens   2
## Schisandra_propinqua     2
## Kadsura_heteroclita      2
```

```
## check to verify tree and data match
name.check(pollen.tree,pollen.data)

## [1] "OK"
```

Next, to use corHMM, we need put our data frame into a special format.[27]

The data frame we make is very simple. It contains one column (denominated Genus.species) that contains the tip labels of all the taxa in our data and a second column (in our case, we'll call it pollen.number) that has a numerically coded[28] discrete state.

```
pollen<-data.frame(Genus.species=rownames(pollen.data),
     pollen.number=pollen.data[,1]-1)
head(pollen)

##               Genus.species pollen.number
## 1              Nuphar_lutea             2
## 2             Nuphar_pumila             1
## 3             Nuphar_advena             1
## 4     Austrobaileya_scandens           1
```

---

[27] We could have done this in our input file, but it's useful to see how to make the data frame within R anyway!

[28] This part is very important: our trait has to be coded as 1, 2, and so on. Since our original data have values of either 2 or 3, for our analysis to work, we first need to calculate pollen.data[,1]-1. If our discrete character was coded as character or factor, we'd need to convert it to numerical format first.

```
## 5    Schisandra_propinqua         1
## 6    Kadsura_heteroclita          1
```

Now, we'll go ahead and fit our model using the *function* corHMM.[29]

```
fit.pollen<-corHMM(pollen.tree,pollen,rate.cat=3,nstarts=10,
    root.p="maddfitz")
```

```
## State distribution in data:
## States: 1 2
## Counts: 354 157
## Beginning thorough optimization search -- performing 10 random
## restarts
## Finished. Inferring ancestral states using marginal
## reconstruction.
```

Some of the arguments we chose to specify here included rate.cat, which is equivalent to ncat in fitHRM: the number of rate categories per state. We chose to use three rate categories because this is the number that Williams et al. (2014) found in their best-fitting hidden-rate model.[30]

We also set nstarts, the number of optimizations, to 10 and root.p, the root prior, to "maddfitz". This is the same root prior probability distribution as is used by fitDiscrete in the *geiger* package that we used in chapter 6 and the same one we set for fitHRM via the argument pi="fitzjohn" (Maddison et al. 2007; FitzJohn et al. 2009).

```
fit.pollen

##
## Fit
## -lnL AIC AICc Rate.cat ntax
## -201.6847 427.3694 427.9959 3 511
##
## Rates
##         (1,R1)      (2,R1)      (1,R2)      (2,R2)
## (1,R1) NA          0.124134084 0.010200449 NA
## (2,R1) 0.391068370 NA          NA          0.010200449
## (1,R2) 0.004257594 NA          NA          0.003066471
## (2,R2) NA          0.004257594 0.000000001 NA
## (1,R3) 0.000000001 NA          0.003457013 NA
## (2,R3) NA          0.000000001 NA          0.003457013
##         (1,R3)      (2,R3)      (1,R3)      (2,R3)
## (1,R1) 0.000000001 NA          0.000000001 NA
## (2,R1) NA          0.000000001 NA          0.000000001
```

---

[29] Yes—the package and function have the same name. In R, this is not too unusual!

[30] Although our results are different because we only used a random subset of the data from Williams et al. (2014).

```
## (1,R2) 0.001279661 NA           0.001279661 NA
## (2,R2) NA           0.001279661 NA           0.001279661
## (1,R3) NA           0.000000001 NA           0.000000001
## (2,R3) 0.000000001 NA           0.000000001 NA
##
## Arrived at a reliable solution
```

One way to think about the fitted model is as a series of different transition process: one for each level of the hidden state (three, in our case) and another for the transitions among levels.

These can be plotted using the corHMM function plotMKmodel as follows (figure 7.9).

```
plotMKmodel(fit.pollen,display="square",text.scale=0.5,
    vertex.scale=0.6,arrow.scale=0.5)
```

Alternatively, we can just pull out the transition (**Q**) matrix of our *full* model, including its hidden states, and plot it in the *same* way that we plotted our M*k* models from chapter 6 (figure 7.10).

```
plot(as.Qmatrix(fit.pollen),show.zeros=FALSE,lwd=1,
    cex.traits=0.7)
```

Last, sometimes we want to do more than simply fit our model to data—we'd also like to map our hidden states onto the tree.

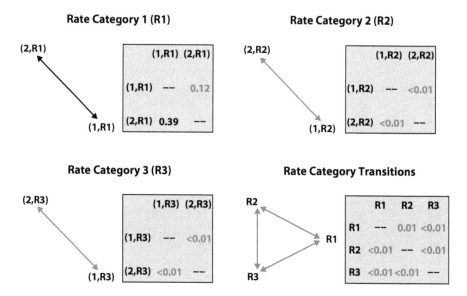

Figure 7.9
Plotted hidden-rates model from corHMM.

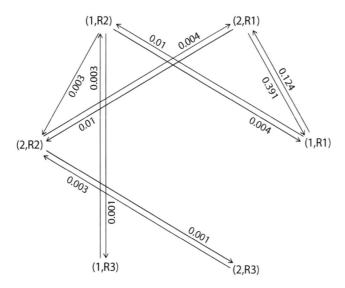

Figure 7.10
Q matrix from the hidden-rates model.

One of the neat things about corHMM is that it lets us do just that. In this case, what we're going to do is plot the *marginal ancestral state reconstructions*[31] onto all of the edges *and* nodes of the phylogeny. The result is in figure 7.11.[32]

Let's see.

```
## create a new matrix containing the tip and internal node
## marginal likelihoods
states<-rbind(fit.pollen$tip.states[pollen.tree$tip.label,],
    fit.pollen$states)
rownames(states)<-1:max(pollen.tree$edge)
## normalize each row to sum to 1.0
states<-t(apply(states,1,function(x) x/sum(x)))
## set the colors for plotting
reds<-c("#ec9488","#eb5a46","#933b27")
blues<-c("#8bbdd9","#0079bf","#094c72")
COLS<-c(reds[1],blues[1],reds[2],blues[2],reds[3],blues[3])
## plot the tree nstates times, using transparency colors
for(i in 1:ncol(states)){
    tree<-pollen.tree
    edge.col<-rep(NA,nrow(tree$edge))
    for(j in 1:nrow(tree$edge)){
        edge.col[j]<-make.transparent(COLS[i],
            mean(states[tree$edge[j,],i]))
        tree<-paintBranches(tree,tree$edge[j,2],
```

---

[31] To be discussed in chapter 8.

[32] There's also a built-in function within the *corHMM* package called plotRECON that graphs much the same information without as much scripting. We like our version, but plotRECON is inarguably easier to use with this object type.

Complex discrete models

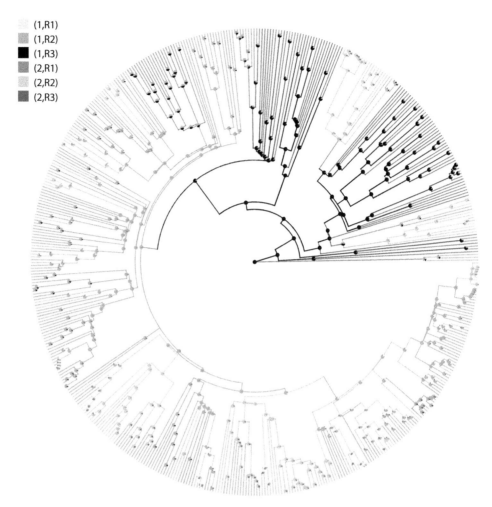

(1,R1)
(1,R2)
(1,R3)
(2,R1)
(2,R2)
(2,R3)

Figure 7.11
Mapped hidden-rates model.

```
            as.character(j))
    }
    cols<-setNames(edge.col,1:nrow(tree$edge))
    plot(tree,type="fan",colors=cols,ftype="off",
        lwd=1,add=(i!=1))
}
## add node labels using pies to show marginal probabilities
## at each node
par(fg="transparent")
nodelabels(pie=fit.pollen$states,piecol=COLS,cex=0.2)
par(fg="black")
## finally, add a legend
legend("topleft",rownames(fit.pollen$solution)[c(1,3,5,2,4,6)],
    pch=15,col=COLS[c(1,3,5,2,4,6)],pt.cex=2,bty="n")
```

The way we did this was by using `paintBranches` in the *phytools* package to map different states onto the tree based on the marginal reconstructions in our `"corhmm"` object.

We set the transparency level of our colors in each tree based on the probabilities from the reconstruction and then graphed the trees on top of each other. Finally, we added the probabilities from the reconstructions at all the nodes.

Try to work through the code and figure out each of these steps.

7.5 A polymorphic trait model

To this point in both chapters 6 and in this chapter, we've always assumed that our discrete character has only one state per lineage.

Sometimes, however, different individuals within a species have different values for the trait. When these two or more states are common, it makes sense to treat this phenotypic trait as *polymorphic*.

In this section, we will see how to fit a polymorphic discrete state trait evolution model. This model is new to this book, so we encourage readers to try it out and see what happens!

The polymorphic trait evolution model we'll use is pretty simple. It merely assumes that polymorphism is an intermediate condition between each pair of monomorphic states.

That is to say, for a character with two monomorphic conditions, a and b, to transition from a to b, we imagine that any lineage should first pass through the intermediate state of $a + b$.

7.5.1 Structure of a polymorphic trait model

Everything works nicely for a binary trait; however, when our discrete character can assume more than two values, for instance, a, b, and c, things begin to get more complicated.

First, we're faced with the question of whether to treat evolution of our character as *ordered* or *unordered*.

If evolution is *ordered*, then to change from a to c, a lineage must (minimally) evolve $a \rightarrow a+b \rightarrow b \rightarrow b+c \rightarrow c$ or[33] (and this is very important) $a \rightarrow a+b \rightarrow a+b+c \rightarrow b+c \rightarrow c$. Both of this, two possibilities must be part of our fitted model.

If evolution is *unordered*, then a transition from a to c should be able to occur just as it did in our binary trait: $a \rightarrow a+c \rightarrow c$.

It often makes sense to treat *meristic*[34] traits as ordered. Figure 7.12 shows the conceptual structure of ordered three-state (figure 7.12a) and four-state (figure 7.12b) models.

```
## split the plot into two panels
par(mfrow=c(1,2))
## graph an ordered polymorphic trait model with
## three monomorphic conditions
graph.polyMk(k=3,model="ARD",ordered=TRUE,
    cex.traits=0.8,xlim=c(-1,1),ylim=c(-1,1))
mtext("(a)",line=0,adj=0)
## graph an ordered polymorphic trait model with
## four monomorphic conditions
```

---

[33] If having more than two states is allowed.
[34] That is, counted.

```
graph.polyMk(k=4,model="ARD",ordered=TRUE,
    cex.traits=0.8,xlim=c(-1,1),ylim=c(-1,1),asp=1)
mtext("(b)",line=0,adj=0)
```

Second, just as for monomorphic discrete character models that we learned in chapter 6, we must choose how to parameterize our fitted models. That is, we need to decide if we're to assume that all types of transitions occur at the same rate (i.e., the ER model of chapter 6), that different transitions occur at different rates (the SYM model), that all types of transitions occur at different rates (the ARD model), or if evolution proceeds under some other scenario.

This important decision is made more difficult because of the potential for rapid expansion of model complexity. This occurs because the *types* of transitions rise *much more rapidly* than with the square of the number of monomorphic states (as it did with our simpler Mk models of the previous chapter).

We can see this by inspecting figure 7.13. The number of arrows between states or state combinations in our graph is equal to the maximum number of parameters in the most complex model for trait evolution given each number of character states. For two states, there are three such arrows (figure 7.13a). For three states there are 18, for four states 56, and for five states up to 150 different transition rates between character states or state combinations (figure 7.13).

```
## split our figure into four panels
par(mfrow=c(2,2))
## graph a representation of our polymorphic trait
## evolution for two, three, four, and five
## character states
for(i in 1:4){
    graph.polyMk(k=i+1,model="ARD",states=letters[1:(i+1)],
        cex.traits=0.5,xlim=c(-1.2,1.2),ylim=c(-1,1),
        asp=NULL)
    mtext(paste("(",letters[i],")",sep=""),line=1,adj=0)
}
```

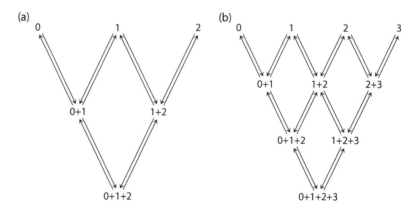

Figure 7.12
Structure of ordered polymorphic character models as implemented in the *phytools* function fitpolyMk. (a) A three-state character. (b) A four-state character.

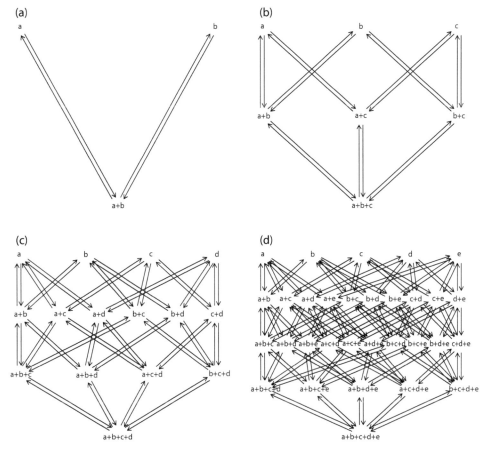

Figure 7.13
Unordered polymorphic trait evolution models for (a) two, (b) three, (c) four, and (d) five levels of the discrete character.

If it's hard to count the arrow in figure 7.13d, that's kind of the point. Normally, it will not be possible to fit a model to the data that typify phylogenetic comparative studies if that model has 150 parameters to be estimated!

7.5.2 The transient model

Given the rapidly escalating scale of complexity for polymorphic discrete trait evolution models, it makes sense to think of biologically sensible ways to *simplify* our fitted model.

We propose a model that we're calling the *transient* model in which we imagine that polymorphism is an inherently less stable condition than monomorphism. Under this model, we suppose that the polymorphism is acquired at one (constant) rate and then lost at another (presumably faster) constant rate.

This transient model makes sense even if many of our species are polymorphic because it helps explain why[35] higher degrees of polymorphism (i.e., lineages with four, five, or more

---

[35] And is consistent with an empirical observation that.

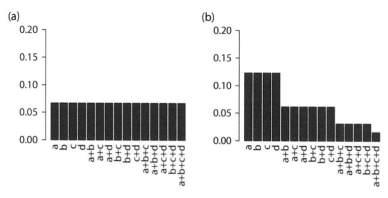

Figure 7.14
Equilibrium distributions for a polymorphic trait evolution model under (a) equal rates of transitions between all states and state combinations and (b) the transient model, in which the rate of loss of polymorphism is assumed to be (in this case, four times) higher than its acquisition.

states for a polymorphic trait) tend to be rarer than species with one, two, or three conditions for a trait.

We can see this by comparing the *equilibrium distributions*[36] between an equal-rates (ER) polymorphic model with four states and a transient four-state model in which we assume that the rate of loss of polymorphism is four times higher than its rate of gain (figure 7.14).

```
library(expm)
## split our plotting area
par(mfrow=c(1,2))
## create a design Q matrix for polymorphic trait
## evolution with four character states
## but with equal transition rates between all
## polymorphic conditions
Q.ER<-graph.polyMk(k=4,model="ER",states=letters[1:4],
    plot=FALSE)
diag(Q.ER)<--rowSums(Q.ER)
## create a barplot showing the expected frequencies
## under an equal-rates model
barplot(rep(1/nrow(Q.ER),nrow(Q.ER))%*%expm(Q.ER*1000),
    las=2,ylim=c(0,0.2),cex.axis=0.8,cex.names=0.8)
mtext("(a)",line=1,adj=0)
## create a design Q matrix for the transient model
Q.transient<-graph.polyMk(k=4,model="transient",
    states=letters[1:4],plot=FALSE)
Q.transient[Q.transient==2]<-0.5
diag(Q.transient)<--rowSums(Q.transient)
```

---

[36]The equilibrium distribution of an Mk model is just the relative frequency distribution of character states that we'd expect if we allowed evolution to proceed independently in a large number of lineages for a long time.

```
## graph the equilibrium frequencies under the
## transient model
barplot(rep(1/nrow(Q.transient),nrow(Q.transient))%*%
    expm(Q.transient*1000),las=2,ylim=c(0,0.2),
    cex.axis=0.8,cex.names=0.8)
mtext("(b)",line=1,adj=0)
```

What this analysis shows us is that under the transient model, lineages with higher and higher degrees of polymorphism are expected to be rarer and rarer—without imposing any particular "cap" on the number of states that are permitted or specifying different transition rates of increasing or decreasing polymorphism that depend on the number of states already possessed by a lineage.

7.5.3 Fitting a polymorphic trait model to data: An empirical example

To conclude this section, let's fit the polymorphic trait evolution model to some data.

For this part of the exercise, we'll use a phylogeny[37] and data set for flatworms from Benítez-Alvarez et al. (2020). The two files we'll use are planaria.timetree.nex and planaria.csv, both of which can be obtained from the book website.

We can start by reading the tree and data from file:

```
planaria.tree<-read.nexus(file="planaria.timetree.nex")
print(planaria.tree,printlen=3)

##
## Phylogenetic tree with 28 tips and 27 internal nodes.
##
## Tip labels:
##    Kronborgia_isopodicola, Urastoma_cyprinae, Phagocata_vitta, ...
##
## Rooted; includes branch lengths.

planaria.habitat<-read.csv(file="planaria.csv",
    row.names=1,stringsAsFactors=TRUE)
planaria.habitat

##                              Habitat
## Kronborgia_isopodicola        Marine
## Urastoma_cyprinae             Marine
## Phagocata_vitta            Freshwater
## Crenobia_alpina            Freshwater
## Cephaloflexa_bergi         Freshwater
## Geoplana_quagga            Freshwater
## Girardia_sp.               Freshwater
## Dugesia_gonocephala        Freshwater
```

---

[37] We modified the phylogeny slightly because the authors analyzed a non-ultrametric tree, which we don't generally recommend for neontological data.

```
## Schmidtea_polycroa                      Freshwater
## Rhodax_sp.2                             Freshwater
## Opisthobursa_mexicana                   Freshwater
## Novomitchellia_bursaelongata           Freshwater
## Hausera_hauseri                         Freshwater
## Kawakatsua_plumila                      Freshwater
## Obrimoposthia_wandeli           Freshwater+Marine
## Miroplana_shenzhensis                   Freshwater
## Procerodes_littoralis           Freshwater+Marine
## Uteriporus_sp.                             Marine
## Bdelloura_candida                          Marine
## Paucumara_falcata                       Freshwater
## Ectoplana_limuli                           Marine
## Sluysia_triapertura             Freshwater+Marine
## Nerpa_fistulata                            Marine
## Oregoniplana_geniculata                    Marine
## Palombiella_stephensoni                    Marine
## Pentacoelum_kazukolinda                 Freshwater
## Sabussowia_dioica                          Marine
## Cercyra_hastata                            Marine
```

We can see that the data consist of a two-state habitat character (Freshwater and Marine) with the polymorphic condition (Freshwater+Marine).

Let's plot our tree and data.

Here, instead of plotting habitat as a three-state character, let's show the polymorphism using a bicolor label, as follows. The result is shown in figure 7.15.

```
## plot planaria tree
plotTree(planaria.tree,ftype="i",fsize=0.8,offset=0.5)
## extra habitat as a vector
habitat<-setNames(planaria.habitat$Habitat,
    rownames(planaria.habitat))
## split each element of the vector containing + character
xx<-strsplit(as.character(habitat),split="+",fixed=TRUE)
## convert to matrix
pp<-matrix(0,length(habitat),2,dimnames=list(names(habitat),
    c("Freshwater","Marine")))
for(i in 1:nrow(pp)) pp[i,xx[[i]]]<-1/length(xx[[i]])
## graph pie tip labels
tiplabels(pie=pp,piecol=c("white","black"),cex=0.6)
## add legend
legend("topleft",c("Freshwater","Marine"),pch=21,
    pt.bg=c("white","black"),pt.cex=2,bty="n")
```

Because our character has only two monomorphic conditions, we don't have to worry about whether to fit an ordered or unordered model—they're both the same model!

Let's thus go ahead and fit four unordered models to these data: ER, SYM, ARD, and transient.

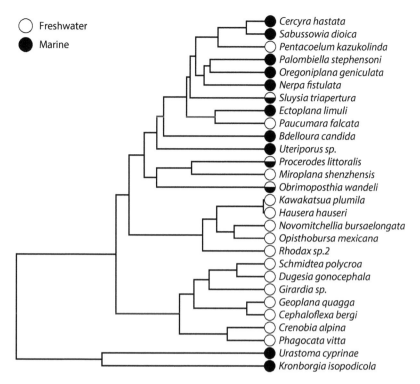

Figure 7.15
Phylogeny of flatworm taxa with habitat mapped onto the tips of the tree.

We'll do this using the *phytools* function `fitpolyMk`.[38]

```
## fit each of our for polymorphic trait evolution
## models using fitpolyMk
planaria.ER<-fitpolyMk(planaria.tree,habitat,
    model="ER",quiet=TRUE)
planaria.SYM<-fitpolyMk(planaria.tree,habitat,
    model="SYM",quiet=TRUE)
planaria.ARD<-fitpolyMk(planaria.tree,habitat,
    model="ARD",quiet=TRUE)
planaria.transient<-fitpolyMk(planaria.tree,habitat,
    model="transient")
```

```
##
## This is the design matrix of the fitted model.
## Does it make sense?
##
##                    Freshwater Marine
```

---

[38]The `fitpolyMk` argument `quiet`, which defaults to `quiet=FALSE`, tells R whether or not to print the design matrix for the fitted model to the terminal. We leave as FALSE on for the last model just so readers can see what this looks like.

```
## Freshwater                              0        0
## Marine                                  0        0
## Freshwater+Marine                       1        1
##                              Freshwater+Marine
## Freshwater                                    2
## Marine                                        2
## Freshwater+Marine                             0
```

Now, let's put together a table to compare our different models:

```
data.frame(model=c("ER","SYM","ARD","transient"),
    logLik=c(logLik(planaria.ER),logLik(planaria.SYM),
    logLik(planaria.ARD),logLik(planaria.transient)),
    k=c(attr(logLik(planaria.ER),"df"),
    attr(logLik(planaria.SYM),"df"),
    attr(logLik(planaria.ARD),"df"),
    attr(logLik(planaria.transient),"df")),
    AIC=aic<-c(AIC(planaria.ER),AIC(planaria.SYM),
    AIC(planaria.ARD),AIC(planaria.transient)),
    weight=unclass(aic.w(aic)))
```

```
##           model    logLik k      AIC      weight
## 1            ER -24.37419 1 50.74838 0.12548888
## 2           SYM -23.86102 2 51.72205 0.07712136
## 3           ARD -20.68220 4 49.36439 0.25068805
## 4     transient -21.90250 2 47.80500 0.54670171
```

This shows that the best-supported model, taking into account the number of estimated parameters, is indeed the transient model.

Let's print out this model and plot it (figure 7.16).

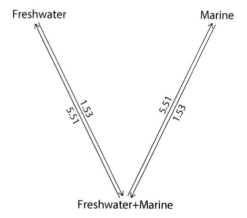

Figure 7.16
Fitted "transient" polymorphic trait model for the flatworm habitat preference evolution.

```
planaria.transient

## Object of class "fitpolyMk".
##
## Evolution was modeled as 'unordered' using the "transient" model.
##
## Fitted (or set) value of Q:
##                      Freshwater    Marine
## Freshwater           -1.533179  0.000000
## Marine                0.000000 -1.533179
## Freshwater+Marine     5.510608  5.510608
##                      Freshwater+Marine
## Freshwater                    1.533179
## Marine                        1.533179
## Freshwater+Marine           -11.021216
##
## Fitted (or set) value of pi:
##        Freshwater            Marine
##          0.333333          0.333333
## Freshwater+Marine
##          0.333333
##
## Log-likelihood: -21.902502
##
## Optimization method used was "nlminb"

plot(planaria.transient,signif=2,mar=rep(1.1,4),
    cex.traits=0.6,cex.rates=0.4)
```

We see that in the best-fitting transient model, the rate *away* from polymorphism is more than three times higher than the rate of transition *to* polymorphism, which is kind of satisfying.

7.6 The threshold model for studying discrete and continuous character traits

In both this chapter and the one that preceded, we've seen an incredibly wide variety of different models for the evolution of discrete characters.

All of these models, though, are fundamentally based on the same underlying premise about the evolution of discrete traits. That is, they assume that evolution proceeds via a continuous-time Markov chain.

In this final section of the chapter, we'll mention one last model for discrete character evolution; however, it's one that *is not* based on a continuous-time Markov chain. This model is called the *threshold model*, and it's derived from evolutionary quantitative genetics[39] and was first applied to the problem of analyzing interspecific data on a phylogeny by Felsenstein (2005, 2012; also see Revell 2014a).

---

[39] In fact, it was originally developed by Sewall Wright (1934).

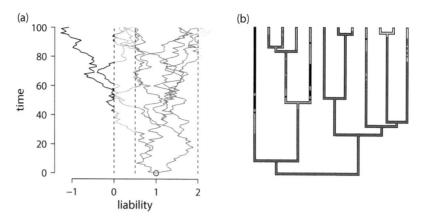

Figure 7.17

Evolution under the threshold model. (a) The evolution of liabilities through time. Thresholds are shown by the vertical dashed lines. (b) The resultant discrete character history mapped onto the tree.

According to the threshold model, the state of our discrete character is determined by the value of an unobserved continuous trait (normally called "liability") along with one or more thresholds. Whenever liability crosses the threshold, our discrete character changes state (Felsenstein 2005, 2012; Revell 2014a).

Figure 7.17 shows an example of evolution under the threshold model. In panel (a) of the figure,[40] we see liability evolution through time by Brownian motion evolution. The thresholds between discrete states are represented by way of the vertical dashed lines. In panel (b), we see how liability evolution translates to changes in the threshold trait on the tree.

```
library(RColorBrewer)
## split plot and set margins
par(mfrow=c(1,2),mar=c(4.1,4.1,2.1,1.1))
## simulate and graph Brownian liability evolution
## with discrete character changes under the
## threshold model
X<-bmPlot(tree<-pbtree(b=0.018,n=12,t=100,
    type="discrete",quiet=TRUE),
    type="threshold",thresholds=c(0,0.5,2),
    anc=1,sig2=1/100,ngen=max(nodeHeights(tree)),
    return.tree=TRUE,bty="n",
    colors=brewer.pal(4,"RdYlBu"))
mtext("(a)",line=1,adj=0)
## rotate the nodes of our simulated tree to
## (as much as possible) match the rank order of
## simulated liabilities
```

---

[40] To create the color scheme for this plot, we used the R package *RColorBrewer* (Neuwirth 2014) that creates custom color palettes for graphing. We haven't explicitly called *RColorBrewer* yet, but it is a *phytools* dependency, so you should already have it installed.

```
tt<-minRotate(as.phylo(X$tree),X$x,
    print=FALSE)$tip.label
## plot an outline for our phylogenetic tree
plotTree(X$tree,lwd=4,ftype="off",
    tips=setNames(1:Ntip(X$tree),tt),
    direction="upwards",mar=c(4.1,1.1,2.1,4.1))
## add the discrete character history
plot(X$tree,lwd=2,ftype="off",colors=X$colors,
    tips=setNames(1:Ntip(X$tree),tt),
    direction="upwards",mar=c(4.1,1.1,2.1,4.1),
    add=TRUE)
mtext(" (b) ",line=1,adj=0)
```

If we compare the result of figure 7.17a with that of figure 6.1 of the previous chapter, it should be quite evident that the threshold model has properties quite different from the standard M*k* model.

In particular, the rate of change between states is visibly heterogeneous across the tree.[41]

This is because when a lineage is near a threshold, the discrete character changes frequently between states. By contrast, if a lineage is far from any threshold, the character may not change at all.

For some characters, it's easy to imagine that this continuously varying rate heterogeneity is more likely to capture the real evolutionary process of our data.

One of the most significant advantages of the threshold model, however, is that it also provides us with an extremely convenient framework within which to model *correlated* evolution between discrete characters—something that we learned in section 7.2 can be hard to define for discrete traits. In the threshold model, the correlation between traits is merely the evolutionary correlation of their liabilities!

7.6.1 Correlated evolution under the threshold model: An empirical example using bony fish

There are a few applications of the threshold model in software.

For instance, Felsenstein (2005, 2012) applied the correlational threshold model in his stand-alone *PHYLIP* software.[42]

In R, the correlational threshold model has been implemented in the *phytools* function threshBayes, which[43] uses Bayesian inference.

Let's try fitting the correlational threshold model to the same two binary characters that we analyzed using the Pagel (1994) method earlier in this chapter (Benun Sutton and Wilson 2019).

Remember, the tree and data for this analysis are in the files bonyfish.tre and bonyfish.csv.[44]

---

[41] We can actually approximate this heterogeneity quite well using the hidden-rates model of a previous section in this chapter. In a sense, the threshold model could be seen as a limiting case of the hidden-rates model as the number of hidden rates goes to ∞!

[42] If you have *PHYLIP* installed locally, it's even possible to run *PHYLIP* from within R using the *Rphylip* package (Revell and Chamberlain 2014).

[43] As the name would suggest.

[44] Available from the book website.

Complex discrete models

In case anyone just jumped to this section without doing the exercise earlier in the chapter, why don't we just reread these data from file into R.

```
## read tree from file
bonyfish.tree<-read.tree(file="bonyfish.tre")
## read data from file
bonyfish.data<-read.csv(file="bonyfish.csv",row.names=1,
    stringsAsFactors=TRUE)
```

Now we can run our MCMC.

Since our parameter space for this model consists of a liability value for each terminal taxon in the tree for each trait, plus the correlation coefficient between them,[45] we'll run 4,000,000 generations—anticipating that our Metropolis–Hasting MCMC sampler might be a bit inefficient in sampling the posterior distribution.[46]

threshBayes also tries to autotune the acceptance ratio to a value of 0.23.[47]

```
## set the number of generations for the MCMC
ngen<-4e6
## run the MCMC in threshBayes
mcmc.bonyfish<-threshBayes(bonyfish.tree,bonyfish.data,
    type=c("disc","disc"),ngen=ngen,plot=FALSE,
    control=list(print.interval=ngen/10))

## Starting MCMC....
## generation: 400000; mean acceptance rate: 0.23
## generation: 800000; mean acceptance rate: 0.19
## generation: 1200000; mean acceptance rate: 0.35
## generation: 1600000; mean acceptance rate: 0.27
## generation: 2000000; mean acceptance rate: 0.16
## generation: 2400000; mean acceptance rate: 0.23
## generation: 2800000; mean acceptance rate: 0.25
## generation: 3200000; mean acceptance rate: 0.25
## generation: 3600000; mean acceptance rate: 0.3
## generation: 4000000; mean acceptance rate: 0.26
## Done MCMC.

mcmc.bonyfish

##
## Object of class "threshBayes" consisting of a matrix (L) of
## sampled liabilities for the tips of the tree & a second matrix
## (par) with the sample model parameters & correlation.
##
## Mean correlation (r) from the posterior sample is: -0.45502.
```

---

[45] In other words, lots of dimensions!

[46] Even more generations of MCMC may sometimes be necessary in practice—particularly for larger trees!

[47] With apologies to the Bayesian purists—we do this throughout the whole MCMC, not just during burn-in.

```
##
## Ordination of discrete traits:
##
##   Trait 1: group <-> pair
##   Trait 2: male <-> none
```

Some of the arguments that we supply to threshBayes include type, in which we indicate the data type, continuous or discrete, for each character of our input data frame or matrix; ngen, where we specify the number of generations of MCMC to run; and control, which is a list of control parameters of the MCMC. Here, we only set the print interval to the display buffer, but under other circumstances, we may want to use this argument to adjust the prior probability distributions of our model or the proposal distributions for the MCMC.

Just plotting our result shows a multipanel figure with the likelihood profile (that is, the likelihood for each generation of the MCMC), the acceptance rate of the MCMC, and the profile plot of the correlation coefficient (r), as this is the parameter that we're probably most interested in (figure 7.18).

```
plot(mcmc.bonyfish)
```

Since we're likely to be most interested in the *correlation* between[48] our two discrete characters, why don't we also plot the posterior density of r in our model.

This can be done pretty easily too. We see the results in figure 7.19.

```
## set margins
par(mar=c(5.1,4.1,2.1,2.1))
## plot posterior density
plot(density(mcmc.bonyfish),cex.lab=0.8,
    cex.axis=0.7)
```

Likewise, it's straightforward to compute a 95 percent high-probability density (HPD) interval around the correlation.

We'll do this by first extracting the post–burn-in sample of values of r (assuming a 20 percent burn-in), assigning the class attribute "mcmc" to our vector, and then using the function HPDinterval from the very handy R package *coda*.[49]

If you haven't used *coda* (Plummer et al. 2006) before, you should install it now.

```
## load the coda package
library(coda)
## extract our post burn-in sample for r
r.mcmc<-tail(mcmc.bonyfish$par$r,
    0.8*nrow(mcmc.bonyfish$par))
## set the class to "mcmc"
```

---

[48] The liabilities of.

[49] *coda* does lots of other cool things too—like evaluate convergence to the posterior distribution and compute effective sample sizes.

(a) log–likelihood trace

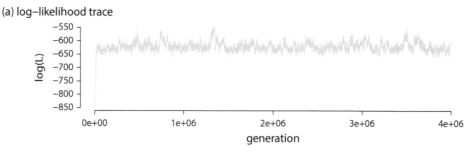

(b) mean acceptance rate (sliding window: bw = 400)

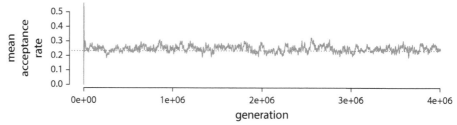

(c) trace of the correlation coefficient, r

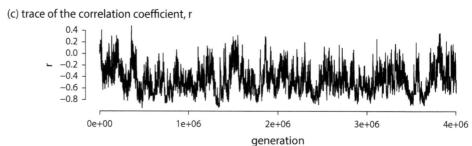

Figure 7.18
Profile plots from a Bayesian MCMC analysis of the threshold model for spawning mode and paternal care evolution in bony fishes. (a) The likelihood profile. (b) Acceptance rates from the MCMC averaged across variables in the model and on a bandwidth as indicated in the figure panel. (c) Profile of the correlation coefficient, *r*.

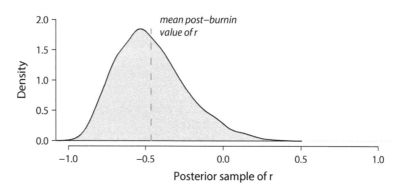

Figure 7.19
Posterior density of the correlation between spawning mode and paternal care in bony fishes.

```
class(r.mcmc) <- "mcmc"
## compute HPD interval for the correlation
HPDinterval(r.mcmc)

##            lower        upper
## var1 -0.8605848 -0.03631072
## attr(,"Probability")
## [1] 0.9500016
```

This tells us that our 95 percent HPD interval for *r*, although broad, does not include zero—although just barely!

What does it mean that the mean correlation coefficient from the posterior distribution is *negative*?

To interpret the direction of the correlation coefficient, we need to know the *ordination* of our binary traits. By ordination, we mean which state for each binary trait is *up*[50] and which is down. By default, `threshBayes` will ordinate our traits in alphabetical order.

In our case, this would mean that the state `"pair"` (spawning) and `"none"` (paternal care) are both *up*, and thus a negative evolutionary correlation implies that higher values of liability for group spawning tend to coevolve with lower values of liability for paternal care. (Or, equivalently, that higher values of liability for pair spawning tend to evolve with higher values of liability toward male parental care!)

We can double-check the ordination of each of our binary traits just by printing the fitted model object to the screen.

```
print(mcmc.bonyfish)

##
## Object of class "threshBayes" consisting of a matrix (L) of
## sampled liabilities for the tips of the tree & a second matrix
## (par) with the sample model parameters & correlation.
##
## Mean correlation (r) from the posterior sample is: -0.45502.
##
## Ordination of discrete traits:
##
##   Trait 1: group <-> pair
##   Trait 2: male <-> none
```

Once again, this result shows us that lineages evolving *pair spawning* (in trait 1) tend to evolve *male parental care* (for trait 2). How does this finding concord with what we found using Pagel's (1994) model in figure 7.2?

7.6.2 Analyzing discrete and continuous traits using the threshold model

Finally, an interesting aspect of the threshold model (identified by Felsenstein 2012) is that it also creates a natural framework within which to evaluate the evolutionary correlation between discrete and continuous traits.

---

[50]So to speak.

Complex discrete models

In this case, the correlation coefficient is now defined as the correlation between liabilities for the discrete trait and the numerical values of the continuous character in our analysis.

To see how this works, we'll use an example of the evolution of viviparity (from oviparity) in the South American lizard family Liolaemidae.

These data are from a study by Esquerréet al. (2019) in which the authors investigated the role played by the uplift of the Andes mountains in the biogeography and reproductive biology of this group.

To follow along, you'll need to download the files `Liolaemidae.MCC.nex` and `Liolaemidae.data.csv` from the book website.

```
## read tree
Liolaemidae.tree<-read.nexus(file="Liolaemidae.MCC.nex")
print(Liolaemidae.tree,printlen=2)
```

```
##
## Phylogenetic tree with 258 tips and 257 internal nodes.
##
## Tip labels:
##    Ctenoblepharys_adspersa, Liolaemus_abaucan, ...
##
## Rooted; includes branch lengths.
```

```
## read data
Liolaemidae.data<-read.csv(file="Liolaemidae.data.csv",
    row.names=1,stringsAsFactors=TRUE)
head(Liolaemidae.data)
```

```
##                          parity_mode max_altitude
## Ctenoblepharys_adspersa            O          750
## Liolaemus_abaucan                  O         2600
## Liolaemus_albiceps                 V         4020
## Liolaemus_andinus                  V         4900
## Liolaemus_annectens                V         4688
## Liolaemus_anomalus                 O         1400
##                          temperature
## Ctenoblepharys_adspersa        23.05
## Liolaemus_abaucan              20.20
## Liolaemus_albiceps             12.38
## Liolaemus_andinus              11.40
## Liolaemus_annectens             5.10
## Liolaemus_anomalus             23.78
```

Let's repeat the analysis we undertook with the data from bony fishes, above, but this time using viviparity and temperature.

In this case, we'll test the hypothesis[51] that the evolution of squamate viviparity[52] is affected by environmental temperature—in this case the mean temperature of the warmest month (Esquerréet al. 2019).

---

[51] As others have done in the past.

[52] Sometimes described as *ovoviviparity* or *aplacental viviparity*.

It's generally thought that lower environmental temperatures should favor the evolution of viviparity—because viviparity improves the ability of the mother to ensure favorable egg incubation conditions via behavioral thermoregulation (Shine and Bull 1979; Blackburn 1982).

This seems like a perfect hypothesis to examine using the threshold model!

This time, we'll only run our MCMC for 1,000,000 generations (figure 7.20)—but not because we should, but because this is a bigger tree and data set, so to run it for more generations would simply take too long!

```
## set number of generations
ngen<-1e6
## run MCMC
mcmc.Liolaemidae<-threshBayes(Liolaemidae.tree,
    Liolaemidae.data[,c(1,3)],
    type=c("disc","cont"),ngen=ngen,plot=FALSE,
    control=list(print.interval=ngen/10))

## Starting MCMC....
## generation: 100000; mean acceptance rate: 0.25
## generation: 200000; mean acceptance rate: 0.31
## generation: 300000; mean acceptance rate: 0.32
## generation: 400000; mean acceptance rate: 0.3
## generation: 500000; mean acceptance rate: 0.22
## generation: 600000; mean acceptance rate: 0.23
## generation: 700000; mean acceptance rate: 0.21
## generation: 800000; mean acceptance rate: 0.22
## generation: 900000; mean acceptance rate: 0.21
## generation: 1000000; mean acceptance rate: 0.49
## Done MCMC.

mcmc.Liolaemidae

##
## Object of class "threshBayes" consisting of a matrix (L) of
## sampled liabilities for the tips of the tree & a second matrix
## (par) with the sample model parameters & correlation.
##
## Mean correlation (r) from the posterior sample is: -0.51692.
##
## Ordination of discrete traits:
##
##   Trait 1: O <-> V

plot(mcmc.Liolaemidae)
```

Many of our arguments are similar to the prior example, except for type, which here is set to type=c("disc","cont") to indicate that the first of our two traits is discretely coded, while the second is *continuous*.

Now let's get an HPD interval for the correlation coefficient and plot our posterior distribution (figure 7.21).

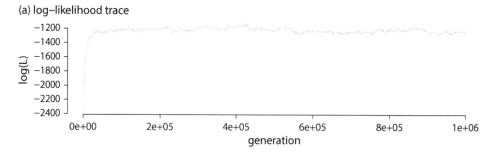

(a) log–likelihood trace

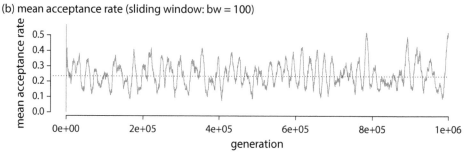

(b) mean acceptance rate (sliding window: bw = 100)

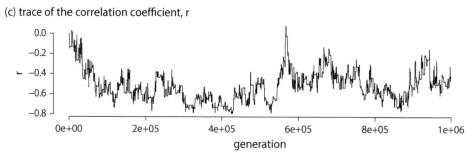

(c) trace of the correlation coefficient, r

Figure 7.20
Profile plots from a Bayesian MCMC analysis of the threshold model for oviparity versus viviparity and environmental temperature in liolaemid lizards. Panels are as in figure 7.18.

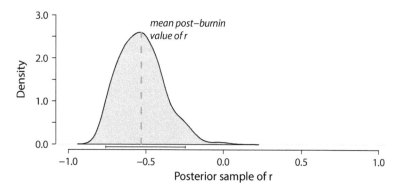

Figure 7.21
Posterior density of the correlation between oviparity and mean temperature of the warmest month in Liolaemidae.

For fun, we'll draw a line with whiskers just above the *x*-axis to show our 95 percent HPD for *r*.

```
## pull out the post burn-in sample and
## compute HPD
r.mcmc<-tail(mcmc.Liolaemidae$par$r,
    0.8*nrow(mcmc.Liolaemidae$par))
class(r.mcmc)<-"mcmc"
hpd.r<-HPDinterval(r.mcmc)
hpd.r
```

```
##          lower       upper
## var1 -0.7597935 -0.245344
## attr(,"Probability")
## [1] 0.9500062
```

```
## plot posterior density
par(mar=c(5.1,4.1,2.1,2.1))
plot(density(mcmc.Liolaemidae),bty="n",
    cex.lab=0.9,cex.axis=0.7)
## add whiskers to show HPD
h<-0-par()$usr[3]
lines(x=hpd.r,y=rep(-h/2,2))
lines(x=rep(hpd.r[1],2),y=c(-0.3,-0.7)*h)
lines(x=rep(hpd.r[2],2),y=c(-0.3,-0.7)*h)
```

Since we see that the ordination of our discrete state is oviparity → viviparity, and the correlation coefficient is *negative*, we know that viviparity is associated with *lower* environmental temperatures.

This matches what we predicted a priori.

7.7 Practice problems

7.1 Ramm et al. (2020) asked if the evolution of defensive tail spines in squamate reptiles could be related to microhabitat utilization. Using a phylogeny of squamate reptiles (lizard_tree.nex from the book website) and a data set of morphology and microhabitat use (lizard_spines.csv), apply Pagel's (1994) method to test the hypothesis that the evolution of tail spines is associated with shifts from non-*saxicolous*[53] to saxicolous microhabitat use. As in prior exercises, you'll need to check that the tree and data set match before beginning your analysis. Make sure to plot both your phylogeny and data, as well as the results from your fitted model. This analysis follows Ramm et al. (2020), but with a subsampled data set.

7.2 Working with the same phylogeny and data set as in the previous question, also test the hypothesis that microhabitat state affects the process of tail spine evolution but *not* the converse. Compare this model to the one you fit for practice problem 7.1, in which both

---

[53] Saxicolous is a fancy way of saying "rock-dwelling."

microhabitat evolution affected tail spine evolution as well as the converse, along with one in which the two characters evolved independently from each other. Use both likelihood ratio tests and AIC to compare among your different fitted models.

7.3 Halali et al. (2020) investigated habitat evolution in Mycalesina butterflies. Many butterfly species can be found exclusively in forested, forest-fringe, or open habitats, but numerous, more generalist species are found in two or all three habitat types. Use `fitpolyMk` to fit a polymorphic trait evolution model to the evolution of habitat type in Mycalesina assuming both ordered character evolution (that is, forest ↔ forest fringe ↔ open) or unordered polymorphic trait evolution. The two files you need (`Mycalesina_phylogeny.nex` and `Mycalesina_habitat.csv`) are modified from Halali et al. (2020) and can be downloaded from the book website.

8

Reconstructing ancestral states

8.1 Introduction

Many phylogenetic comparative methods purport to give us some insight about the evolutionary past.

For instance, in chapters 2, 3, and 7, we learned how to ask if the evolutionary change of one trait (be it continuous or discrete) tended to precipitate a concurrent or subsequent change in a second character.

In later chapters of this book, we'll learn how to investigate if the evolution of a key trait increases speciation, decreases extinction, or both.

But perhaps the most direct question we can ask about ancestral species is simply, what were they like? The endeavor of answering this question using a phylogenetic tree is called *ancestral state reconstruction* (reviewed in Felsenstein 2004; Harmon 2019), and for better or for worse, the estimation of ancestral traits has long been, and continues to be, an important goal in phylogenetic comparative biology.

When estimating or *reconstructing*[1] ancestral states, the objective is typically to estimate with some measure of confidence the condition of a hypothetical ancestral taxon in the past.

This is usually done using the data for present-day taxa[2] combined with a phylogenetic tree. The hypothetical ancestral taxa usually correspond to the internal nodes of the phylogeny.

In this chapter, we'll:

1. Learn how to estimate ancestral states for continuous characters using maximum likelihood, as well as how to compute and interpret 95 percent confidence intervals around these estimates.
2. Examine the statistical properties of maximum likelihood estimation of ancestral states for a continuous trait using numerical simulation: both when our assumed model is correct as well as when it is wrong.

---

[1] We prefer to use the term *estimation* over *reconstruction*, because the former emphasizes its statistical (and thus probabilistic) nature. Nevertheless, reconstruction is widely used, so we employ both words more or less interchangeably here.

[2] Although it's possible—and can be very useful—to include data from fossils.

3. Learn about reconstructing ancestral states for discrete characters, as well as the distinction between joint and marginal estimation.
4. Explore a discrete character method called stochastic character mapping.
5. Finally, consider why the method of maximum parsimony should probably not be used for ancestral state reconstruction on the phylogenies.

8.2 Ancestral states for continuous characters

So far in this book, we've seen the analysis of the evolution of continuous characters on the phylogeny (in chapters 2, 3, 4, and 5), as well as the analysis of discrete trait evolution (in chapters 6 and 7).

It should thus be of no surprise that we can estimate ancestral states on a phylogeny for both of these types of characters.

We'll start with continuously valued character traits. These are characters that are (logically) measured on a continuous or *metric* scale and might include traits, such as body size, mass, and limb length, but also ecological traits, such as average perch height, dietary niche width,[3] and so on.

If we'd like to estimate ancestral states for the internal nodes of our phylogeny, we'll need to make some assumptions about the process[4] under which we suppose that our character trait evolved on the tree.

Although we could assume any of a range of different evolutionary models, by far the most widely used model for ancestral state reconstruction on the phylogeny for continuous traits is an important model that we've already learned in this book: Brownian motion.

8.2.1 Estimating ancestral states under Brownian motion

To estimate ancestral states under Brownian motion, our tactic will be to identify the set of states that have the highest probability under our assumed model (Felsenstein 2004). As we first discussed in chapter 4, these states will be, by definition, our maximum likelihood estimates (MLEs; see Harmon 2019).

Ancestral character estimation is implemented in a variety of different R functions. The most commonly used is ace[5] from the same *ape* R package that we've used in prior chapters.

Here, instead of using ace, we'll start out by using the *phytools* function fastAnc;[6] however, all methods that reconstruct ancestral states under Brownian motion using likelihood should give the same results.

Why don't we start out by estimating ancestral body size of elopomorph eels! This group includes species such as *Anguilla anguilla* (the European eel) and the moray eels of the family Muraenidae (such as the spotted moray, *Gymnothorax moringa*).

To do this, we'll use a data set from Collar et al. (2014) in the form of two files that can be obtained from the book webpage[7]: elopomorph.tre and elopomorph.csv. These files contain an estimated phylogeny and a trait data set for feeding mode and maximum body size in elopomorph eels, respectively.

As soon as you have these files, we can proceed to load the *phytools* package and then continue by reading our tree and data from file as follows.

---

[3] If measured numerically.

[4] That is, we need to assume a model of evolution.

[5] An acronym for *ancestral character estimation*.

[6] Simply because it gives us some information not supplied in the output of ace.

[7] http://www.phytools.org/Rbook/.

```
## load libraries
library(phytools)
## read tree from file
eel.tree<-read.tree("elopomorph.tre")
print(eel.tree,printlen=2)

##
## Phylogenetic tree with 61 tips and 60 internal nodes.
##
## Tip labels:
##     Moringua_edwardsi, Kaupichthys_nuchalis, ...
##
## Rooted; includes branch lengths.

## read data
eel.data<-read.csv("elopomorph.csv",row.names=1,
    stringsAsFactors=TRUE)
head(eel.data)

##                     feed_mode Max_TL_cm
## Albula_vulpes         suction       104
## Anguilla_anguilla     suction        50
## Anguilla_bicolor      suction       120
## Anguilla_japonica     suction       150
## Anguilla_rostrata     suction       152
## Ariosoma_anago        suction        60
```

As our next step, let's extract total body length (Max_TL_cm) as a new vector from our trait data frame as follows. We'll also transform the character to a log-scale using log.

```
## extract total body length and log-transform
lnTL<-setNames(log(eel.data$Max_TL_cm),rownames(eel.data))
head(lnTL)

##     Albula_vulpes Anguilla_anguilla
##          4.644391          3.912023
##  Anguilla_bicolor Anguilla_japonica
##          4.787492          5.010635
## Anguilla_rostrata    Ariosoma_anago
##          5.023881          4.094345
```

At this point, we're already ready to estimate ancestral states.

Remember, when we do this using maximum likelihood (ML), as we intend to, what we're doing is choosing the values for internal nodes of the tree that maximize the probability of obtaining the data that we see at the tips of the tree (Felsenstein 2004; Harmon 2019).[8]

This seems sensible.

---

[8]The way we describe them here, then, these are the joint ancestral state estimates—we'll return to this concept later in the chapter!

In addition to finding the MLEs for internal nodes, `fastAnc` will also compute both variances and the 95 percent confidence intervals for each node.

This is very important since, as we'll see in a moment, ancestral states tend to be estimated with lots of uncertainty (Losos 1999).

```
## estimate ancestral states using fastAnc
fit.lnTL<-fastAnc(eel.tree,lnTL,vars=TRUE,CI=TRUE)
```

Next, let's print[9] our result.

```
print(fit.lnTL,printlen=10)

## Ancestral character estimates using fastAnc:
##          62         63         64         65         66
##    4.212848  4.106868  4.164451  4.169512  4.171423
##          67         68         69         70         71
##    4.080113  4.082773  3.998125  4.247916  4.268798
##
##    . . . .
##
## Variances on ancestral states:
##          62         63         64         65         66
##    0.31021  0.211767  0.170578  0.107321  0.093304
##          67         68         69         70         71
##    0.110589  0.111161  0.009653  0.112782  0.093701
##
##    . . . .
##
## Lower & upper 95% CIs:
##          lower      upper
## 62  3.121197  5.304498
## 63  3.204911  5.008824
## 64  3.354951  4.973952
## 65  3.527418  4.811605
## 66  3.572727  4.770118
## 67  3.428317   4.73191
## 68  3.429294  4.736252
## 69  3.805553  4.190697
## 70  3.589689  4.906143
## 71   3.66883  4.868766
##          . . . .      . . . .
```

In our printout, it's helpful to keep in mind that the numbers that we see as either element or row names correspond to the node indices of our input `"phylo"` object, just as we learned in prior chapters (particularly chapters 1 and 2).

---

[9]By specifying `printlen=10`, we're choosing to print out the ML states, variances, and 95 percent confidence intervals for only the first ten nodes of the tree. If we'd just entered the name of the object into the command line, R would've printed the states for all the nodes of the phylogeny.

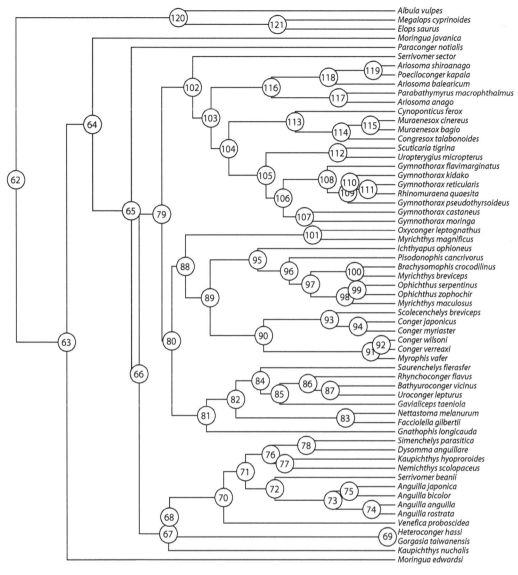

Figure 8.1
A phylogeny of elopomorph eel species with node indices of the R "phylo" object indicated. The same indices are used in ancestral state reconstruction.

Figure 8.1 shows these numerical indices mapped onto the nodes of our elopomorph tree.[10]

```
## plot eel phylogeny using plotTree
plotTree(eel.tree,ftype="i",fsize=0.5,lwd=1)
## add node labels for reference
```

---

[10] As in chapter 1, we could have done this using ape::nodelabels, although that function has the annoying habit of resizing our circles depending on the number of digits in the corresponding node index.

```
labelnodes(1:eel.tree$Nnode+Ntip(eel.tree),
    1:eel.tree$Nnode+Ntip(eel.tree),
    interactive=FALSE,cex=0.5)
```

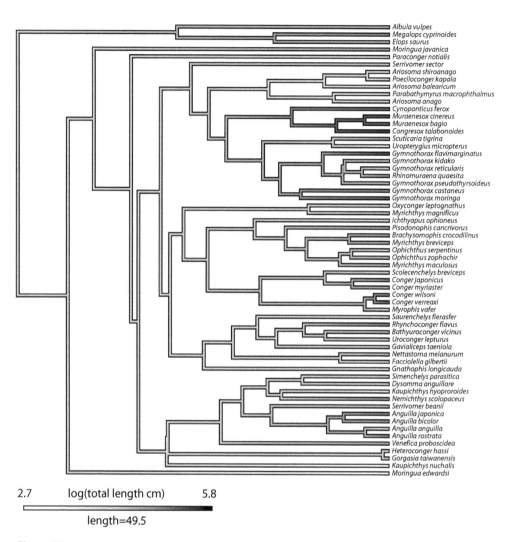

Figure 8.2
A projection of maximum body length (on a log-scale) onto the tree of elopomorph eels. The color gradient shows observed (at the tips) or estimated trait values.

8.2.2 Plotting reconstructed ancestral states on the tree

In addition to merely printing the values of the trait at internal nodes of the tree, it's also not particularly difficult to graph our estimated ancestral states on the tree.

Still working with our elopomorph body size example, let's do this using a `phytools` function called `contMap`.[11] The result is shown in figure 8.2.

---

[11] Short for *continuous character mapping.*

```
## compute "contMap" object
eel.contMap<-contMap(eel.tree,lnTL,
    plot=FALSE,lims=c(2.7,5.8))
## change the color gradient to a custom gradient
eel.contMap<-setMap(eel.contMap,
    c("white","orange","black"))
## plot "contMap" object
plot(eel.contMap,sig=2,fsize=c(0.4,0.7),
    lwd=c(2,3),leg.txt="log(total length cm)")
```

In this code, we first computed a "contMap" object, which is a tree with a continuous character ancestral state reconstruction along the branches and nodes of the tree.

Next, we used setMap to change the color gradient of the map from the default (which is a rainbow scale) to a heat color (white to orange to black) gradient.

Finally, we plotted our object—setting various options for font size and style.

It's also possible to add error bars at the internal nodes of tree showing the *uncertainty* that is associated with each node ancestral state estimate.

These can be difficult to read for larger trees, so let's just extract one clade of this tree and then replot it with error bars at the nodes. This graph is shown in figure 8.3.

```
## identify the tips descended from node 102
tips<-extract.clade(eel.tree,102)$tip.label
tips

##  [1] "Gymnothorax_moringa"
##  [2] "Gymnothorax_castaneus"
##  [3] "Gymnothorax_pseudothyrsoideus"
##  [4] "Rhinomuraena_quaesita"
##  [5] "Gymnothorax_reticularis"
##  [6] "Gymnothorax_kidako"
##  [7] "Gymnothorax_flavimarginatus"
##  [8] "Uropterygius_micropterus"
##  [9] "Scuticaria_tigrina"
## [10] "Congresox_talabonoides"
## [11] "Muraenesox_bagio"
## [12] "Muraenesox_cinereus"
## [13] "Cynoponticus_ferox"
## [14] "Ariosoma_anago"
## [15] "Parabathymyrus_macrophthalmus"
## [16] "Ariosoma_balearicum"
## [17] "Poeciloconger_kapala"
## [18] "Ariosoma_shiroanago"
## [19] "Serrivomer_sector"

## prune "contMap" object to retain only these tips
pruned.contMap<-keep.tip.contMap(eel.contMap,tips)
## plot object
```

```
plot(pruned.contMap,xlim=c(-2,90),lwd=c(3,4),
    fsize=c(0.7,0.8))
## add error bars
errorbar.contMap(pruned.contMap,lwd=8)
```

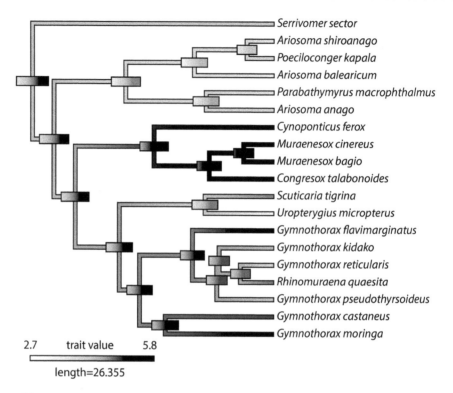

Figure 8.3
Eel "contMap" plot with error bars showing the uncertainty associated with the estimated values for the trait at internal nodes of the tree.

In figure 8.3, the horizontal bars at each node show (via the color gradient) the 95 percent confidence intervals[12] of each reconstruction. We can already tell that in some cases the uncertainty is quite large.

8.3 Properties of ancestral state estimation for continuous traits

The next thing that we'll do is explore some of the properties of ancestral state reconstruction of continuous traits in general, starting with our Brownian model.

To do this, though, we'll first need to simulate some data.

For simulating a phylogeny, we can use the *phytools* function pbtree,[13] as we've done before, and to simulate data under Brownian motion, we'll use the function fastBM.

---

[12] In other words, a measure of the uncertainty of each estimated state.

[13] There are lots of phylogenetic tree simulators in other software. The R package *TreeSim* by Tanja Stadler (2011, 2019) is perhaps the most powerful. For our purposes at the moment, phytools::pbtree will work just fine.

We already encountered `fastBM` in chapter 2. Here, instead of merely simulating trait values at the tips of the tree, we're going to use it to simulate states at *all* of the internal nodes of our phylogeny as well.[14]

```
## simulate a tree & some data
tree<-pbtree(n=26,scale=1,tip.label=LETTERS)
## simulate with ancestral states
x<-fastBM(tree,internal=TRUE)
```

Our vector x contains the states for both the internal nodes and the tips of the tree.

For our analysis, we want to split this object, x, into two separate vectors: one containing the tip states and a second with the states for all internal nodes. This is not difficult because the tip taxa in our vector have names that correspond to all the species in the tree, whereas our internal node values are numbered, once again, according to our input `"phylo"` object node indices.

```
## ancestral states
a<-x[1:tree$Nnode+Ntip(tree)]
## tip data
x<-x[tree$tip.label]
```

Now, as above, let's proceed to estimate ancestral states for our vector x just as we did for the eel body length data, using the function `fastAnc`.

```
## estimate ancestral states for simulated data
fit<-fastAnc(tree,x,CI=TRUE)
print(fit,printlen=6)

## Ancestral character estimates using fastAnc:
##        27        28        29        30        31
##   0.298908  0.174493  0.144861 -0.086368 -0.364955
##        32
##  -0.381101 ....
##
## Lower & upper 95% CIs:
##       lower     upper
## 27 -0.479894  1.077709
## 28 -0.462764   0.81175
## 29  -0.49133  0.781052
## 30 -0.731548  0.558811
## 31 -1.000656  0.270745
## 32 -1.017425  0.255224
##           ....      ....
```

Since we simulated these data (and in contrast to every empirical case, including our eel data above), we know the right answers—that is to say, the *true* states at the nodes of our tree.

---

[14]The latter we accomplish by setting the argument `internal=TRUE`.

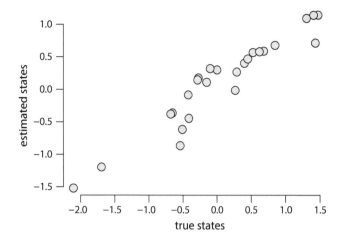

Figure 8.4
Simulated ancestral states compared to reconstructed states from maximum likelihood.

Let's take advantage of this fact and *compare* these reconstructions to our known true values of x.

```
par(mar=c(5.1,4.1,2.1,2.1))
plot(a,fit$ace,xlab="true states",ylab="estimated states",
    bty="n",bg="gray",cex=1.5,pch=21,las=1,cex.axis=0.8)
```

Straightaway, we should see (figure 8.4) that the true values from simulation are quite highly correlated with our MLE ancestral states.

That's a good start.

But remember, we're doing statistical estimation, so perhaps an even more important consideration[15] is not merely whether our estimates are correlated with the generating values but the type I error rate of our estimation procedure. That is to say, whether or not a hypothesis test based on our estimate would reject our known true values at a rate that exceeds the nominal α value of the test, say, 0.05.

We could evaluate this in more than one way—for instance, by computing the difference between the estimate and its known value over the standard error.[16]

However, perhaps the easiest thing to do is simply measure the fraction of times that our 95 percent confidence interval for each estimate includes the (known, in our case) true value of the parameter.

On average, this should be around 95 percent of the time.

Instead of simply computing this, let's plot its measure by representing our estimates as vertical lines that run from the bottom to the top of each confidence interval, instead of as simple x, y points.

We should find that these vertical lines cross the 1:1 line[17] about 19 times out of 20—that is to say, 95 percent of the time!

---

[15] At least, that is, from the traditional frequentist perspective.
[16] Which should be distributed as a t-statistic.
[17] In which the parameter and its estimate are the same.

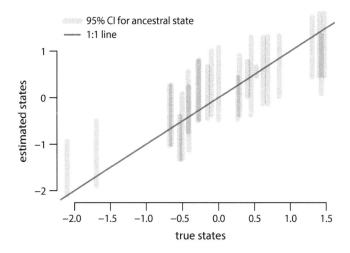

Figure 8.5
Simulated ancestral states compared to reconstructed 95 percent confidence intervals from maximum likelihood.

```
## set margins
par(mar=c(5.1,4.1,2.1,2.1))
## plot true vs. estimated states
plot(a,fit$ace,xlab="true states",
    ylab="estimated states",bty="n",
    ylim=range(fit$CI95),col="transparent",
    las=1,cex.axis=0.8)
## add 1:1 line
lines(range(c(x,a)),range(c(x,a)),
    col="red",lwd=2) ## 1:1 line
## use a for loop to add vertical lines showing
## each confidence interval
for(i in 1:tree$Nnode)
    lines(rep(a[i],2),fit$CI95[i,],lwd=5,
        col=make.transparent("blue",0.25),
        lend=0)
## create a legend
legend(x="topleft",legend=c("95% CI for ancestral state",
    "1:1 line"),col=c(make.transparent("blue",0.25),
    "red"),lty=c("solid","solid"),lwd=c(5,2),cex=0.7,
    bty="n")
```

In this case,[18] it looks like most, but not all, of our confidence intervals touch or include the 1:1 line, just as we'd expect (figure 8.5).

If we want to attach a number to this, we can just compute the fraction of CIs in our simulation in which the true ancestral state falls within the 95 percent CI for that state.

---

[18]Of course, this is a stochastic numerical simulation that uses random-number generation, so readers will all obtain slightly different results!

```
withinCI<-((a>=fit$CI95[,1]) & (a<=fit$CI95[,2]))
table(withinCI)

## withinCI
## FALSE   TRUE
##     2     23

mean(withinCI)

## [1] 0.92
```

If you repeat this exercise yourself, sometimes you should expect to obtain values greater than 0.95 and sometimes lower than 0.95, as we did.

Our expectation, in fact, is that after a large number of such simulations, we should (on average) find that 95 percent of the confidence intervals included the true value of the state at each node.

Rather than simply assert this, however, let's test it!

The code below does exactly that—by creating a custom function that simulates a tree, simulates data for the nodes and tips of that tree, computes 95 percent confidence intervals around the ML ancestral state estimates at each node, and then counts the fraction of times in 100 replicate simulations that this 95 percent CI includes the true value of the ancestral state.

```
## custom function that conducts a simulation, estimates
## ancestral states, & returns the fraction on 95% CI
foo<-function(){
    tree<-pbtree(n=100)
    x<-fastBM(tree,internal=TRUE)
    fit<-fastAnc(tree,x[1:length(tree$tip.label)],
        CI=TRUE)
    withinCI<-((x[1:tree$Nnode+length(tree$tip.label)]>=
        fit$CI95[,1]) &
        (x[1:tree$Nnode+length(tree$tip.label)]<=
        fit$CI95[,2]))
    mean(withinCI)
}
## conduct 100 simulations
pp<-replicate(100,foo())
mean(pp)

## [1] 0.9479798
```

This value is very close to 0.95, just as hoped.

Our simulation demonstrates that, when our model of evolution is correct, the confidence intervals around our ML ancestral states will tend to include the generating value $(1 - \alpha) \times 100$ percent of the time. In other words, when our model is correct, ancestral state estimation as a statistical method seems to work just fine (Martins 1999; Webster and Purvis 2002)!

8.3.1 What happens when the model is wrong?

Unfortunately, although ancestral state estimation using ML performs very well as a statistical method when we have the model of evolution correct, if we've got it wrong,[19] estimation can be much worse.

To show this, let's imagine that instead of evolution by Brownian motion, our trait has evolved by a slightly different process: Brownian motion with a trend.

Brownian motion with a trend is a very simple modification of Brownian motion, but in which the character value tends to change gradually through time (Hunt 2006).

Why don't we simulate under this model and then see how well we do at reconstructing ancestral states?[20]

```
## simulate a vector of data under a trend model
y<-fastBM(tree,mu=2,internal=TRUE)
```

Since we know the true internal states for our data, let's create a projection of the tree into morphospace[21] using the function phenogram from *phytools*.

```
## set margins
par(mar=c(5.1,4.1,2.1,2.1))
## plot traitgram with known states
phenogram(tree,y,fsize=0.6,
    color=make.transparent("blue",0.5),
    spread.cost=c(1,0),cex.axis=0.8,
    las=1)
```

These data look fairly Brownian, but with a clear trend toward greater and greater values for the trait through time (figure 8.6).

Now, as promised, let's estimate ancestral states while *incorrectly* assuming a Brownian model of evolutionary change.

We can then proceed to compare our estimates to the known true ancestral values at internal nodes of the tree for our trait.

```
## estimate ancestral states
fit.trend<-fastAnc(tree,y[tree$tip.label],CI=TRUE)
## set margins
par(mar=c(5.1,4.1,2.1,2.1))
## create plot showing true states vs. estimated
## states
```

---

[19] All models are wrong by nature: or, as George Box famously said, "All models are wrong, but some are useful." Our question should not be if the model of Brownian motion is right or wrong but whether or not it is *badly* wrong. In practice, this is harder to assess.

[20] For now, we'll *reconstruct* states using the incorrect model: untrended Brownian motion. Although under some circumstances, such as when our tree has fossils, we can reconstruct ancestral states for a trended model (Slater et al. 2012), this is more often not the case with the neontological data that typify most comparative studies.

[21] Called a *traitgram* (Evans et al. 2009).

```
plot(a<-y[1:tree$Nnode+Ntip(tree)],fit.trend$ace,
    xlab="true states",
    ylab="estimated states",bty="n",
    ylim=range(fit.trend$CI95),col="transparent",
    cex.axis=0.8,las=1)
## add 1:1 line & lines showing CI of each estimate
lines(range(y),range(y),lty="dashed",
    col="red") ## 1:1 line
for(i in 1:tree$Nnode)
    lines(rep(a[i],2),fit.trend$CI95[i,],lwd=5,
        col=make.transparent("blue",0.25),
        lend=0)
points(a,fit.trend$ace,bg="grey",cex=1.5,pch=21)
## plot legend
legend(x="bottomright",
    legend=c("estimated ancestral state",
    "95% CI for ancestral state",
    "1:1 line"),cex=0.7,col=c("black",
    make.transparent("blue",0.25),
    "red"),pch=c(21,NA,NA),pt.bg=c("grey",NA,NA),
    pt.cex=c(1.5,NA,NA),bty="n",
    lty=c(NA,"solid","dashed"),lwd=c(NA,5,1))
```

This result (figure 8.7) shows that even though the 95 percent CIs for many of the ancestral states in the tree are quite broad, when our model is wrong (as it is in this case), these 95 percent CIs nonetheless frequently *fail* to include the true value for the ancestral state (see Oakley and Cunningham 2000 for an empirical example).

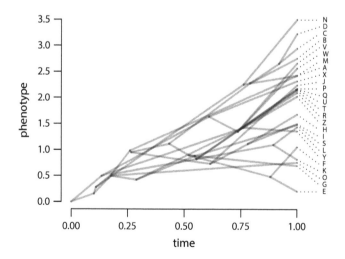

Figure 8.6
Projection of a simulated tree into phenotype space in which the data have been simulated by Brownian evolution with a trend.

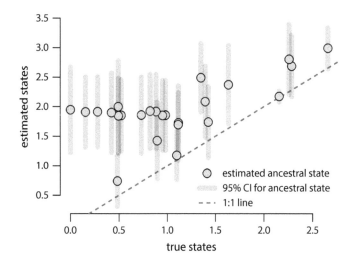

Figure 8.7
True compared to estimated states when the generating model of evolution of the trait was Brownian evolution with a trend.

Rather than leave ancestral state reconstruction of continuous traits on such a dim note, we'll grant that the trend model (although trivially simple to imagine and simulate) is among the worst for ancestral state reconstruction, at least without fossils (Slater et al. 2012), and many other model violations may be less pathological.[22]

8.4 Discrete characters

Next, we'll consider ancestral state reconstruction for discrete character traits.

Most commonly, the estimation of ancestral character states for discretely valued traits uses a continuous-time Markov chain model commonly known as the extended M*k* model (Lewis 2001; Harmon 2019).

This is *exactly* the same model that we learned about in chapter 6 of this book. Now, however, instead of merely fitting the parameters of the transition process to our data, we'll simultaneously estimate both the matrix that represents the process of transitions between states, **Q**, as well as the ancestral conditions at all the internal nodes of the tree.

For this part of the chapter, we'll use the same phylogenetic tree of eels that we used to study continuous trait ancestral state reconstruction but combine it with some data for feeding mode.

In our data, feeding mode is classified very simply into one of two binary categories: biting versus suction feeding (Collar et al. 2014).[23]

As we noted earlier, the data and tree files for this part of the chapter can be downloaded from the book website.

We already read our tree and data from file. Now let's plot the binary discrete character at the tips of the tree[24] (figure 8.8). We'll do this using the *phytools* function `plotTree.datamatrix`.

---

[22] Although we nonetheless recommend that more research be undertaken on this subject.

[23] There is no requirement that we reconstruct ancestral states only for binary traits. Everything we learn here about binary ancestral state estimation applies equally to a multi-state character.

[24] We think this is always a good idea—a topic that we'll discuss in greater detail in chapter 12!

```
## extract feeding mode as a vector
feed.mode<-setNames(eel.data[,1],rownames(eel.data))
## set colors for plotting
cols<-setNames(c("red","lightblue"),levels(feed.mode))
## plot the tree & data
plotTree.datamatrix(eel.tree,as.data.frame(feed.mode),
    colors=list(cols),header=FALSE,fsize=0.45)
## add legend
legend("topright",legend=levels(feed.mode),pch=22,
    pt.cex=1.5,pt.bg=cols,bty="n",cex=0.8)
```

8.4.1 Choosing a character model

To do ancestral state reconstruction for a discrete trait, we need to start by choosing a model.

We learned about this in chapter 6, but it's equally important for ancestral state reconstruction, if not more so.

For a two-state character, there are four possible models: an equal-rates (ER) model, an all-rates-different (ARD) model, and two irreversible models—one that allows only 0 to 1[25] changes (but not the reverse) and the other that allows only 1 to 0 changes (but not the reverse).[26]

As a review of what we learned in chapter 6, let's begin by comparing all four of these models.

We'll identify which model is best supported by our data and then proceed to use this model for ancestral state estimation.[27]

```
## fit ER model
fitER<-fitMk(eel.tree,feed.mode,model="ER")
## fit ARD model
fitARD<-fitMk(eel.tree,feed.mode,model="ARD")
## fit bite->suction model
fit01<-fitMk(eel.tree,feed.mode,
    model=matrix(c(0,1,0,0),2,2,byrow=TRUE))
## fit suction->bite model
fit10<-fitMk(eel.tree,feed.mode,
    model=matrix(c(0,0,1,0),2,2,byrow=TRUE))
## extract AIC values for each model
aic<-c(AIC(fitER),AIC(fitARD),AIC(fit01),AIC(fit10))
## print summary table
data.frame(model=c("ER","ARD","bite->suction",
    "suction->bite"),
    logL=c(logLik(fitER),logLik(fitARD),
    logLik(fit01),logLik(fit10)),
    AIC=aic,delta.AIC=aic-min(aic))
##          model       logL      AIC delta.AIC
## 1           ER -37.03307 76.06614  0.000000
```

---

[25]Or, in our case, *bite* to *suction*.

[26]Since there are only two states of the character, our ER and SYM models are the same.

[27]This time, we'll use the function fitMk from the *phytools* package to fit our M*k* model instead of *geiger*'s fitDiscrete; however, using fitDiscrete would've turned out almost the same.

```
## 2              ARD -37.00365 78.00730 1.941156
## 3 bite->suction -38.64057 79.28114 3.214999
## 4 suction->bite -40.50138 83.00276 6.936613
```

This analysis doesn't show a huge difference between models;[28] however, nor does it indicate any real justification for using a model more complex than the simple equal-rates (ER) model. Consequently, we'll use this model for the rest of our analyses.

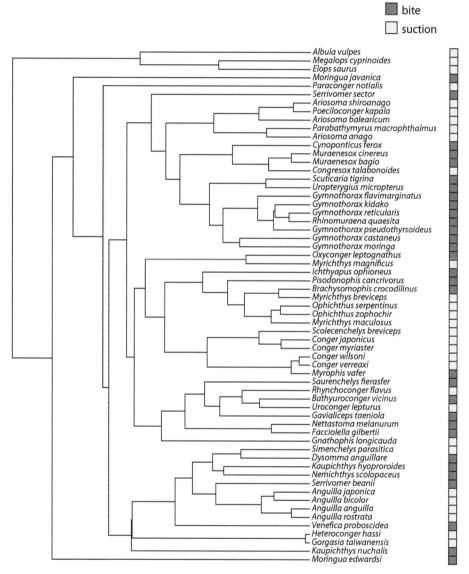

Figure 8.8
Phylogeny of elopomorph eels with feeding mode shown at the tips of the phylogeny.

---

[28] Except for the model that permits only transitions from suction feeding to biting, but not the reverse, which explains our data very *poorly*.

8.4.2 Joint vs. marginal ancestral state reconstruction

In undertaking ancestral state estimation for a discrete trait using maximum likelihood, the first thing we need to do is decide whether we want to do *joint* or *marginal* reconstruction (Pagel 1999b; Yang 2006).

In joint ancestral state reconstruction, what we do is ask which set of character values at all the internal nodes maximizes the probability of obtaining the values for the character that we've observed at the tips of the tree, given our model.

These states will constitute our maximum likelihood *joint ancestral character state reconstructions* (Pagel 1999b; Yang 2006).

By contrast, in marginal ancestral state reconstruction, we imagine traversing the tree node by node. At each node, we ask which state has the highest likelihood, integrating over all possible states at all the other nodes of the phylogeny.

If we do this, we will have found the maximum likelihood *marginal ancestral state reconstructions* (Pagel 1999b; Yang 2006).[29]

Note that since we must compute the likelihoods for each possible character state at each node, the ratio of the likelihood over the sum of the likelihoods of each state gives us a value between 0 and 1 for each state, often called the *marginal scaled likelihoods*.[30]

Marginal ancestral state reconstruction is much more popular than joint reconstruction in phylogenetic comparative biology, because it is the type of reconstruction we should do if we want to make specific, probabilistic statements about the values at particular nodes of the tree (Pagel 1999b).

Nonetheless, both types of analysis can be undertaken in R, and we'll start with joint reconstruction here.

8.5 Joint ancestral state reconstruction

To do joint ancestral state reconstruction in R, we'll use the package *corHMM* (Beaulieu et al. 2020).

We used *corHMM* in chapter 7, so you already have it installed. If so, it can be loaded as follows.

```
library(corHMM)
```

Just as we did in chapter 7, for *corHMM* we need to first create a special data frame containing our species names and trait data. Not only that, but we must recode our character data in numerical integer format: 0, 1, 2, and so on.[31]

```
## create new data frame for corHMM
eel.data<-data.frame(Genus_sp=names(feed.mode),
    feed.mode=as.numeric(feed.mode)-1)
```

---

[29] For continuous characters, these two different approaches lead to exactly the same set of ancestral states, but for discrete traits they differ!

[30] Perhaps unsurprisingly, these scaled likelihoods are also a special kind of probability called an *empirical Bayesian posterior probability* (Yang 2006).

[31] This is actually surprisingly easy to do with factors because as.numeric called on a factor vector returns a numeric vector with values 1, 2, 3, and so on, corresponding to the first, second, third, and so on levels of the factor. If we subtract 1 from this vector, we'll get a numerical vector that starts at 0 instead!

Let's look at the top part of our data frame `eel.data` using `head`.

```
head(eel.data,n=10)
```

```
##                        Genus_sp feed.mode
## 1            Albula_vulpes         1
## 2       Anguilla_anguilla         1
## 3        Anguilla_bicolor         1
## 4       Anguilla_japonica         1
## 5      Anguilla_rostrata         1
## 6          Ariosoma_anago         1
## 7      Ariosoma_balearicum         1
## 8      Ariosoma_shiroanago         1
## 9    Bathyuroconger_vicinus        0
## 10 Brachysomophis_crocodilinus        0
```

This looks correctly formatted, so let's go ahead and fit our model.

```
## estimate joint ancestral states using corHMM
fit.joint<-corHMM(eel.tree,eel.data,node.states="joint",
    rate.cat=1,rate.mat=matrix(c(NA,1,1,NA),2,2))
```

```
## State distribution in data:
## States:  1    2
## Counts:  31   30
## Beginning thorough optimization search -- performing 0 random
## restarts
## Finished. Inferring ancestral states using joint
## reconstruction.
```

```
fit.joint
```

```
##
## Fit
##       -lnL      AIC      AICc Rate.cat ntax
##   -37.03307 76.06614 76.13394        1   61
##
## Legend
##   1   2
## "0" "1"
##
## Rates
##             (1,R1)      (2,R1)
## (1,R1)          NA 0.01582924
## (2,R1) 0.01582924          NA
##
## Arrived at a reliable solution
```

Our call to the function `corHMM` includes a number of arguments. `rate.cat` is for the hidden rate model that we learned in chapter 7. By setting `rate.cat=1`, we are indicating

bite

suction

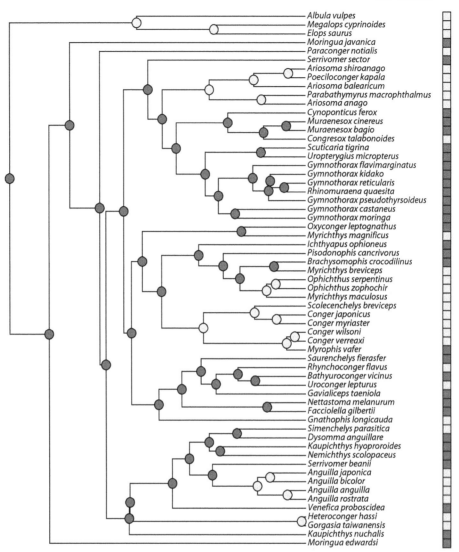

Figure 8.9

Phylogeny of elopomorph eels with joint ancestral state estimates at the nodes.

that our model has no hidden rates. `rate.mat` is a design matrix for our model. The way it has been specified in this case corresponds to the ER model that we decided, based on model selection, was the most appropriate for our data in this example.

After optimization, the ML joint ancestral states are hidden inside the object as node labels of the phylogeny. Let's pull them out and add them to our tree from before (figure 8.9).

```
## plot the tree & data
plotTree.datamatrix(eel.tree,as.data.frame(feed.mode),
```

```
    colors=list(cols),header=FALSE,fsize=0.45)
## add legend
legend("topright",legend=levels(feed.mode),pch=22,
    pt.cex=1.5,pt.bg=cols,bty="n",cex=0.8)
## add node labels showing ancestral states
nodelabels(pie=to.matrix(
    levels(feed.mode)[fit.joint$phy$node.label],
    levels(feed.mode)),piecol=cols,cex=0.4)
```

Note that although we have found the ML ancestral states under our model, our reconstructions do not in any way show the uncertainty inherent in their estimation.

Although this is still a statistical estimation procedure (because we could always propose an alternative set of states and compare that hypothesis to the ML solution), the best way to measure the uncertainty about the specific values for ancestral states (and, likewise, the best way to test hypotheses or ask questions about particular ancestral states) is with marginal ancestral state reconstruction.[32]

8.6 Marginal ancestral state reconstruction

In this section, we'll now see how to undertake marginal ancestral state reconstruction on the phylogeny.

This can also be done using the function corHMM (from the *corHMM* package) (Beaulieu et al. 2020), just as we did in the last section.[33]

Let's do just that. Once again, we'll assume an ER model because that model was best supported by our data in the model comparison exercise that we undertook earlier.

```
## estimate marginal ancestral states under a ER model
fit.marginal<-corHMM(eel.tree,eel.data,node.states="marginal",
    rate.cat=1,rate.mat=matrix(c(NA,1,1,NA),2,2))

## State distribution in data:
## States:  1   2
## Counts:  31  30
## Beginning thorough optimization search -- performing 0 random
## restarts
## Finished. Inferring ancestral states using marginal
## reconstruction.

fit.marginal

##
## Fit
##        -lnL      AIC      AICc Rate.cat ntax
##   -37.03307 76.06614 76.13394        1   61
```

---

[32]Or with another procedure called *stochastic character mapping*, which we'll see later in the chapter.

[33]More often, marginal ancestral state reconstruction in R is conducted using the function ace in the *ape* package.

```
##
## Legend
##   1   2
## "0" "1"
##
## Rates
##                (1,R1)        (2,R1)
## (1,R1)            NA 0.01582874
## (2,R1) 0.01582874           NA
##
## Arrived at a reliable solution
```

Sensibly, our marginal ancestral states are stored in a matrix of our fitted model object called `states`.

Let's take a look at the first few lines of that matrix using `head`.

```
head(fit.marginal$states)

##           (1,R1)     (2,R1)
## [1,]  0.5479961  0.4520039
## [2,]  0.6242188  0.3757812
## [3,]  0.6560719  0.3439281
## [4,]  0.6903942  0.3096058
## [5,]  0.7066283  0.2933717
## [6,]  0.7157887  0.2842113
```

The correct interpretation of this matrix is that it gives the posterior probabilities that each state is in each node—that is, conditioning on an assumption that our fitted model is correct—just as it was for continuous traits.

It's quite straightforward to overlay these posterior probabilities on the tree. The result is shown in figure 8.10.

```
## plot the tree & data
plotTree.datamatrix(eel.tree,as.data.frame(feed.mode),
    colors=list(cols),header=FALSE,fsize=0.45)
## add legend
legend("topright",legend=levels(feed.mode),pch=22,
    pt.cex=1.5,pt.bg=cols,bty="n",cex=0.8)
## add node labels showing marginal ancestral states
nodelabels(pie=fit.marginal$states,piecol=cols,
    cex=0.5)
```

The root node of the tree is almost completely unknown.[34] On the other hand, it is possible to make some statements about more recent nodes in the tree with much greater confidence (figure 8.10).

------

[34] That is to say, the probabilities that it is in each of the two possible states are effectively equal.

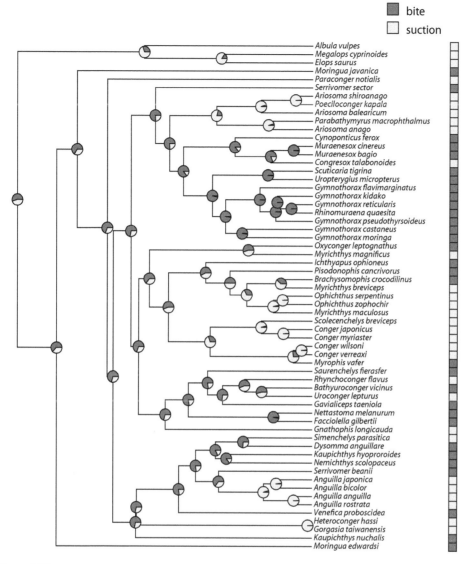

bite
suction

Figure 8.10
Phylogeny of elopomorph eels with marginal ancestral state estimates at the nodes.

Finally, the findings seem largely consistent with our data—the ancestral states for clades that seem to be fixed for one state or the other tend to show a high probability of sharing that state (figure 8.10). This makes sense.

8.7 Stochastic character mapping

An alternative tactic to the one outlined above is to use an MCMC approach to sample character histories from their posterior probability distribution.

This technique is called stochastic character mapping (Huelsenbeck et al. 2003). The model of evolution for our trait is the same, but in this case, we're going to sample unambiguous

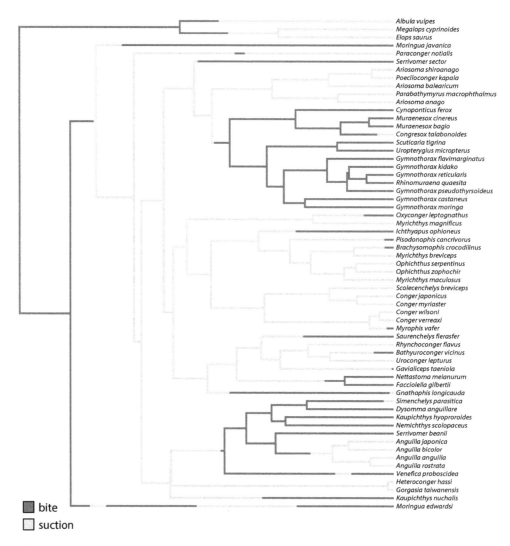

Albula vulpes
Megalops cyprinoides
Elops saurus
Moringua javanica
Paraconger notialis
Serrivomer sector
Ariosoma shiroanago
Poeciloconger kapala
Ariosoma balearicum
Parabathymyrus macrophthalmus
Ariosoma anago
Cynoponticus ferox
Muraenesox cinereus
Muraenesox bagio
Congresox talabonoides
Scuticaria tigrina
Uropterygius micropterus
Gymnothorax flavimarginatus
Gymnothorax kidako
Gymnothorax reticularis
Rhinomuraena quaesita
Gymnothorax pseudothyrsoideus
Gymnothorax castaneus
Gymnothorax moringa
Oxyconger leptognathus
Myrichthys magnificus
Ichthyapus ophioneus
Pisodonophis cancrivorus
Brachysomophis crocodilinus
Myrichthys breviceps
Ophichthus serpentinus
Ophichthus zophochir
Myrichthys maculosus
Scolecenchelys breviceps
Conger japonicus
Conger myriaster
Conger wilsoni
Conger verreaxi
Myrophis vafer
Saurenchelys fierasfer
Rhynchoconger flavus
Bathyuroconger vicinus
Uroconger lepturus
Gavialiceps taeniola
Nettastoma melanurum
Facciolella gilbertii
Gnathophis longicauda
Simenchelys parasitica
Dysomma anguillare
Kaupichthys hyoproroides
Nemichthys scolopaceus
Serrivomer beanii
Anguilla japonica
Anguilla bicolor
Anguilla anguilla
Anguilla rostrata
Venefica proboscidea
Heteroconger hassi
Gorgasia taiwanensis
Kaupichthys nuchalis
Moringua edwardsi

■ bite
□ suction

Figure 8.11
A single stochastic map of feeding mode on the tree of elophomorph eels.

histories for our discrete character's evolution on the tree—rather than a probability distribution for the character at nodes.

For instance, given the data simulated above, we can generate a *single* stochastic character map as follows (figure 8.11).

```
## generate one stochastic character history
mtree<-make.simmap(eel.tree,feed.mode,model="ER")

## make.simmap is sampling character histories conditioned on
## the transition matrix
##
```

```
## Q =
##               bite       suction
## bite    -0.01582783   0.01582783
## suction  0.01582783  -0.01582783
## (estimated using likelihood);
## and (mean) root node prior probabilities
## pi =
##     bite suction
##      0.5     0.5

## Done.
```

```
## plot single stochastic map
plot(mtree,cols,fsize=0.4,ftype="i",lwd=2,offset=0.4,
    ylim=c(-1,Ntip(eel.tree)))
## add legend
legend("bottomleft",legend=levels(feed.mode),pch=22,
    pt.cex=1.5,pt.bg=cols,bty="n",cex=0.8)
```

As any astute reader could probably guess, a single stochastic character map does not mean a whole lot in isolation—we need to look at the whole *distribution* from a sample of stochastic maps (Huelsenbeck et al. 2003).

For instance, the following code generates 1,000 stochastic character maps.

Here (and in contrast to when we produced a single stochastic character map in the previous example), rather than fixing the single transition rate of the ER model, *q*, to its ML value, we'll use Bayesian MCMC to sample *q* from its posterior distribution.[35]

```
## generate 1,000 stochastic character maps in which
## the transition rate is sampled from its posterior
## distribution
mtrees<-make.simmap(eel.tree,feed.mode,model="ER",
    nsim=1000,Q="mcmc",vQ=0.01,
    prior=list(use.empirical=TRUE),samplefreq=10)
```

```
## Running MCMC burn-in. Please wait....
## Running 10000 generations of MCMC, sampling every 10 generations.
## Please wait....
##
## make.simmap is simulating with a sample of Q from
## the posterior distribution
##
## Mean Q from the posterior is
## Q =
##               bite       suction
## bite    -0.02600077   0.02600077
```

---

[35] In fact, this is one of the great advantages of stochastic mapping, as we'll discuss a bit more later.

```
## suction   0.02600077 -0.02600077
## and (mean) root node prior probabilities
## pi =
##    bite suction
##     0.5     0.5

## Done.
```

```
mtrees
```

```
## 1000 phylogenetic trees with mapped discrete characters
```

The first thing we might do now is to plot the posterior density for our transition rates given in **Q** (figure 8.12). In our case, we are using the ER model, so we only have one rate to plot.

```
## set plot margins
par(mar=c(5.1,4.1,2.1,2.1))
## create a plot of the posterior density from stochastic
## mapping
plot(d<-density(sapply(mtrees,function(x) x$Q[1,2]),
    bw=0.005),bty="n",main="",xlab="q",xlim=c(0,0.5),
    ylab="Posterior density from MCMC",las=1,
    cex.axis=0.8)
polygon(d,col=make.transparent("blue",0.25))
## add line indicating ML solution for the same parameter
abline(v=fit.marginal$solution[1,2])
text(x=fit.marginal$solution[1,2],y=max(d$y),"MLE(q)",
    pos=4)
```

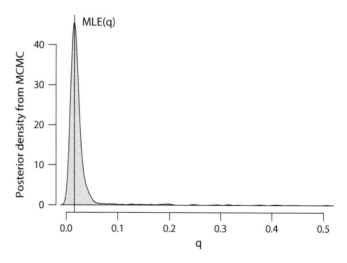

Figure 8.12
Posterior distribution of the transition rate between states, *q*, for the evolution of feeding mode on the phylogeny of elopomorph eels.

We can see that the posterior density for q is centered closely on the MLE of the transition rate between states under our fitted ER model (figure 8.12), which makes a lot of sense—but by sampling this rate from its posterior distribution, our ancestral state estimation now takes into account (Huelsenbeck et al. 2003)[36] the uncertainty that we might feel about q.

So now we have a set of stochastic maps, but what can we do with them?

Well, a simple thing to do is just to look at all of them. In fact, this is completely allowed but, in most cases, will seem a bit overwhelming.

Here, for instance, are our 100 stochastic maps (sampled evenly from our set of 1,000) plotted in a simple grid (figure 8.13).

```
## create a 10 x 10 grid of plot cells
par(mfrow=c(10,10))
## graph 100 stochastic map trees, sampled evenly from
## our set of 1,000
null<-sapply(mtrees[seq(10,1000,by=10)],
    plot,colors=cols,lwd=1,ftype="off")
```

From this plot, it's evident that we are confident about the states at some nodes while much less so about others, but trying to extract any other details is virtually impossible (figure 8.13).

Fortunately, *phytools* makes it fairly easy to summarize a set of stochastic maps in much more meaningful ways.

For instance, we can estimate the number of changes of each type, the proportion of time spent in each state, and the posterior probabilities that each internal node is in each state, under our model.

To do this, we'll use a *phytools* summary method for the object class that we're dealing with: *phytools'* "multiSimmap" object. The result is shown in figure 8.14.

```
## compute posterior probabilities at nodes
pd<-summary(mtrees)
pd

## 1000 trees with a mapped discrete character with states:
##   bite, suction
##
## trees have 51.633 changes between states on average
##
## changes are of the following types:
##      bite,suction suction,bite
## x->y    27.107        24.526
##
## mean total time spent in each state is:
##            bite      suction     total
## raw  1112.8247134 871.2765283 1984.101
## prop    0.5608709   0.4391291    1.000
```

---

[36]To a certain extent. It still ignores uncertainty about the *model* of evolution per se.

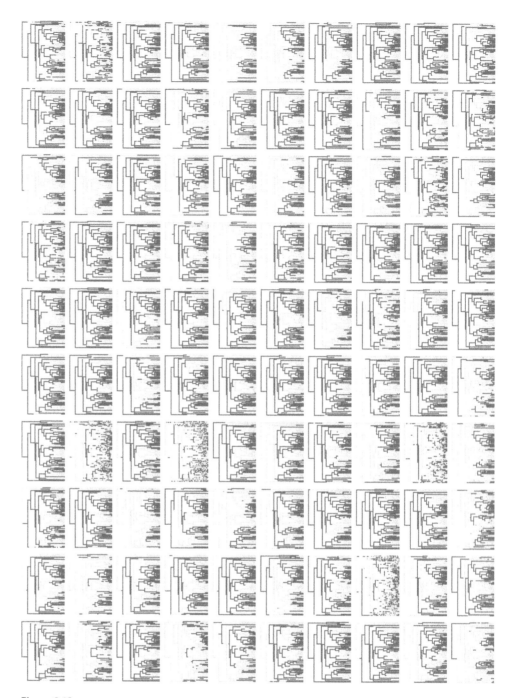

Figure 8.13
One hundred stochastic character maps.

```
## create a plot showing PP at all nodes of the tree
plot(pd,colors=cols,fsize=0.4,ftype="i",lwd=2,
    offset=0.4,ylim=c(-1,Ntip(eel.tree)),
    cex=c(0.5,0.3))
## add a legend
legend("bottomleft",legend=levels(feed.mode),pch=22,
    pt.cex=1.5,pt.bg=cols,bty="n",cex=0.8)
```

Now, many readers are probably wondering how the posterior probabilities from stochastic mapping compare with marginal ancestral states.

In fact, if we had *fixed* **Q** to its MLE,[37] the quantities would converge exactly—given enough stochastic simulations.

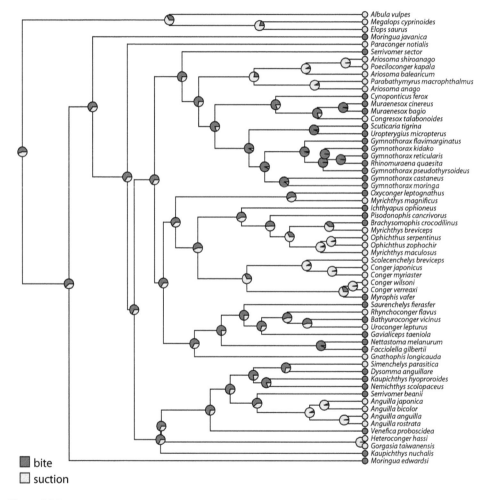

Figure 8.14
Posterior probabilities at nodes from 100 stochastic character maps.

---

[37] Actually, this is what's done by default in make.simmap!

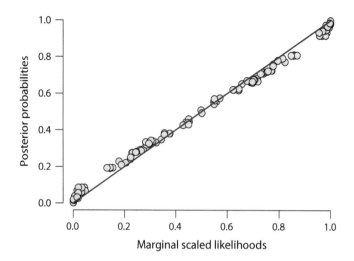

Figure 8.15
Marginal ancestral states compared to posterior probabilities from stochastic mapping for feeding mode on the phylogeny of elopomorph eels.

However, in our case, we chose to sample **Q** from its posterior distribution. As such, we might expect posterior probabilities to "regress to the mean" compared to the marginal reconstructions.

That is to say, compared to marginal reconstruction, high-scaled likelihoods should go down, and low ones (that is, values close to 0) should likewise go up.

This is indeed what we find (figure 8.15)—although the effect is quite subtle in our case. It would be considerably more marked if we had more uncertainty about **Q**.

```
## set margins
par(mar=c(5.1,4.1,2.1,2.1))
## graph marginal ancestral states and posterior
## probabilities from stochastic mapping
plot(fit.marginal$states,pd$ace[1:eel.tree$Nnode],pch=21,
    cex=1.2,bg="grey",xlab="Marginal scaled likelihoods",
    ylab="Posterior probabilities",
    bty="n",las=1,cex.axis=0.8)
lines(c(0,1),c(0,1),col="blue",lwd=2)
```

Last of all, for binary discrete characters, we can also use a method in *phytools* called densityMap to visualize the posterior probability of being in each state across all the edges and nodes of the tree. We can see what this looks like in figure 8.16.

```
## create a "densityMap" object
eel.densityMap<-densityMap(mtrees,
    states=levels(feed.mode)[2:1],plot=FALSE)

## sorry - this might take a while; please be patient
```

```
## update color gradient
eel.densityMap<-setMap(eel.densityMap,cols[2:1])
## plot it, adjusting the plotting parameters
plot(eel.densityMap,fsize=c(0.3,0.7),lwd=c(3,4))
```

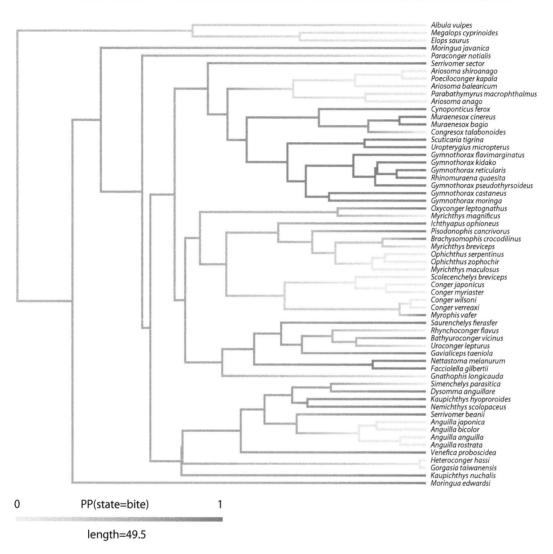

Figure 8.16
Posterior density map from stochastic mapping of feeding mode on the tree of elopomorph eels.

8.8 What about parsimony?

For decades, the most popular method for reconstructing discretely valued ancestral states on
the tree was a procedure called maximum parsimony (reviewed in Felsenstein 2004).[38]

---

[38] The principle of maximum parsimony in phylogenetics is to minimize the number of changes of the character on the tree. Thus, a maximum parsimony ancestral state reconstruction is one in which the number of
changes is minimized.

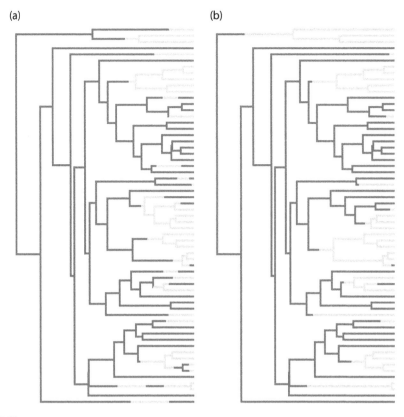

(a) (b)

Figure 8.17
(a) A stochastic character map with the transition rate of the ER model fixed to its ML value. (b) A stochastic character map in which we have forced evolution to be much lower than its empirical rate. This makes (b) a parsimony reconstruction.

Parsimony was an important method for phylogeny reconstruction and still has many uses in phylogenetic analysis. We do not, however, recommend using parsimony for ancestral state reconstruction. This is for a couple of different reasons.

First, parsimony is not a statistical method. That is to say, if we find a reconstruction of our character that is maximally parsimonious, we have no way to compare this reconstruction to another that is less so.

Second, parsimony implicitly assumes that the rate of evolution is slow,[39] even if strongly contrary evidence exists in our data that the rate of evolution is relatively fast.

To show this in a relatively simple way, we'll *force* a very slow[40] rate of evolution of feeding mode on our elopomorph eel data (Collar et al. 2014) using stochastic character mapping and show that this causes the stochastic maps to exhibit a number of changes precisely *equal* to the parsimony score for the trait on our tree[41] (figure 8.17).

---

[39] Actually, it has also been postulated that parsimony implicitly assumes a different rate of evolution for every branch of the tree discretely-valued a model denominated the *no common mechanism* model by Tuffley and Steel (1997).

[40] Relative to the MLE of our rate.

[41] Another way of putting it is that stochastic character mapping returns a parsimony inference about ancestral states—if we force the rate of evolution to be very low.

```
## set Q to its MLE
fixed.Q<-mtree$Q
mtree.mle<-make.simmap(eel.tree,feed.mode,
    Q=fixed.Q)

## make.simmap is sampling character histories conditioned on
## the transition matrix
##
## Q =
##                 bite        suction
## bite      -0.01582783  0.01582783
## suction    0.01582783 -0.01582783
## (specified by the user);
## and (mean) root node prior probabilities
## pi =
##     bite suction
##      0.5     0.5
## Done.

## set Q to a value 1/1000 smaller
mtree.slow<-make.simmap(eel.tree,feed.mode,
    Q=0.001*fixed.Q)

## make.simmap is sampling character histories conditioned on
## the transition matrix
##
## Q =
##                   bite           suction
## bite      -1.582783e-05  1.582783e-05
## suction    1.582783e-05 -1.582783e-05
## (specified by the user);
## and (mean) root node prior probabilities
## pi =
##     bite suction
##      0.5     0.5
## Done.

## plot the results
par(mfrow=c(1,2))
plot(mtree.mle,cols,ftype="off")
plot(mtree.slow,cols,ftype="off")
```

If the character state changes infrequently,[42] then ML and parsimony will tend to be in agreement.

---

[42] That is, if the rate of evolution *is* low.

However, if the rate of evolution is relatively high, then parsimony can give a misleading picture about evolution on our tree, particularly about our *confidence* in the reconstructed ancestral states.

8.9 Practice problems

8.1 Investigate your reconstructed ancestral character states for the continuous character in the eel data set, log-transformed total length ("lnTL"). In particular, determine how many ancestral eels we can confidently reconstruct as longer than 80 cm. That is, how many nodes have a confidence interval for the ancestral state that is larger than, and excludes, 80 cm?

8.2 Create a new discrete character, bigOrSmall, for the anoles using the data sets from chapter 1 (anole.data.csv and Anolis.tre). Assign a species to the state "big" if lnSVL > 4 and "small" otherwise. Estimate ancestral states for your new character under an appropriate model. Finally, compare the ancestral states for your new character with those considering size as a continuous character. Are the anoles reconstructed as "big" also reconstructed to have a large lnSVL?

8.3 Explore the accuracy of ancestral state reconstruction when data are simulated under an OU model but reconstructed using Brownian motion. What happens when we get the model wrong in this way? Do enough simulations—and follow-up analyses—so that you can see the general pattern!

Analysis of diversification with phylogenies

The processes of speciation, where one lineage divides into two or more descendants, and extinction, in which species and clades die off and are lost from the phylogenetic tree, are a fundamental part of most macroevolutionary theories.

We might consider whether our group of interest has a higher speciation rate than its close relatives, for example (Slowinski and Guyer 1989). Alternatively, we may want to try to see the impact that extinction has had on our phylogenetic tree (Nee, Holmes et al. 1994).

Phylogenetic comparative methods focused on the processes that underlie diversification can potentially help address these sorts of questions, as well as many others (reviewed in Nee 2006).

In the next two chapters (this one and chapter 10), we'll change gears a bit and consider some phylogenetic methods that use the tree alone to measure the course of evolutionary diversification.

Then, in chapter 11, we'll look to combine the analysis of diversification with that of phenotypic evolution, to see how we might better understand how species' traits can influence the processes that underlie the production and maintenance of diversity.

Thus, in this chapter:

1. We'll create a popular graphical visualization of diversification called a *lineage-through-time* or LTT plot.
2. We'll see how to conduct the γ-test of Pybus and Harvey (2000) for constant-rate speciation, including under circumstances in which the species of our tree are incompletely sampled.
3. We'll learn how to fit a model called the "birth-death model" for speciation and extinction on the phylogeny using the *phytools* package.
4. Finally, we'll introduce a powerful R package called *diversitree* (FitzJohn 2012) and see how we can use *diversitree* to both fit a birth-death model with likelihood (as in item 3) but also using Bayesian MCMC. This package will become more important in chapters 10 and 11 of the book.

Two of the very simplest methods of diversification analysis using phylogenetic trees are a graphical method called a lineage-through-time (LTT) plot and the γ summary statistic of Pybus and Harvey (2000).

An LTT plot is a simple graph of the accumulation of lineages through time from the base of our tree to the present day. We can compare this pattern of accumulation to some point of reference and see, visually, how diversification in our tree compares to what we might expect under different models.

For the γ method (Pybus and Harvey 2000), we convert this same pattern of lineage accumulation (precisely what we graphed in our LTT plot) to a summary statistic called γ that has a known distribution under a constant-rate process of stochastic speciation called *pure birth*.[1] We can then determine whether the pattern in our LTT plot is consistent or not with such a process.

Let's take a closer look at each of these methods in turn.

9.2.1 Lineage-through-time plots

The lineage-through-time or LTT plot is a popular graphical method in the analysis of diversification on trees. The LTT method we'll use is implemented as part of the *phytools* R package that we've used in every chapter of the book to this point.

The concept of an LTT plot is quite simple, really.

We merely count the number of lineages as they accumulate from the root of the tree toward the present day and then create a graph showing this accumulation as a function of time.

Figure 9.1 shows in panel (a) a simple simulated phylogeny, with every speciation evident indicated using a vertical dotted line, while panel (b) shows the same events recorded in an LTT plot.

The following code is complicated, but it creates a figure that connects the pattern of branching in the tree to our LTT plot. Subsequently in this chapter, we'll show you a simpler way to make an LTT plot for a real phylogenetic tree.

```
## load the phytools package
library(phytools)
## simulate a pure-birth (Yule) tree using pbtree
tree<-pbtree(n=12,scale=100)
## split our plotting area in two
par(mfrow=c(2,1))
## graph our phylogeny
plotTree(tree,ftype="off",mar=c(4.1,4.1,2.1,1.1))
## compute the lineages through time using ltt
obj<-ltt(tree,plot=FALSE)
## draw vertical lines at each lineage accumulation
## event
```

---

[1] Pure birth is also known as a Yule process, so named for the British statistician George Udny Yule. In phylogenetics, a *Yule tree* thus just means a tree generated by a constant-rate speciation process, without extinction, and consequently has nothing at all to do with Christmas!

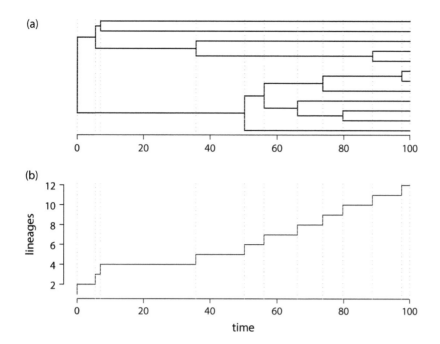

Figure 9.1
(a) Example simulated phylogeny with twelve species. (b) LTT plot. Vertical blue lines in both (a) and (b) show the temporal positions of events on the tree.

```
abline(v=obj$times,lty="dotted",
    col=make.transparent("blue",0.5))
## add a horizontal axis and plot label
axis(1,cex.axis=0.8)
mtext("(a)",line=1,at=-10)
## create a second plot graphing our LTT
plot(obj,mar=c(5.1,4.1,2.1,1.1),bty="n",
    log.lineages=FALSE,las=1,cex.axis=0.8)
## add the same vertical lines as in panel a)
abline(v=obj$times,lty="dotted",
    col=make.transparent("blue",0.5))
## label our plot
mtext("(b)",line=1,at=-10)
```

Theory tells us that under a scenario of constant-rate speciation (with no extinction, also called a pure-birth model), lineages should accumulate as an exponential function of time since the root of the tree.

As such, we expect that a curve giving the number of lineages through time[2] should be "log-linear": in other words, linear on a semilogarithmic scale, assuming a process of diversification that involves only speciation and in which speciation occurs at a constant rate (Nee 2006).

---

[2] That is, our *lineage-through-time* plot—the thing we're plotting.

Obviously, very few studies undertake to compute an LTT plot for a phylogeny with only twelve taxa as we have done in figure 9.1.

To explore a more realistic case study of diversification analysis, we'll use a phylogeny of darters (a fish group) that's rather larger and derives from a publication by Near et al. (2011).

In their study, the authors sampled 201 of 216 species from the group (~93 percent of taxa; Near et al. 2011). For diversification analyses, it is often important to correct for such incomplete sampling. We'll discuss the issue of incomplete sampling in more depth later in the chapter.

The tree file of this exercise (`etheostoma_percina_chrono.tre`) is available from the book website,[3] just like all the files used in previous chapters.

To proceed, why don't we go ahead and load our darter tree from file in the normal way.

```
darter.tree<-read.tree("etheostoma_percina_chrono.tre")
```

Now, let's visualize the tree and make an LTT plot.

Our phylogeny is relatively large (>200 species), which can make right-facing phylograms difficult to read. As such, instead we'll use a circular or "fan-style" graph to plot it, just as we've done for larger trees in earlier chapters of this book.

Normally, in fan-style trees, all 360° of the circle is used to plot the phylogeny. Here, seeing as how our tree is calibrated in units of millions of years, let's leave a little space in our plot to add a time axis.

To do that, we can use the `phytools::plotTree` argument `part`, which we'll set to `0.88`. This argument tells the function that we want to plot our tree through 88 percent of the arc of the circle[4] (figure 9.2).[5]

```
## plot our tree in fan style
plotTree(darter.tree,ftype="i",
    fsize=0.4,type="fan",lwd=1,part=0.88)
## compute the total height of the tree
h<-max(nodeHeights(darter.tree))
## graph a temporal axis without labeling
obj<-axis(1,pos=-2,at=h-c(0,5,10,15,20),
    cex.axis=0.5,labels=FALSE)
## add labels, but going backwards from the
## present day
text(obj,rep(-5,length(obj)),h-obj,
    cex=0.6)
## add a text label to the axis
text(mean(obj),-8,"time (mybp)",
    cex=0.8)
```

---

[3] http://www.phytools.org/Rbook/.

[4] We'll set `part=0.88` here, but 0.88 is not a magic number. We just determined that was the right fraction via trial and error!

[5] Another complication that we run into when creating this plot is that the horizontal dimension of our plotted tree runs from $-h$ to h, in which h is the total length of our tree—but what we really want is our axis to start with zero in the present day and then run backward! The code chunk illustrates one way in which this can be accomplished.

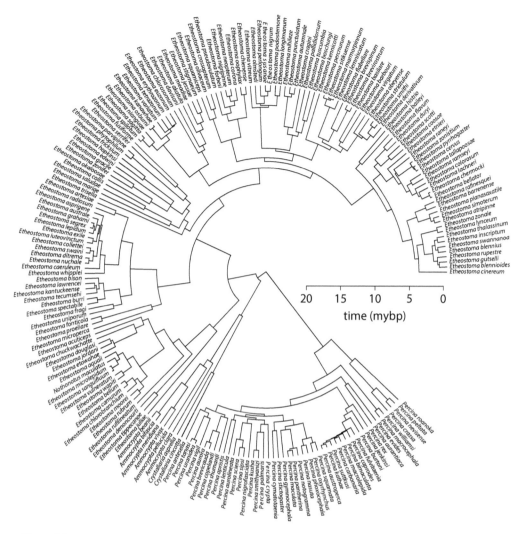

Figure 9.2
Time-calibrated phylogeny of darters. Phylogeny modified from Near et al. (2011).

Now that we've successfully read and plotted our tree, let's proceed to use the function `ltt` from *phytools* to generate an LTT plot.

This time, and in contrast to figure 9.1 (but following the most common convention for LTT plots), we'll graph our plot on a semilogarithmic scale (figure 9.3).[6]

```
## compute "ltt" object
darter.ltt<-ltt(darter.tree, plot=FALSE)
## modify the figure margins
```

---

[6]Astute readers may notice that we set `log.lineages=FALSE` and `log="y"`. This has the effect of transforming our plotting axis y values, so that the labels on that axis still represent numbers of lineages rather than log(lineages), which we think makes the graph easier to read.

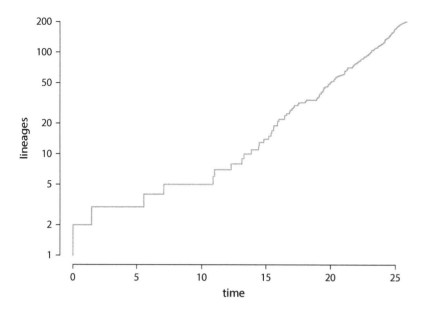

Figure 9.3
Lineage through time plot for darters. The vertical axis of the plot is shown on the logarithmic scale.
Phylogeny from Near et al. (2011).

```
par(mar=c(5.1,4.1,2.1,2.1))
## plot "ltt" object
plot(darter.ltt,log.lineages=FALSE,log="y",
    col="blue",lwd=2,bty="n",las=1,
    cex.axis=0.8)
```

9.2.2 The γ statistic

Another useful simple measure of diversification through time is given by the γ statistic of
Pybus and Harvey (2000).

By design, γ has a *standard normal distribution*[7] under a null hypothesis of constant-rate,
pure-birth speciation (Pybus and Harvey 2000).

When we computed our "ltt" object, we should have already calculated γ, so let's look at
it now by printing out the object we created: darter.ltt.

```
print(darter.ltt)

## Object of class "ltt" containing:
##
## (1) A phylogenetic tree with 201 tips and 198 internal
##     nodes.
##
```

---

[7]That is, it is normally distributed with $\mu = 0$ and $\sigma = 1$.

```
## (2) Vectors containing the number of lineages (ltt) and
##     branching times (times) on the tree.
##
## (3) A value for Pybus & Harvey's "gamma" statistic of
##     gamma = NA, p-value = NA.
```

This is weird—for some reason, we're not getting values for γ, just NA. What is the problem?

Well, it turns out that our problem was that our tree is not fully bifurcating, so we need to resolve polytomies to continue.

To do that, we'll use the handy *ape* function `multi2di`.[8]

```
## resolve polytomies using multi2di
darter.tree<-multi2di(darter.tree)
## recompute "ltt" object
darter.ltt<-ltt(darter.tree,plot=FALSE)
darter.ltt
```

```
## Object of class "ltt" containing:
##
## (1) A phylogenetic tree with 201 tips and 200 internal
##     nodes.
##
## (2) Vectors containing the number of lineages (ltt) and
##     branching times (times) on the tree.
##
## (3) A value for Pybus & Harvey's "gamma" statistic of
##     gamma = 0.2007, p-value = 0.841.
```

This result tells us that our value of γ is about 0.2, and since γ has a standard normal distribution under the null, this value of γ is sufficiently close to zero that we *fail* to reject the null hypothesis of constant-rate pure birth on our darter phylogeny.

We can conclude that our LTT plot is not any different from what we would expect under a pure-birth process.

Note that if you plot the LTT, you will also see that the process of resolving polytomies had no effect on the LTT plot. This is because we inserted branches with zero length, so that the accumulation of lineages through time is totally unchanged!

9.2.3 The MCCR test for γ with incomplete sampling

One implicit assumption that we made when analyzing γ in this way is that the taxa of our tree have been completely sampled. That is to say, we've assumed that our tree includes all the descendants of the hypothetical ancestral taxon represented by the root of the tree.

In addition to developing the γ method, Pybus and Harvey (2000) also proposed a method, referred to as the MCCR[9] test, that can be used to take incomplete sampling into consideration.

---

[8] The converse operation—collapsing zero-length internal branch lengths into multichotomies—can be done using `ape::di2multi`.

[9] An abbreviation of *Monte Carlo constant rates*.

This approach is relatively simple because it merely involves simulating trees under the null model of constant-rate speciation, randomly subsampling the simulated trees to contain the same sampling fraction as our empirical tree, and then computing a test distribution for γ from these simulated, randomly subsampled phylogenies.

Let's try it using the function `mccr` from *phytools*. In `mccr`, we can specify our sampling fraction using the argument `rho`. Remember that Near et al. (2011) purport to have sampled 201 of the 216 extant species in the clade.

```
darter.mccr<-mccr(darter.ltt,rho=201/216,
    nsim=500)
darter.mccr

## Object of class "mccr" consisting of:
##
## (1) A value for Pybus & Harvey's "gamma" statistic of
##     gamma = 0.2007.
##
## (2) A two-tailed p-value from the MCCR test of 0.632.
##
## (3) A simulated null-distribution of gamma from 500
##     simulations.
```

Finally, we can even plot our null distribution and observed value of γ.[10]

```
par(mar=c(5.1,4.1,2.1,2.1))
plot(darter.mccr,main="",las=1,cex.axis=0.8)
```

Both the result and the plot of figure 9.4 say the same thing. Once again, we fail to reject the null and conclude that our LTT plot is not any different from what we would expect under a pure-birth model.

9.3 Estimating speciation and extinction rates from a reconstructed phylogeny

So far, our methods for studying diversification have been a step away from the actual process that is responsible for producing the structure of our estimated tree: speciation and extinction of lineages.

We can get a bit closer to this process using a method developed nearly thirty years ago by Nee et al. (1992). Although many other approaches have since been proposed, the method of Nee et al. (1992) continues to form the basis for the overwhelming majority of diversification models analyzed using reconstructed trees.

Although the mathematical details are a bit beyond the scope of this book,[11] we think that it would nonetheless be of value to take a moment to try and understand how the rates of speciation and (especially) extinction can be estimated from a reconstructed tree that contains only extant taxa.

---

[10] This mainly serves to confirm that our result is still nonsignificant; however, it also shows us how the distribution of γ is recentered by missing taxa.

[11] See Harmon (2019) for a nice explanation.

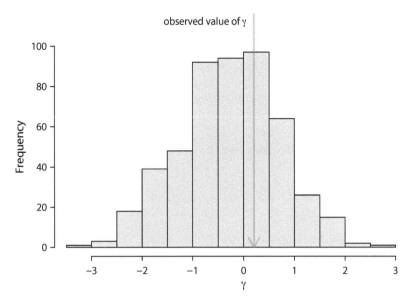

Figure 9.4
Null distribution for the MCCR test based on 201 of 216 species sampled in the phylogeny of figure 9.2.

9.3.1 The pull of the present

In previous sections of this chapter, we learned that if diversification occurs *purely* by speciation, then we should expect that the accumulation of lineages through time will occur exponentially—which linearizes on a semilogarithmic scale.

We can go even further and note that the *slope* of this log-linear accumulation of lineages through time has an expected value equal to the speciation rate (Nee et al. 1992; Nee, May, et al. 1994).

Interestingly, however, when diversification occurs by both speciation *and* extinction, the accumulation of lineages through time is no longer log-linear, even if speciation and extinction take place with unvarying rates through time. Instead, lineage accumulation will tend to curve *upward* toward the tips of the tree—a phenomenon that's usually referred to as the "pull of the present" (Nee et al. 1992; Nee, Holmes, et al. 1994).[12]

In essence, the magnitude of this upward curve[13] is what tells us the rate of extinction compared to speciation in our tree.

An increase in slope toward the present day is the expected signature of extinction, and the higher the extinction rate is compared to speciation, the more dramatic we should expect this curve to be!

Let's do a simple simulation to illustrate the concept.

In the following code, we'll first use pbtree to simulate a pure-birth, speciation-only tree.

Remember, under pure birth, we expect species to accumulate log-linearly with time and furthermore that the slope of this linearized accumulation curve should be equal to our speciation rate!

---

[12]The pull of the present is best understood as being the effect of recently speciated lineages that have not yet had sufficient time to go extinct!

[13]Combined with some complicated mathematics, of course; see Nee (2006) for a review.

Next, we'll proceed to simulate a birth-death tree with a rather high rate of extinction.[14] Then we can compute an LTT plot for this tree—including its extinct lineages—using ltt.

Finally, we'll go ahead and prune all the extinct tips out of our tree and create an LTT plot for this "reconstructed" phylogeny.[15] Here's where we should expect to see the so-called pull of the present.

The result from this three-part experiment is given in figure 9.5.

```
## first simulate tree with no extinction
tree.noExtinction<-pbtree(b=0.039,n=100,t=100,
    method="direct")

## simulating with both taxa-stop (n) & time-stop (t) using
## 'direct' sampling. this is experimental

## next simulate tree with extinction
tree.withExtinction<-pbtree(b=0.195,d=0.156,
    n=100,t=100,method="direct")

## simulating with both taxa-stop (n) & time-stop (t) using
## 'direct' sampling. this is experimental

## prune extinct lineages to recreate the
## "reconstructed" phylogeny
tree.reconstructed<-drop.tip(tree.withExtinction,
    getExtinct(tree.withExtinction))
## add a root edge (to accommodate lineages pruned
## before the first divergence event in the
## reconstructed tree)
tree.reconstructed$root.edge<-100-
    max(nodeHeights(tree.reconstructed))
## convert root edge to an unbranching node
tree.reconstructed<-rootedge.to.singleton(
    tree.reconstructed)
## create "ltt" object from each tree
ltt.noE<-ltt(tree.noExtinction,plot=FALSE)
ltt.wE<-ltt(tree.withExtinction,plot=FALSE)
ltt.recon<-ltt(tree.reconstructed,plot=FALSE)
## graph the LTTs
par(lend=1,mar=c(5.1,4.1,2.1,2.1))
plot(ltt.noE,bty="n",log.lineages=FALSE,log="y",
    lwd=2,xlim=c(0,110),las=1,cex.axis=0.8)
plot(ltt.wE,log.lineages=FALSE,lty="dotted",
    col="black",lwd=2,add=TRUE)
```

---

[14]We set extinction to be 80 percent of the rate of speciation. Extinction rates are rarely estimated to be as high as this from reconstructed phylogenies. On the other hand, the paleontological record may tell a different story.

[15]We call it the reconstructed tree because this is the tree we would hope to reconstruct from our neontological data.

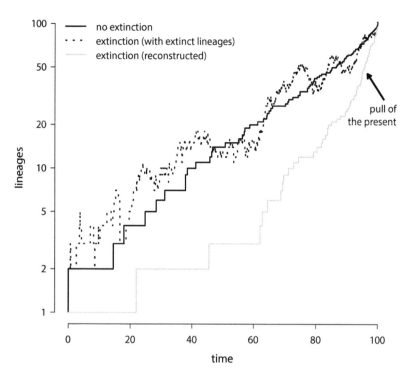

Figure 9.5
Lineage through times for simulated trees. The solid black line shows the number of lineages through time under constant-rate pure birth. The dotted line and the solid gray curves show the number of lineages on a birth-death phylogeny both with and without extinct lineages included, respectively. The latter curve clearly exhibits the "pull of the present."

```
plot(ltt.recon,log.lineages=FALSE,
    lwd=2,add=TRUE,col="darkgray")
## add a legend
legend(x="topleft",lty=c("solid","dotted","solid"),
    lwd=2,col=c("black","black","darkgray"),
    legend=c("no extinction",
    "extinction (with extinct lineages)",
    "extinction (reconstructed)"),
    bty="n",cex=0.7)
## add arrow and text to point to the "pull of
## the present"
arrows(102,29.9,96.6,45.2,length=0.05,lwd=2)
text(x=106,y=28,"pull of\nthe present",adj=c(1,1),
    cex=0.7)
```

In these simulated LTT plots, we can clearly see the pull of the present for the reconstructed tree with extinction (figure 9.5). It's precisely this effect that we'll exploit to estimate speciation and extinction rates from a reconstructed phylogenetic tree!

9.3.2 Fitting the birth-death model to an empirical phylogeny

Fitting the birth-death model using maximum likelihood[16] (Nee, May, et al. 1994) has been implemented in the *phytools* function `fit.bd`, and it's pretty easy to use, so let's try it.[17]

```
bd.model<-fit.bd(darter.tree)
bd.model

##
## Fitted birth-death model:
##
## ML(b/lambda) = 0.2362
## ML(d/mu) = 0.01495
## log(L) = 370.8757
##
## Assumed sampling fraction (rho) = 1
##
## R thinks it has converged.
```

When we print out our fitted model, we can see both of the estimated model parameters, given here as b and d, along with the (maximum) likelihood of the model fit.

This model printout also tells us that we've assumed a sampling fraction (given by ρ) of `rho = 1`, which corresponds to 100 percent sampling.

In fact, we know that our tree includes only 201 of 216 species in the group.

Luckily, we can take this incomplete sampling fraction into consideration quite easily, following Stadler (2013).

```
## compute sampling fraction
sampling.f<-201/216
sampling.f

## [1] 0.9305556

## re-fit out model, setting rho equal to
## our computed sampling fraction
bd.model<-fit.bd(darter.tree,
    rho=sampling.f)
bd.model

##
## Fitted birth-death model:
##
## ML(b/lambda) = 0.2538
```

---

[16] An estimation technique that we've discussed in prior sections of this book, beginning in chapter 4.

[17] Note that the rates of speciation and extinction are normally represented using the Greek characters λ and μ, respectively.

```
## ML(d/mu) = 0.03258
## log(L) = 370.8757
##
## Assumed sampling fraction (rho) = 0.9306
##
## R thinks it has converged.
```

We now have ML estimates of birth or speciation (λ) and death or extinction (μ) that account for incomplete sampling of the species in our phylogeny. Notice that the estimated extinction rate, μ, increases when we take incomplete sampling into account. Why do you think that is?

To try to understand why, let's go back again to figure 9.5.

Logically, we might suppose, even entirely random missing taxa should probably tend to have the effect of "flattening out" the pull of the present—the upward curvature of the curve of lineage accumulation through time that we need to estimate extinction.

Let's prove this idea by returning to our reconstructed phylogeny, tree.reconstructed, and randomly pruning 50 percent of lineages from the tree. We can do this by using the *ape* package function drop.tip, as we have done before, this time along with the *base* R function sample to randomly choose tips to prune out of our tree.

```
## collapse singleton nodes
tree.reconstructed<-collapse.singles(tree.reconstructed)
## drop 50 random species
tree.missing<-drop.tip(tree.reconstructed,
    sample(tree.reconstructed$tip.label,50))
## compute "ltt" object without plotting
ltt.recon<-ltt(tree.reconstructed,plot=FALSE)
ltt.missing<-ltt(tree.missing,plot=FALSE)
## set margins and line-ending style
par(mar=c(5.1,4.1,2.1,2.1),lend=2)
## plot our "ltt" objects
plot(ltt.recon,bty="n",log.lineages=FALSE,log="y",
    lwd=2,col="darkgray",las=1,cex.axis=0.8)
plot(ltt.missing,log.lineages=FALSE,lty="dotted",
    lwd=2,add=TRUE)
## add a legend
legend(x="topleft",lty=c("solid","dotted"),
    lwd=c(2,2),col=c("darkgray","black"),
    legend=c("reconstructed phylogeny",
    "phylogeny with missing taxa"),
    bty="n",cex=0.8)
```

Even though we dropped 50 percent of the taxa of the tree, our two LTT plots are exactly superimposed throughout much of the history of the clade (figure 9.6).

They only begin to diverge toward the present when the LTT plot for the *full* tree begins to curve upward while the LTT for the tree with the missing taxa stays more or less straight.

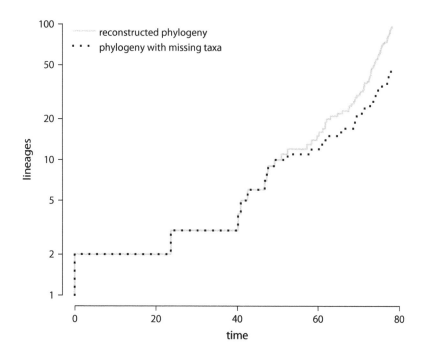

Figure 9.6

Lineage-through-time plot for a reconstructed phylogeny (gray) and the same phylogeny with random missing taxa (black dotted).

Remember, it's the pull of the present that contains information about extinction. As such, we can conclude that not accounting for missing taxa will tend to obscure the pattern expected from extinction.

It should be very clear from this analysis that it's essential to properly account for any missing taxa when fitting diversification models to phylogenetic trees.

9.3.3 Comparing alternative models for diversification

While taking into account incomplete sampling, as we've done, we can also fit and compare alternative models for diversification using likelihood ratio tests and/or the Akaike information criterion (AIC).

This allows us, for instance, to compare models with (birth-death) and without (pure birth, or "Yule") extinction in our tree.

Let's go ahead and compare a pure-birth model to the birth-death model we fitted earlier, as follows.

```
## fit Yule model
yule.model<-fit.yule(darter.tree,
    rho=sampling.f)
yule.model

##
## Fitted Yule model:
```

```
##
## ML(b/lambda) = 0.2375
## log(L) = 370.7178
##
## Assumed sampling fraction (rho) = 0.9306
##
## R thinks it has converged.

## compute AICs for Yule and birth-death
## models
AIC(yule.model,bd.model)

##              df       AIC
## yule.model   1 -739.4357
## bd.model     2 -737.7514
```

Since we should generally prefer the model with the *lower* AIC score, this result suggests that (controlling for the number of parameters), a pure-birth model better explains our tree than does a birth-death model.

Given that our models are *nested*,[18] if we'd prefer we can also easily conduct a likelihood ratio (LR) test.

For this, we'll need the package *lmtest* (Zeileis and Hothorn 2002).

```
library(lmtest)
lrtest(yule.model,bd.model)

## Likelihood ratio test
##
## Model 1: yule.model
## Model 2: bd.model
##    #Df LogLik Df  Chisq Pr(>Chisq)
## 1   1 370.72
## 2   2 370.88  1 0.3157     0.5742
```

Both the AIC comparison and the likelihood ratio test agree with our results from the γ test. We again conclude that our tree is not any different from what we might expect from a pure-birth process of lineage accumulation.

9.4 The effect of incomplete sampling on diversification rates

As we saw in previous sections, random missing taxa tend to remove the pull of the present from our lineage accumulation curve and will thus cause the rate of extinction to be underestimated.

In our darter phylogeny from Near et al. (2011), we supposed that the true phylogeny of darters contained a total of 216 real species—201 of which were represented in our empirical phylogeny.

---

[18] That is, pure birth is a special case of birth-death in which the death rate, μ, is set to zero.

For fun, let's imagine that instead of 216 species of darter, there were really a huge number of undiscovered species and the group actually should include 400 lineages.[19]

What effect do you think this will have on our estimated rate of extinction?

Let's find out.

```
## fit a birth-death model but assuming a true
## N of 400
bd.model2<-fit.bd(darter.tree,
    rho=201/400)
bd.model2
```

```
##
## Fitted birth-death model:
##
## ML(b/lambda) = 0.47
## ML(d/mu) = 0.2488
## log(L) = 370.8757
##
## Assumed sampling fraction (rho) = 0.5025
##
## R thinks it has converged.
```

When we compare this to our previous result, it looks like our estimated extinction rate has increased by nearly eightfold. Likewise, if we *now* compare our birth-death model to the simpler Yule model, will we similarly find a different result?

```
yule.model2<-fit.yule(darter.tree,
    rho=201/400)
yule.model2
```

```
##
## Fitted Yule model:
##
## ML(b/lambda) = 0.313
## log(L) = 365.6883
##
## Assumed sampling fraction (rho) = 0.5025
##
## R thinks it has converged.
```

```
lrtest(yule.model2,bd.model2)
```

```
## Likelihood ratio test
##
## Model 1: yule.model2
## Model 2: bd.model2
```

---

[19] Don't get too excited, ichthyologists.

```
##    #Df LogLik Df  Chisq Pr(>Chisq)
## 1    1 365.69
## 2    2 370.88  1 10.375    0.001277 **
## ---
## Signif. codes:
## 0 '***' 0.001 '**' 0.01 '*' 0.05 '.' 0.1 ' ' 1
```

The answer is inarguably *yes*: this time, we *strongly* reject a null hypothesis of constant-rate pure birth.

We see from this experiment that incorrectly estimating the *true* species richness of our clade can have a very large effect on not only parameter estimates but also hypothesis tests about our fitted models!

Let's go a step further and conduct a small experiment to look at the effect of ρ (our sampling fraction) on the estimation of λ and μ across a broad range of possible values of ρ—let's say from 20 to 100 percent.[20]

In our darter phylogeny, 20 percent sampling corresponds to a genuine species diversity in the group of 1,000 species—far beyond what anyone would imagine. By contrast, 100 percent sampling supposes that the fifteen species ostensibly *missing* from our phylogeny are not, in fact, real taxa at all!

```
## generate a sequence of values of rho
rho<-seq(0.2,1,by=0.01)
## fit our birth-death model given each
## rho value
fits<-lapply(rho,function(x,t) fit.bd(t,rho=x),
    t=darter.tree)
## organize these results into a table
BD<-t(sapply(fits,function(x) c(x$b,x$d)))
## set our plot margins and line-ending style
par(mar=c(5.1,4.1,2.1,2.1),lend=2)
## graph the birth-rate, as a function of rho
plot(rho,BD[,1],type="l",ylim=c(0,max(BD)),
    xlab=expression(paste("Assumed ",rho)),
    ylab=expression(paste("Estimated ",lambda,
    " or ",mu)),col="blue",lwd=2,bty="l",
    las=1,cex.axis=0.6,cex.lab=0.8)
## add a line showing the death rate
lines(rho,BD[,2],col="darkgreen",lwd=2,
    lty="dotted")
## create a legend
legend("topright",legend=c(expression(lambda),
    expression(mu)),lwd=2,col=c("blue","darkgreen"),
    lty=c("solid","dotted"),cex=0.8)
```

---

[20] This is one of the greatest strengths of R—particularly as you become more and more proficient at scripting. Imagine doing this analysis using a point-and-click software!

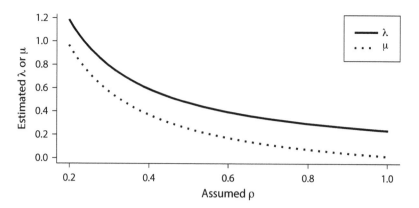

Figure 9.7
Estimated speciation and extinction rates for various assumed sampling fractions, using the phylogeny of darters from Near et al. (2011).

We see that both the estimated speciation rate (λ) and extinction rate (μ) are affected by our sampling fraction (figure 9.7). This again suggests that it is critical to account for incomplete sampling when carrying out diversification analyses (Chang et al. 2020; Sun et al. 2020), and not only that, but that getting this fraction wrong can have a rather large effect on the estimated parameters of our fitted model!

9.5 Likelihood surface for a birth-death model

In addition to what we've already seen, it may be somewhat useful to dig a bit deeper and have a look at the likelihood surface for a birth-death model (Nee, May, et al. 1994).

This can be valuable for a couple of reasons. First, the likelihood surface for our birth-death model has a fairly characteristic shape, and that shape gives us insight into our ability to estimate speciation and extinction rates.

Second, visualizing likelihood surfaces is a good way to check to make sure your own ML analyses[21] are arriving at the correct answer!

`fit.bd` exports a likelihood function that makes it pretty easy for us to go ahead and plot a likelihood surface for our model.

This can be done using the `contour` function in R as follows.[22]

```
## set the number of grid cells in our plot
ngrid<-100
## set the values of b & d that we'll use
## to compute the likelihood
```

---

[21] Normally, likelihood optimization in R and other computer software relies on heuristic numerical optimization algorithms rather than exhaustive search. This means that they are generally not guaranteed to get the correct, maximum likelihood solution! Fortunately, particularly for optimization problems with relatively few dimensions, the heuristic algorithms in R tend to be pretty good.

[22] R has lots of other three-dimensional plotting functions too—some of which can be used to create very neat-looking surfaces. `contour` is just the easiest for our purposes here.

```
b<-seq(0.24,0.27,length.out=ngrid)
d<-seq(0.02,0.05,length.out=ngrid)
## create an empty matrix
logL<-matrix(NA,ngrid,ngrid)
## use a for loop to cover our grid
## and compute likelihoods for each pair of
## b and d
for(i in 1:ngrid) for(j in 1:length(d))
    logL[i,j]<-bd.model$lik(c(b[i],d[j]))
## graph our contour plot
par(mar=c(5.1,4.1,2.1,2.1))
contour(x=b,y=d,logL,nlevels=50,
    xlab=expression(lambda),
    ylab=expression(mu),bty="l",axes=FALSE)
## add axes
axis(1,at=seq(0.24,0.27,by=0.01),cex.axis=0.8)
axis(2,at=seq(0.02,0.05,by=0.01),cex.axis=0.8,
    las=1)
## add a point showing our MLE
points(bd.model$b,bd.model$d,cex=1.5,pch=4,
    col="blue",lwd=2)
```

The shape of the likelihood surface in this plot (figure 9.8) is fairly typical of many diversification models—and this is consequential to estimation.

In particular, the long ridge that we see for different combinations of simultaneously increasing λ and μ means that these values will tend to have similar likelihoods (and thus be difficult to distinguish statistically) in our model.

9.6 Analyzing diversification using *diversitree*

Diversification models are also implemented in several other R packages, including *ape* and *diversitree* (FitzJohn 2012).

We'll learn more about *diversitree* in the next two chapters, but in this section, we'll also demonstrate how it can be used to fit and compare birth-death models using likelihood as well as with Bayesian MCMC.

The following quickly shows how we would fit the same birth-death model using *diversitree*. It involves two steps—first we create the likelihood function (with make.bd), and then we fit it (using find.mle). Remember, if you haven't used *diversitree* before, you will need to install it[23] before proceeding!

```
## load diversitree
library(diversitree)
## make birth-death likelihood function
```

---

[23] Which can be done using install.packages.

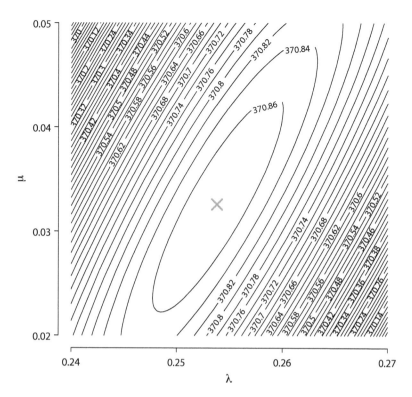

Figure 9.8
Likelihood surface for our birth-death model using the darter phylogeny of Near et al. (2011).

```
bd<-make.bd(darter.tree,
    sampling.f=sampling.f)
## find MLE using optimization function
fitted.bd<-find.mle(bd,x.init=c(0.1,0.05),
    method="optim",lower=0)
fitted.bd
```

```
## $par
##     lambda         mu
## 0.25381855 0.03257651
##
## $lnLik
## [1] 370.8757
##
## $counts
## function gradient
##       13       13
##
## $convergence
## [1] 0
##
```

```
## $message
## [1] "CONVERGENCE: REL_REDUCTION_OF_F <= FACTR*EPSMCH"
##
## $optim.method
## [1] "L-BFGS-B"
##
## $method
## [1] "optim"
##
## $func.class
## [1] "bd"       "dtlik"     "function"
##
## attr(,"func")
## Constant rate birth-death likelihood function:
##    * Parameter vector takes 2 elements:
##       - lambda, mu
##    * Function takes arguments (with defaults)
##       - pars: Parameter vector
##       - condition.surv [TRUE]: Condition
##       -              likelihood on survival?
##       - intermediates [FALSE]: Also return
##       -              intermediate values?
##    * Phylogeny with 201 tips and 200 nodes
##       - Taxa: Etheostoma_cinereum, ...
##    * Reference:
##       - Nee et al. (1994) doi:10.1098/rstb.1994.0068
## R definition:
## function (pars, condition.surv = TRUE, intermediates = FALSE)
## attr(,"class")
## [1] "fit.mle.bd" "fit.mle"
```

Even though the R syntax we used to get there is quite different, we should see that the fitted parameter estimates are highly similar, if not identical, to those obtained from fit.bd.

Likewise, let's also make a pure-birth (Yule) model and compare them, just as we did using *phytools*.

```
## make pure-birth likelihood function
yule<-make.yule(darter.tree,
    sampling.f=sampling.f)
## find MLE using optimization function
fitted.yule<-find.mle(yule,x.init=0.1,method="optim",
    lower=0)
fitted.yule

## $par
##    lambda
## 0.2374687
```

```
##
## $lnLik
## [1] 370.7178
##
## $counts
## function gradient
##       10       10
##
## $convergence
## [1] 0
##
## $message
## [1] "CONVERGENCE: REL_REDUCTION_OF_F <= FACTR*EPSMCH"
##
## $optim.method
## [1] "L-BFGS-B"
##
## $method
## [1] "optim"
##
## $func.class
## [1] "yule"      "bd"         "dtlik"      "function"
##
## attr(,"func")
## Yule (pure birth) likelihood function:
##    * Parameter vector takes 1 elements:
##       - lambda
##    * Function takes arguments (with defaults)
##       - pars: Parameter vector
##       - condition.surv [TRUE]: Condition
##       -              likelihood on survival?
##    * Phylogeny with 201 tips and 200 nodes
##       - Taxa: Etheostoma_cinereum, ...
##    * Reference:
##       - Nee et al. (1994) doi:10.1098/rstb.1994.0068
## R definition:
## function (pars, condition.surv = TRUE)
## attr(,"class")
## [1] "fit.mle.bd" "fit.mle"
```

```
anova(fitted.yule,fitted.bd)
```

```
##           Df lnLik     AIC    ChiSq Pr(>|Chi|)
## minimal  1 370.72 -739.44
## model 1  2 370.88 -737.75 0.31567     0.5742
```

If *diversitree* could only do this, it wouldn't be too interesting, would it?

Fortunately, *diversitree* is much more flexible. For instance, it also permits us to easily pass the likelihood function that we just created[24] to an MCMC sampler. This in turn will allow us to calculate posterior distributions of our model parameters, represented here as lambda and mu.

Let's do just that.[25]

```
## run MCMC using diversitree::mcmc
samples<-diversitree::mcmc(bd,c(1,0),nsteps=1e+05,
    lower=c(0,0),upper=c(Inf,Inf),w=c(0.05,0.05),
    fail.value=-Inf,print.every=10000)

## 10000: {0.2843, 0.0725} -> 370.51971
## 20000: {0.3198, 0.1381} -> 369.32367
## 30000: {0.2463, 0.0481} -> 370.41591
## 40000: {0.2940, 0.0649} -> 369.91835
## 50000: {0.3019, 0.0764} -> 369.70222
## 60000: {0.2923, 0.0924} -> 370.32497
## 70000: {0.2790, 0.0615} -> 370.60117
## 80000: {0.2770, 0.1103} -> 369.43987
## 90000: {0.3304, 0.1422} -> 368.97892
## 100000: {0.2625, 0.0313} -> 370.73033
```

samples is a matrix containing our posterior sample of λ and μ. Let's look at just the first bit of it as follows:

```
head(samples,20)

##     i   lambda           mu          p
## 1   1 0.5954741 0.217032291 317.2437
## 2   2 0.2051849 0.169243264 339.4852
## 3   3 0.3700257 0.152214204 366.0379
## 4   4 0.3575484 0.161431901 367.5393
## 5   5 0.3470252 0.132087981 367.6754
## 6   6 0.3315445 0.147374893 368.9064
## 7   7 0.3105099 0.072455762 368.8117
## 8   8 0.2538445 0.049108170 370.7099
## 9   9 0.2637679 0.038986929 370.7968
## 10 10 0.2261093 0.004285941 370.4163
## 11 11 0.2623881 0.070935758 370.4349
## 12 12 0.2839447 0.087021236 370.4610
## 13 13 0.2911339 0.074586862 370.2908
## 14 14 0.2690738 0.053197990 370.7831
```

---

[24]Using make.bd or make.yule.

[25]Readers might note that here we called the *diversitree* function mcmc using the function call diversitree::mcmc. That's because more than one package we have loaded in our R session contains a function called mcmc. Consequently, we need to specify diversitree::mcmc so that R will know which of these functions to use!

```
## 15 15 0.2578440 0.058707966 370.6184
## 16 16 0.2681394 0.048903781 370.7830
## 17 17 0.2651933 0.051798630 370.8200
## 18 18 0.2615416 0.091630131 369.4260
## 19 19 0.2852074 0.011471047 367.9658
## 20 20 0.2203870 0.048033532 367.6421
```

Convergence to the posterior distribution is pretty fast for this model,[26] so let's just chuck out the first 5 percent of our samples (5,000 generations, in this case) as burn-in.

```
samples<-samples[-(1:5000),]
```

There are numerous different ways we might go about analyzing our posterior sample. For instance, the arithmetic mean of the sample can be used as an estimate of the parameter. We might also compute a 95 percent (or other interval) high posterior density interval.

The former we can do easily with *base* R. For the latter, let's use the R package *coda*.[27]

```
## load coda package
library(coda)
## extract posterior sample of lambda & mu
lambda<-samples$lambda
mu<-samples$mu
## set object class to "mcmc"
class(lambda)<-class(mu)<-"mcmc"
## create data frame with our estimates and
## HPD for each parameter
object<-data.frame(lambda=c(mean(lambda),
    HPDinterval(lambda)),mu=c(mean(lambda),
    HPDinterval(mu)))
rownames(object)<-c("estimate","2.5% HPD",
    "97.5% HPD")
## round our object to print it
print(round(object,6))
```

```
##             lambda        mu
## estimate    0.272868  0.272868
## 2.5% HPD    0.218377  0.000000
## 97.5% HPD   0.335179  0.147813
```

We can also visualize the posterior sample pretty easily in R, which we show in figure 9.9. To do this, we'll use the *diversitree* function `profiles.plot`, which creates a very nice graph

---

[26]We're ignoring lots of nuance here, but there are some good R packages such as *coda* (Plummer et al. 2006) that can be used for Bayesian MCMC diagnostics—including helping us to figure out whether or not we've converged to the posterior distribution.

[27]Which readers should *load* using `library` but shouldn't need to install since it's a dependency of *phytools*.

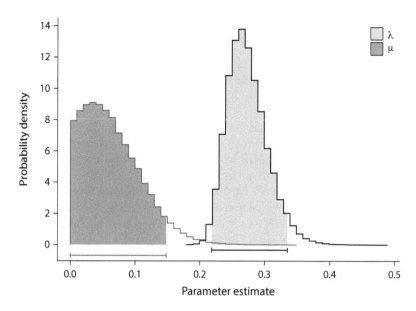

Figure 9.9
Posterior sample for birth-death model parameters from *diversitree*. Phylogeny of darters from Near et al. (2011).

showing both our estimate of the posterior distribution and (with the whisker lines in the margins of the plot) our high probability density intervals.

```
## set plot margins
par(mar=c(5.1,4.1,2.1,2.1))
## create posterior density plot
profiles.plot(samples[c("lambda","mu")],
    col.line=c("red","blue"),bty="l",las=1,
    cex.axis=0.8)
## add legend
legend(x="topright",c(expression(lambda),
    expression(mu)),pch=22,
    pt.bg=make.transparent(c("red","blue"),0.5),
    pt.cex=c(2,2),bty="n",cex=0.8)
```

The Bayesian analysis is in broad agreement with other analyses of the darter tree, in that we can't rule out the possibility that extinction rates (μ) are not statistically distinguishable from zero for this clade.

This is just a very small sample of what *diversitree* can do. We'll see a lot more of this package later in this book.

In summary, we can use a variety of functions in R to fit pure-birth and birth-death models to phylogenetic trees and to compare the fit of these models to make conclusions about the effects of speciation, extinction, and sampling on the shape of our LTT plot.

9.1 Imagine that you discovered a hidden valley in Tennessee that contains 96 new species of darters, none of which are included in your analysis. Ignoring problems with nonrandom sampling, repeat the γ MCCR test of Pybus and Harvey (2000), accounting for your new knowledge of this undersampling.

9.2 You may have noticed that one effect of accounting for missing taxa in the tree is that our chances of estimating a nonzero extinction rate increase. Can you figure out *exactly* how many undiscovered species of darters would need to be missing from our tree in order for our analysis to reject a pure-birth model in favor of birth-death, given an α level of 0.05?

9.3 Download the file 10kTrees_Primates.nex from the book website. This file contains 100 primate trees sampled from the posterior distribution of a Bayesian analysis and was obtained from the "10kTrees Webserver."[28] These trees are (in theory) time-calibrated in millions of years. Use *phytools* or *diversitree* to estimate speciation and extinction rates for all of the trees from the sample. Compute the mean speciation and extinction rates, as well as the variability in rate among trees. Take into account incomplete sampling fraction. Some debate exists around how many species of primate are currently extant. Various sources claim that the mammalian order Primates may include from 376 to 522 species. Set the sampling fraction to assume these different values for total species diversity of the order. What effect do these different assumed diversities have on our estimated speciation and extinction rates?

---

[28] https://10ktrees.nunn-lab.org/.

Time- and density-dependent diversification

10

The constant-rate birth and death model (Nee et al. 1992) has been extremely useful in efforts to understand how diversification tends to proceed through the tree of life. Nonetheless, in real phylogenies, we know that evolutionary diversification is often much more complicated than this very simple model assumes.

In particular, it seems quite reasonable to suppose that both the rate of new species formation and the rate of extinction may vary through time, among clades, as a function of species' traits, as well as in any number of other interesting manners (Jablonski 2008; Rabosky and McCune 2010). A wide range of comparative methods have been developed that are specifically designed to examine these more complicated scenarios of diversification and in turn help us better characterize the processes that underlie the diversity of life on earth.

In this chapter, we will:

1. Learn about a model in which the rates of speciation, extinction, or both vary through time.
2. See how to fit this time-varying speciation and extinction rate model in R using the *diversitree* package.
3. Learn how to compare between different models.
4. Finally, examine an alternative density-dependent model of diversification in which the rate of new species formation depends on the number of extant lineages.

Before we begin, it's worth mentioning that attempts to reconstruct the dynamics of speciation through time should account for a recent study by Louca and Pennell (2020). This article showed that, for any given phylogenetic tree, there is an infinite set of time-varying diversification models that all fit the data equally, and maximally, well. This set, called a congruence class, includes the models we'll estimate here—as well as many others in which speciation, extinction, or both changes through time in complex ways. So what does that mean for your data analysis? Well, nobody really knows for sure—but it stands to reason that your inferred results are only as good as your prior knowledge that the model you're fitting is appropriate for the data and question at hand. We recommend that users worried about this issue calculate the "pulled diversification rate" (Louca and Pennell 2020) as a way to summarize the congruence class.

10.2 Time-varying diversification

The first model we'll consider is one in which the rate of speciation and/or extinction is allowed to change through time from the root of the tree to the present day.

This class of model may be of interest in a wide variety of contexts.

For instance, many theories of adaptive radiation posit that when a lineage experiences new ecological opportunity, perhaps following a mass extinction or after colonizing a new area, the rate of diversification will be initially very high as diversification occurs to fill available niches (Schluter 2000). Subsequently, as niches are filled, the speciation rate may be expected to gradually decline through time (Schluter 2000).

Alternatively, under some scenarios, such as long-term environmental change, we might be interested in gradual increases or decreases in the extinction rate (e.g., Sepkoski 1981).

10.2.1 The exponential model

Among time-varying diversification models,[1] the most common functional form is an *exponential curve* (Morlon et al. 2010).[2]

Figure 10.1 illustrates the speciation (λ) and extinction (μ) rates through time under scenarios of constant rate (panel a), declining speciation (panel b), increasing extinction (panel c), or both (panel d).[3]

```
## set sequence of time intervals
t<-seq(0,35.4,length.out=100)
## set plotting parameters
par(mfrow=c(2,2),bty="n",las=1)
## panel a) constant speciation and extinction
## compute speciation/extinction through time
b<-rep(3,length(t))
d<-rep(1,length(t))
## plot curve
plot(t,b,type="l",col="black",lwd=3,
    ylim=c(0,5),xlab="time",
```

---

[1] That is, models in which the speciation rate, the extinction rate, or both is assumed to vary as a continuous function of time since the root of the tree.

[2] We're not sure why exponential curves are so popular. An exponential function can monotonically increase or decrease with quite different curvatures, depending on how it's parameterized, so that may be the reason. Also, the exponential functions popular in comparative methods are never negative, thus avoiding the problem of negative speciation or extinction rates.

[3] We chose the total amount of time, 35.4 million years, to match the total depth of the tree in an empirical example we'll see in the next section.

```
    ylab=expression(paste("rate (",lambda," or ",mu,")")))
lines(t,d,col="lightgray",lwd=3)
mtext("(a)",line=1,adj=0)
## panel b) declining speciation, constant extinction
## compute speciation or extinction through time
l=3
a=0.1
b<-l*exp(-a*t)
d<-rep(1,length(t))
## plot curve
plot(t,b,type="l",col="black",lwd=3,ylim=c(0,5),
    xlab="time",
    ylab=expression(paste("rate (",lambda," or ",mu,")")))
lines(t,d,col="lightgray",lwd=3)
legend("topright",lwd=3,col=c("black","lightgray"),
    legend=c(expression(paste("speciation (",lambda,")")),
    expression(paste("extinction (",mu,")"))))
mtext("(b)",line=1,adj=0)
## panel c) increasing extinction, constant speciation
## compute speciation/extinction through time
l=1
a=-0.05
b<-rep(3,length(t))
d<-l*exp(-a*t)
## plot curve
plot(t,b,type="l",col="black",lwd=3,ylim=c(0,5),
    xlab="time",
    ylab=expression(paste("rate (",lambda," or ",mu,")")))
lines(t,d,col="lightgray",lwd=3)
mtext("(c)",line=1,adj=0)
## panel d) both speciation & extinction change through time
## compute speciation/extinction through time
lambda.l=3
lambda.a=0.1
mu.l=1
mu.a=-0.05
b<-lambda.l*exp(-lambda.a*t)
d<-mu.l*exp(-mu.a*t)
## plot curve
plot(t,b,type="l",col="black",lwd=3,ylim=c(0,5),
    xlab="time",
    ylab=expression(paste("rate (",lambda," or ",mu,")")))
lines(t,d,col="lightgray",lwd=3)
mtext("(d)",line=1,adj=0)
```

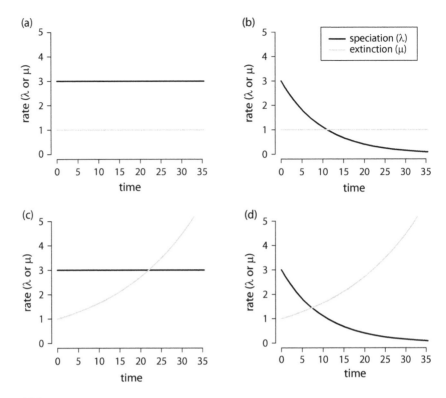

Figure 10.1
Hypothetical models for time-varying diversification. (a) Speciation and extinction are constant through time. (b) Speciation declines through time while extinction is constant. (c) Speciation is constant, but extinction increases through time. (d) Both speciation and extinction change through time.

In figure 10.1, we just picked arbitrary values for the curves that describe speciation or extinction through time. In the next section, we'll see how to estimate these curves from a real phylogeny.

10.3 Fitting time-variable diversification models to data

We can fit all four of the models of figure 10.1, and many others, using the powerful R package *diversitree* (FitzJohn 2012)[4] that we first encountered in chapter 9.

Along with the constant-rate model that we learned about in chapter 9, *diversitree* also contains models in which speciation and extinction can either be constant or vary through time. When speciation and extinction vary, they can follow a variety of different functional forms.

As in previous chapters, in practice, we'll use the statistical estimation procedures of maximum likelihood or Bayesian MCMC to find reasonable parameter estimates given our data—the phylogenetic tree.

---

[4]It's also possible to fit some of the same models with three other useful R packages: *RPANDA* (Morlon et al. 2016), *DDD* (Etienne and Haegeman 2021), and *TreePar* (Stadler 2015). We'll use *DDD* later in the chapter but focus mostly on *diversitree* because in chapter 11 we'll also be using *diversitree* to fit diversification models with the rate of speciation, or extinction can depend on traits.

10.3.1 An empirical example: The phylogeny of Cetacea

Let's try running some analyses on a real phylogeny: the tree of Cetacea, the whales and dolphins (Morlon et al. 2011). To follow along, you'll need to download the file `Cetacea.phy`, which is available, along with all the files we use in this volume, from the book website.[5]

```
## load phytools
library(phytools)
## read whale tree from file
whale.tree<-read.tree(file="Cetacea.phy")
```

We can start by plotting our tree.

Just for fun, let's plot the phylogeny in a such a way that we'll be able to add a horizontal axis showing time before the present day (figure 10.2).

To accomplish this, however, we're going to do something a little bit nutty. We're going to plot our tree in a *left-facing* direction but then flip our x axis so that the upper limit is first and the lower limit is second.[6,7]

```
## compute the dimensions of the plotting area without
## graphing our tree
plotTree(whale.tree,fsize=0.4,direction="leftwards",
    ftype="i",mar=c(5.1,0.1,0.1,0.1),plot=FALSE,lwd=1,
    offset=0.25)
## extract the x-limits for plotting, and flip them
xlim<-get("last_plot.phylo",envir=.PlotPhyloEnv)$x.lim[2:1]
## graph the tree with the axis flipped
plotTree(whale.tree,fsize=0.4,direction="leftwards",ftype="i",
    mar=c(5.1,0.1,0.1,0.1),plot=TRUE,xlim=xlim,lwd=1,
    add=TRUE,offset=0.25)
## add an axis
axis(1,at=seq(0,30,by=10),cex.lab=0.6)
## add an x-axis label
title(xlab="time (mybp)",cex=0.7)
par(font=3)
## add node index labels
whale.tree$node.label<-paste("n",1:whale.tree$Nnode,
    sep="")
labelnodes(whale.tree$node.label,
    1:whale.tree$Nnode+Ntip(whale.tree),
    interactive=FALSE,cex=0.4,bg="lightblue",
    shape="ellipse")
```

OK, you're probably asking yourself—why did we first run `plotTree` in *phytools* without actually plotting the tree at all?[8]

---

[5] http://www.phytools.org/Rbook/.

[6] In R, this is totally legal and will result in our plot being oriented right to left instead of left to right.

[7] Don't worry about the node labels. We're going to use these later!

[8] That is, with `plot=FALSE`.

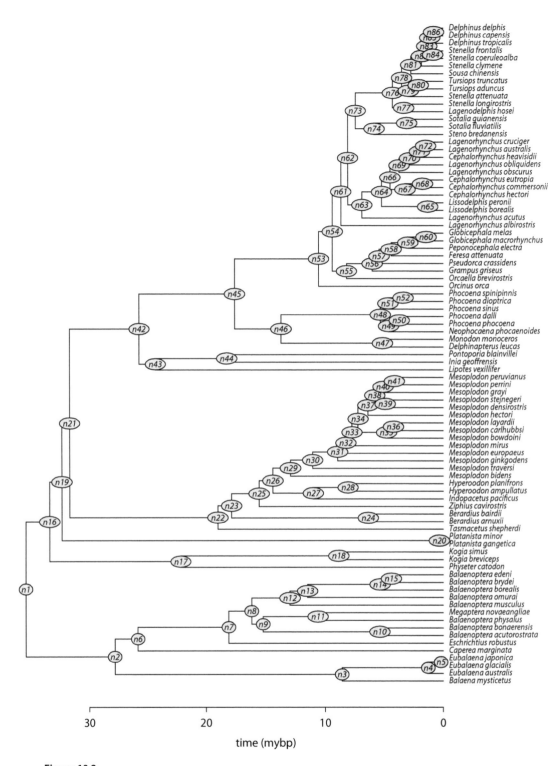

Figure 10.2
Phylogenetic tree of Cetacea.

This occurred because it's actually quite complicated to figure out how much *space* to leave to the right of the tips of a plotted tree in the user coordinates.[9] This calculation is something that's done internally by plotTree.[10] But we *need* these values to be able to flip our *x*-axis and plot our left-facing tree in a rightward direction. This way, we can plot the tree in the desired way without having some of the labels cut off by the figure margin.

Cool, right? We'll learn more about this and other plotting tricks in chapter 13 of this book.

Now that we're done with this important business, let's get down to fitting time-varying diversification models to our tree.

10.3.2 A time-varying speciation model

In *diversitree*, as we learned at the end of chapter 9, fitting a model will *always* involve first *creating* a likelihood function for our model using one function—and then optimizing it, or running an MCMC, with another.[11]

We'll start by doing this for our simple constant-rate birth-death model—in other words, the same model that we fit using *phytools* and then *diversitree* in chapter 9.

For this and subsequent models, we'd like to get our starting parameter values for the speciation (λ) and extinction (μ) rates on the correct order of magnitude. For a quick check, why don't we use the *phytools* function fit.bd to fit a simple birth and death model to our tree? fit.bd does not require starting values for the model parameters for optimization.[12]

```
fit.bd(whale.tree,rho=Ntip(whale.tree)/89)

##
## Fitted birth-death model:
##
## ML(b/lambda) = 0.106
## ML(d/mu) = 0
## log(L) = 23.4321
##
## Assumed sampling fraction (rho) = 0.9775
##
## R thinks it has converged.
```

We set the sampling fraction, here the argument rho, to be rho = Ntip(whale.tree) / 89 because Ntip(whale.tree) gives us the number of tips in our cetacean phylogeny, and we can use 89 as the "true" number of extant whale and dolphin species thought to exist.

---

[9] The coordinate system of our plotted graph.

[10] As well as by other similar plotting functions, such as plot.phylo in *ape*.

[11] We'll likewise always need to assign starting parameter values for optimization. The specific numbers we choose for the starting values don't matter very much for simple models—but will become more and more important as we progress to more sophisticated models in this chapter and in chapter 11. For its most complicated models, *diversitree* even comes with some functions designed to help us choose reasonable starting parameter values—but we're also wise to try a few different sets of values to make sure that we consistently converge to the same ML solution!

[12] At least not supplied by the user! fit.bd, like nearly every method using numerical optimization, needs starting values—it just identifies some reasonable values for this starting condition absent our external input!

This analysis tells us that our ML speciation rate, λ, is about 0.1 and our ML extinction rate, μ, is about 0. Why don't we use starting values for optimization, then, of 0.1 and 0.01 for λ and μ, respectively?

```
## load diversitree
library(diversitree)
## fit our birth-death model, as in chapter 9
bd_model<-make.bd(tree=whale.tree,
    sampling.f=Ntip(whale.tree)/89)
st<-c(0.1,0.01)
bd_mle<-find.mle(func=bd_model,x.init=st)
```

Going through this code chunk line by line, first we load the *diversitree* package (FitzJohn 2012), then we define our likelihood function using make.bd,[13] then we provide our initial values for the speciation and extinction rates,[14] and finally we optimize the parameters of our model using find.mle.

Let's see the result:

```
bd_mle

## $par
##     lambda         mu
## 0.1069670 0.0000006
##
## $lnLik
## [1] 23.4
##
## $counts
## [1] 11
##
## $code
## [1] 2
##
## $gradient
## [1] 7.47 3.95
##
## $method
## [1] "nlm"
##
## $func.class
## [1] "bd"      "dtlik"    "function"
##
```

---

[13] In *diversitree*, we define sampling fraction using the argument sampling.f, but this is equivalent to rho in the *phytools* function fit.bd.

[14] For such a simple model, it isn't too important what values we choose; however, it can be useful to have them on the correct order of magnitude, as we do here using fit.bd.

```
## attr(,"func")
## Constant rate birth-death likelihood function:
##    * Parameter vector takes 2 elements:
##        - lambda, mu
##    * Function takes arguments (with defaults)
##        - pars: Parameter vector
##        - condition.surv [TRUE]: Condition
##                      likelihood on survival?
##        - intermediates [FALSE]: Also return
##                      intermediate values?
##    * Phylogeny with 87 tips and 86 nodes
##        - Taxa: Balaena_mysticetus, ...
##    * Reference:
##        - Nee et al. (1994) doi:10.1098/rstb.1994.0068
## R definition:
## function (pars, condition.surv = TRUE, intermediates = FALSE)
## attr(,"class")
## [1] "fit.mle.bd" "fit.mle"
```

This printout gives us a lot of information,[15] but among this, we see the parameter estimates of our fitted model—the speciation rate (lambda) and extinction rate (mu)—as well as some details about the model we optimized and a summary of the success or failure of the optimization.

Just as we saw in chapter 9,[16] *diversitree* makes it very easy to be frequentist *or* Bayesian.[17] To sample the posterior distribution, we just need to take the likelihood function that we built (bd_model) and pass it to a different function: mcmc.

Let's try it.

```
bd_res_bayes<-mcmc(lik=bd_model,x.init=st,
    nsteps=10000,w=0.01,print.every=1000)

## 1000: {0.1111, 0.0121} -> 23.29430
## 2000: {0.1432, 0.0592} -> 21.91686
## 3000: {0.1009, 0.0077} -> 23.05121
## 4000: {0.1212, 0.0538} -> 21.61637
## 5000: {0.0958, 0.0032} -> 22.84909
## 6000: {0.1118, 0.0034} -> 23.33759
## 7000: {0.1096, 0.0072} -> 23.36152
## 8000: {0.1389, 0.0214} -> 21.60501
## 9000: {0.1212, 0.0303} -> 22.93159
## 10000: {0.1231, 0.0006} -> 22.44633
```

---

[15] Probably more than we'll usually need.

[16] But will nonetheless revisit here. We promise it's important!

[17] Perhaps depending on our mood! Although statisticians may be aghast, this probably describes most biologists.

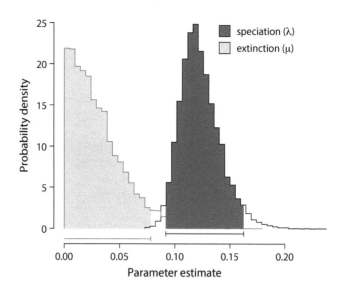

Figure 10.3

Posterior distribution for a Bayesian MCMC analysis of the constant-rate birth-death model for the cetacean phylogeny of figure 10.2.

In this function call, `nsteps` is the number of steps of the MCMC—that is, the number of generations that we intend to sample from the posterior distribution, and w is a *diversitree*-specific tuning parameter of the MCMC.[18]

Let's proceed to eliminate the first 10 percent of samples[19] and then graph the posterior samples of the two different parameters (λ and μ) of our model. The result is in figure 10.3.

```
## remove 1,000 samples for burn-in
bd_res_bayes<-bd_res_bayes[-(1:1000),]
## pull out lambda and mu
postSamples<-bd_res_bayes[,c("lambda","mu")]
## create density plots for lambda and mu
profiles.plot(postSamples,col.line=c("black","gray"),
    las=1,bty="n",cex.axis=0.7,cex.lab=0.8)
legend("topright",c(expression(paste("speciation (",lambda,")")),
    expression(paste("extinction (",mu,")"))),
    cex=0.8,pch=22,pt.cex=1.6,bty="n",
    pt.bg=make.transparent(c("black","gray"),0.5))
```

---

[18]The help page of mcmc provides more information, explaining how w affects the slice sampler of the MCMC—recommending, in short, that users try to pick w by first generating a preliminary sample from the posterior distribution and then using the range of 90 percent of the values in our sample. We won't worry too much about that here.

[19]This is arbitrary. In general, we should minimally visualize the likelihood profile from our MCMC, as we did for the threshold model analysis of chapter 7. In practice, if we are unsure about whether or not our MCMC has converged on the posterior distribution, we can use the R package *coda* (Plummer et al. 2006) to try and check!

Next, we'll fit a model in which we imagine that the rate of speciation, λ, changes through time while the rate of extinction, μ, is constant (Morlon et al. 2010, 2011).

This corresponds to the scenario we illustrated in panel (b) of figure 10.1.

Instead of make.bd, to build our likelihood function, we'll use the *diversitree* function make.bd.t.

make.bd.t is quite similar to make.bd but takes the extra argument functions that we can use to specify the functions of time that we want to use to specify our time-varying speciation and extinction models.

In our case, we'll set functions to c("exp.t", "constant.t"), which corresponds to exponentially changing λ and constant μ. Does that make sense?

```r
## make likelihood function
bvar_model<-make.bd.t(whale.tree,sampling.f=87/89,
  functions=c("exp.t","constant.t"))
## choose starting values of parameters
st<-c(0.1,0.01,0.01)
## optimize likelihood function
bvar_mle<-find.mle(bvar_model,st)
bvar_mle
```

```
## $par
##   lambda.l   lambda.a        mu
## 0.10666540 0.00080028 0.00000399
##
## $lnLik
## [1] 23.4
##
## $counts
## [1] 139
##
## $convergence
## [1] 0
##
## $message
## [1] "success! tolerance satisfied"
##
## $hessian
## NULL
##
## $method
## [1] "subplex"
##
## $func.class
## [1] "bd.t"      "bd"        "dtlik.t"  "dtlik"
## [5] "function"
##
## attr(,"func")
## Constant rate birth-death (time-varying) likelihood function:
```

```
##   * Parameter vector takes 3 elements:
##      - lambda.l, lambda.a, mu
##   * Function takes arguments (with defaults)
##      - pars: Parameter vector
##      - condition.surv [TRUE]: Condition
##   -               likelihood on survival?
##      - intermediates [FALSE]: Also return
##   -               intermediate values?
##   * Phylogeny with 87 tips and 86 nodes
##      - Taxa: Balaena_mysticetus, ...
##   * Reference:
##      - Nee et al. (1994) doi:10.1098/rstb.1994.0068
## R definition:
## function (pars, condition.surv = TRUE, intermediates = FALSE)
## attr(,"class")
## [1] "fit.mle.bd" "fit.mle"
```

In the printout of our fitted model, the first parameter (denominated lambda.l) is the estimated value of λ at the root of the tree. lambda.a is the fitted model parameter of our exponential function. Kind of counterintuitively, because it's *positive*,[20] we know this corresponds to a *decreasing* rate of speciation through time in the fitted model.

Let's go ahead and plot our fitted model in the same way that we did in figure 10.1 just to see what we mean.

```
## extract parameter values from our fitted model
lambda.l<-bvar_mle$par["lambda.l"]
lambda.a<-bvar_mle$par["lambda.a"]
mu<-bvar_mle$par["mu"]
## create exponential or linear functions
b<-lambda.l*exp(-lambda.a*t)
d<-rep(mu,length(t))
## set plotting parameters
par(mar=c(5.1,4.1,1.1,2.1))
## plot our fitted model
plot(t,b,type="l",col="black",lwd=3,ylim=c(0,0.15),
    bty="n",xlab="time",las=1,cex.axis=0.8,
    ylab=expression(paste("rate (",lambda," or ",
    mu,")")))
lines(t,d,col="lightgray",lwd=3)
legend("topright",lwd=3,col=c("black","lightgray"),
    legend=c(expression(paste("speciation (",lambda,
    ")")),expression(paste("extinction (",mu,")"))),
    cex=0.8,bty="n")
```

It's nearly imperceptible, but the curve showing the speciation rate through time is curving *very* slightly downward in our plot (figure 10.4).

---

[20] Although only very slightly so.

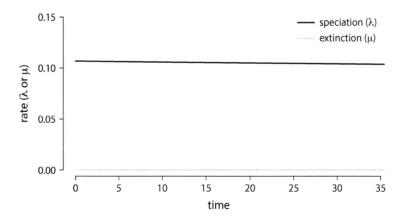

Figure 10.4
Fitted variable-speciation and constant-extinction model to the cetacean phylogeny.

10.3.3 Bayesian MCMC analysis of our time-varying diversification model

Our maximum likelihood analysis gave us a good point estimate about how the rate is changing (or not) through time, but just as we did with our birth-death model, *diversitree* lets us easily undertake a Bayesian MCMC analysis of our variable-rate models.

To do this, we'll once again use the *same* MCMC function, mcmc, but give it the likelihood function in our previous analysis: bvar_model.[21]

```
bvar_res_bayes<-mcmc(lik=bvar_model,
    x.init=bvar_mle$par,nsteps=1000,
    w=0.01,print.every=100)

## 100: {0.1367, 0.0113, 0.0153} -> 22.27948
## 200: {0.1162, 0.0059, 0.0547} -> 19.40163
## 300: {0.1325, 0.0111, 0.0168} -> 22.48347
## 400: {0.1228, 0.0049, 0.0093} -> 23.03892
## 500: {0.1511, 0.0044, 0.0875} -> 19.69471
## 600: {0.1259, 0.0023, 0.0144} -> 22.83982
## 700: {0.1289, 0.0008, 0.0164} -> 22.57296
## 800: {0.1197, 0.0111, 0.0205} -> 22.26425
## 900: {0.1195, 0.0093, 0.0393} -> 20.90854
## 1000: {0.1470, 0.0052, 0.0370} -> 21.65602
```

Here, we only ran our MCMC for 1,000 generations. In practice, this is probably not nearly long enough!

We should also check our MCMC for evidence of convergence to the posterior distribution. This can be done using the R package *coda* (Plummer et al. 2006) that we've mentioned in prior chapters. To keep things simple for now, though, we'll just trim the first 10 percent of generations from our sample and jump straight to plotting the posterior distribution.

---

[21] We're also going to give it our ML solution for the same model as a starting value of the MCMC. For our purposes here, it has the nice effect of starting us off in a region of high posterior density, which should make our MCMC a little bit more efficient.

Now we could show histograms of the posterior distribution of λ and μ like we did in our birth-death model—but, remember, under a variable λ model,[22] the model parameter lambda.1 is pretty difficult to interpret because it is simply the instantaneous value of λ at the root!

It may make *more* sense to visualize the posterior distribution of the rates through time—mimicking what we did with our MLE point estimate in figure 10.4, but for every sample[23] from our posterior distribution!

Let's try it.

```
## trim burn-in
bvar_res_bayes<-bvar_res_bayes[-(1:100),]
## extract parameters
postSamples<-bvar_res_bayes[,c("lambda.1","lambda.a","mu")]
## thin posterior sample
thinnedPosterior<-postSamples[round(seq(1,nrow(postSamples),
    length.out=100)),]
## set plotting parameters and open plot
par(mar=c(5.1,4.1,1.1,2.1))
plot(NULL,xlim=c(0,35.4),ylim=c(0,0.22),bty="n",xlab="time",
    las=1,ylab=expression(paste("rate (",lambda," or ",
    mu,")")),cex.axis=0.8)
legend("topright",lwd=3,col=c("black","lightgray"),
    legend=c(expression(paste("speciation (",lambda,")")),
    expression(paste("extinction (",mu,")"))),
    cex=0.8,bty="n")
## add all curves to our plot
for(i in 1:nrow(thinnedPosterior)){
    lambda.1<-thinnedPosterior[i,"lambda.1"]
    lambda.a<-thinnedPosterior[i,"lambda.a"]
    mu<-thinnedPosterior[i,"mu"]
    b<-lambda.1*exp(-lambda.a*t)
    d<-rep(mu,length(t))
    lines(t,b,col=make.transparent("black",0.5),lwd=1)
    lines(t,d,col=make.transparent("lightgray",0.5),lwd=1)
}
```

This code chunk is not too difficult to understand.

First, we "thin" our posterior sample by picking out 100 of the 900, which works out to one of every nine values, for each parameter in the sample. Next, we set our plotting parameters and add a simple legend. Finally, we use a for loop to iterate across all of the values in our thinned posterior sample, each time re-creating the plot of figure 10.4 for that sample!

We decided to use the *phytools* function make.transparent to plot our lines translucently so it is easier to see their distribution across many samples.

Figure 10.5 shows that across our entire posterior sample, λ usually declines through time but sometimes is nearly flat.

---

[22] Likewise, for the variable μ model, when we get to it.

[23] Or some random subsample thereof.

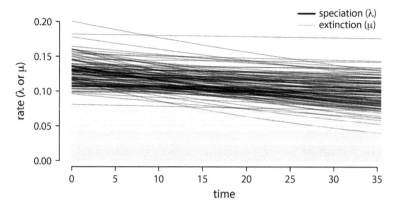

Figure 10.5
Posterior distribution of the rates of speciation and extinction through time in the variable-speciation model.

10.3.4 Fitting variable-extinction and variable-speciation and extinction models

In addition to this time-varying model, we can also fit the other two remaining models of figure 10.1 as well.

The next of these was the variable-extinction model. Just as we did above, let's first fit this model using maximum likelihood. We'll wait until optimizing the variable-speciation *and* variable-extinction model before graphing both of them together.

To make our constant-speciation, variable-extinction model, we'll use the same function (make.bd.t) as we employed in the previous part of the section, but we'll set functions to c("constant.t","exp.t").

```
## make time-varying extinction model
dvar_model<-make.bd.t(whale.tree,sampling.f=87/89,
    functions=c("constant.t","exp.t"))
st<-c(0.1,0.01,0.01)
## optimize model
dvar_mle<-find.mle(dvar_model, st)
dvar_mle
```

```
## $par
##        lambda          mu.l          mu.a
## 0.10598042044 0.00000000702 0.05727598760
##
## $lnLik
## [1] 23.4
##
## $counts
## [1] 184
##
## $convergence
## [1] 0
```

```
##
## $message
## [1] "success! tolerance satisfied"
##
## $hessian
## NULL
##
## $method
## [1] "subplex"
##
## $func.class
## [1] "bd.t"      "bd"         "dtlik.t"  "dtlik"
## [5] "function"
##
## attr(,"func")
## Constant rate birth-death (time-varying) likelihood function:
##    * Parameter vector takes 3 elements:
##        - lambda, mu.l, mu.a
##    * Function takes arguments (with defaults)
##        - pars: Parameter vector
##        - condition.surv [TRUE]: Condition
##        -                 likelihood on survival?
##        - intermediates [FALSE]: Also return
##        -                 intermediate values?
##    * Phylogeny with 87 tips and 86 nodes
##        - Taxa: Balaena_mysticetus, ...
##    * Reference:
##        - Nee et al. (1994) doi:10.1098/rstb.1994.0068
## R definition:
## function (pars, condition.surv = TRUE, intermediates = FALSE)
## attr(,"class")
## [1] "fit.mle.bd" "fit.mle"
```

Once again, because the coefficient mu.a is positive, we know that this fitted model corresponds to one in which the extinction rate, μ, declines through time.

Last, we can fit the model of figure 10.1d, in which both speciation and extinction can change through time.

To do this, we now set functions=c("exp.t","exp.t"), as you've probably guessed.

```
## make time-varying speciation & extinction model
bdvar_model<-make.bd.t(whale.tree,sampling.f=87/89,
    functions=c("exp.t","exp.t"))
st<-c(0.1,-0.01,0.01,0.01)
## optimize model
bdvar_mle<-find.mle(bdvar_model,st)
bdvar_mle
```

```
## $par
##     lambda.l     lambda.a          mu.l
##   0.107076035  0.001108310  0.000000849
##         mu.a
## -0.043289551
##
## $lnLik
## [1] 23.4
##
## $counts
## [1] 617
##
## $convergence
## [1] 0
##
## $message
## [1] "success! tolerance satisfied"
##
## $hessian
## NULL
##
## $method
## [1] "subplex"
##
## $func.class
## [1] "bd.t"      "bd"         "dtlik.t"   "dtlik"
## [5] "function"
##
## attr(,"func")
## Constant rate birth-death (time-varying) likelihood function:
##    * Parameter vector takes 4 elements:
##       - lambda.l, lambda.a, mu.l, mu.a
##    * Function takes arguments (with defaults)
##       - pars: Parameter vector
##       - condition.surv [TRUE]: Condition
##       -             likelihood on survival?
##       - intermediates [FALSE]: Also return
##       -             intermediate values?
##    * Phylogeny with 87 tips and 86 nodes
##       - Taxa: Balaena_mysticetus, ...
##    * Reference:
##       - Nee et al. (1994) doi:10.1098/rstb.1994.0068
## R definition:
## function (pars, condition.surv = TRUE, intermediates = FALSE)
## attr(,"class")
## [1] "fit.mle.bd" "fit.mle"
```

Let's make a plot with both our fitted variable-extinction (but constant-speciation) and variable-speciation and extinction models.

```
## set plotting parameters
par(mfrow=c(2,1),mar=c(5.1,4.1,2.1,2.1))
## extract fitted model coefficients
lambda<-dvar_mle$par["lambda"]
mu.l<-dvar_mle$par["mu.l"]
mu.a<-dvar_mle$par["mu.a"]
b<-rep(lambda,length(t))
d<-mu.l*exp(-mu.a*t)
## plot time-varying extinction model
plot(t,b,type="l",col="black",lwd=3,ylim=c(0,0.15),
    bty="n",xlab="time",las=1,cex.axis=0.8,
    ylab=expression(paste("rate (",lambda," or ",mu,
    ")")))
lines(t,d,col="lightgray",lwd=3)
legend("topright",lwd=3,col=c("black","lightgray"),
    legend=c(expression(paste("speciation (",lambda,
    ")")),expression(paste("extinction (",mu,")"))),
    cex=0.8,bty="n")
mtext("(a)",line=0.5,adj=0)
## extract fitted model coefficients
lambda.l<-bdvar_mle$par["lambda.l"]
lambda.a<-bdvar_mle$par["lambda.a"]
mu.l<-dvar_mle$par["mu.l"]
mu.a<-dvar_mle$par["mu.a"]
b<-lambda.l*exp(-lambda.a*t)
d<-mu.l*exp(-mu.a*t)
## plot time-varying speciation & extinction model
plot(t,b,type="l",col="black",lwd=3,ylim=c(0,0.15),
    bty="n",xlab="time",las=1,cex.axis=0.8,
    ylab=expression(paste("rate (",lambda," or ",mu,
    ")")))
lines(t,d,col="lightgray",lwd=3)
mtext("(b)",line=0.5,adj=0)
```

Just as we did for the variable-speciation model, we could also run a Bayesian MCMC for each of these two latter models, as well as visualize the posterior distributions of the model parameters.[24]

The only thing that remains unresolved is which model fits our data best. Each of the four models is a special case of the most general, variable-speciation and extinction model, and likewise, each of the models has the constant-rate model as a special case. As such, we could compare many of the models against each other using likelihood ratio tests.

To compare all four models at once, however, the most practical solution is to use information theory: in our case, the Akaike information criterion (AIC) that we've seen in prior chapters.

---

[24]We're not going to do it here—because it would take a long time and yield a highly similar result to what we already see from the maximum likelihood analysis.

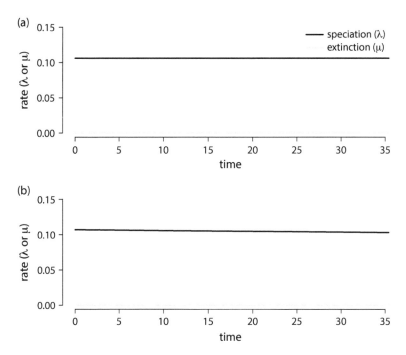

Figure 10.6
(a) Fitted variable-extinction model. (b) Fitted variable-speciation and extinction model.

```
aic<-AIC(bd_mle,bvar_mle,dvar_mle,bdvar_mle)
rownames(aic)<-c("constant-rate","variable-speciation",
    "variable-extinction","variable speciation & extinction")
print(aic)

##                                    df    AIC
## constant-rate                       2 -42.86
## variable-speciation                 3 -40.87
## variable-extinction                 3 -40.86
## variable speciation & extinction    4 -38.87
```

The constant-rate birth-death model has the smallest (most negative) AIC score and is thus the preferred model. There is little support for the other, more complex models—all of which have more parameters than constant-rate birth-death.

To understand how all of the AIC scores are so similar, we can inspect the inferred model (figures 10.4, 10.6).

In all cases, you'll see that, although the reconstructed rates do vary, they don't vary by that much—in other words, the overall model is not all that different from the constant-rate model.

Likewise, in all cases, the likelihoods are similar.

We can infer from these data that our best-supported model has speciation and extinction rates that are constant through time.

Many verbal and mathematical models have suggested that speciation and/or extinction rates should depend not necessarily on time but perhaps on the number of lineages present at any time—that is, that speciation, extinction, or both should be *diversity dependent* (Rabosky and Hurlbert 2015; but see Harmon and Harrison 2015).

Models in which speciation or extinction change through time behave a bit like diversity-dependent models—and, in fact, those models are often used as proxies to detect the phenomenon of diversity dependence. Technically, however, all of the models we've seen so far in the chapter have rates that depend on time, not diversity, and will approximate true diversity-dependent diversification only to the extent that time and diversity are correlated.

Let's figure out how to test diversity dependence more directly (Etienne et al. 2012, 2016; Etienne and Haegeman 2012).

To do this, we'll use the package *DDD* (Etienne and Haegeman 2021). Although we're not going to explore this extensively, *DDD* can fit time-varying models that are nearly identical to the ones we fit using *diversitree*, above.

Here we focus on using *DDD* to fit models of diversity dependence of speciation and extinction rates.

We should start by loading the *DDD* package.[25]

```
library(DDD)
```

The next thing we'll do is compute the *branching times* from our "phylo" object. The branching times are merely the times between branching events on our reconstructed tree.[26]

```
cbt<-branching.times(whale.tree)
cbt

##      n1      n2      n3      n4      n5      n6      n7
## 35.425  27.834   8.577   1.279   0.188  25.843  18.095
##      n8      n9     n10     n11     n12     n13     n14
## 16.169  15.214   5.361  10.603  12.964  11.490   5.343
##     n15     n16     n17     n18     n19     n20     n21
##   4.381  33.416  22.167   8.853  32.377   0.335  31.715
##     n22     n23     n24     n25     n26     n27     n28
## 19.015  17.858   6.343  15.550  14.393  11.006   8.060
##     n29     n30     n31     n32     n33     n34     n35
## 12.916  11.058   8.942   8.235   7.719   7.202   4.775
##     n36     n37     n38     n39     n40     n41     n42
```

---

[25]Which, if you have not done so already, you should install from CRAN now in the typical way.

[26]Almost all diversification methods, including the ones we used early in this chapter and in chapter 9, compute the likelihood from these times—not the phylogeny itself. We supply the phylogeny as input to functions such as fit.bd in *phytools* or birthdeath in *ape* only because this calculation is done internally to the function!

```
## 4.197   6.365   5.786   4.893   5.079   4.116  25.727
##    n43     n44     n45     n46     n47     n48     n49
## 24.334  18.252  17.586  13.696   5.126   5.308   4.608
##    n50     n51     n52     n53     n54     n55     n56
## 3.710   4.664   3.381  10.535   9.373   8.171   5.993
##    n57     n58     n59     n60     n61     n62     n63
## 5.414   4.387   2.977   1.439   8.654   8.066   6.867
##    n64     n65     n66     n67     n68     n69     n70
## 5.198   1.300   4.489   3.213   1.739   3.716   2.818
##    n71     n72     n73     n74     n75     n76     n77
## 1.983   1.469   7.408   5.815   3.061   4.266   3.311
##    n78     n79     n80     n81     n82     n83     n84
## 3.492   2.921   2.087   2.719   1.820   1.368   0.859
##    n85     n86
## 1.109   0.787
```

We can see these are the branching times because if we plot our tree and the branching times (as vertical dotted lines) together,[27] we'll see that each branching time corresponds exactly to a node on our plotted tree (figure 10.7).

```
## use plotTree with plot=FALSE to obtain plotting
## parameters
plotTree(whale.tree,ftype="off",direction="leftwards",
    xlim=c(max(nodeHeights(whale.tree)),0),lwd=1,
    plot=FALSE,mar=c(5.1,1.1,0.1,1.1),
    ylim=c(0,Ntip(whale.tree)))
## draw a set of lines onto our graph
nulo<-sapply(cbt,function(x,N) lines(rep(x,2),y=c(1,N),
    col="lightblue",lty="dotted"),N=Ntip(whale.tree))
## add a temporal axis
axis(1,cex.lab=0.8)
title(xlab="time before present (ma)")
## plot our tree
plotTree(whale.tree,ftype="off",direction="leftwards",
    xlim=c(max(nodeHeights(whale.tree)),0),add=TRUE,
    lwd=1,mar=c(4.1,1.1,0.1,1.1),
    ylim=c(0,Ntip(whale.tree)))
## add points at branching events
points(get("last_plot.phylo",
    envir=.PlotPhyloEnv)$xx[1:whale.tree$Nnode+
    Ntip(whale.tree)],get("last_plot.phylo",
    envir=.PlotPhyloEnv)$yy[1:whale.tree$Nnode+
    Ntip(whale.tree)],pch=21,bg="lightblue")
```

---

[27] We'll use the plot=FALSE trick from earlier in the chapter here again—but this time so we can plot the vertical lines first and then our phylogeny on top.

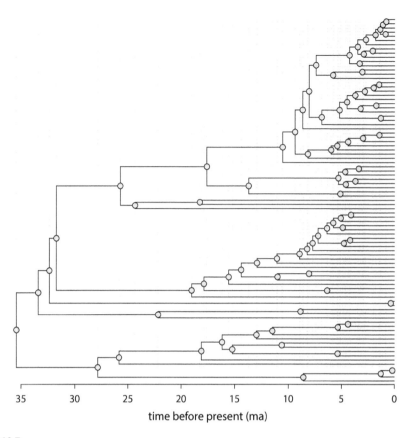

Figure 10.7
Cetacean phylogeny showing the branching times that are used to fit diversification models in *DDD* and other packages.

10.4.1 Model fitting using the *DDD* package

Our next step is to use *DDD* to fit our different models. Just as we did earlier in the chapter, we'll start by fitting a constant-rate model, then we'll fit a variable birth and death rate model, and then, finally, we'll fit our density-dependent model.

To fit our first two models in *DDD*, we're going to use just one function: bd_ML. Which model we fit depends on the values of the different arguments that we supply to the function. Using its default values bd_ML fits the same constant-rate birth-death model that we obtained with fit.bd in the previous chapter.

```
bd_res_ddd<-bd_ML(brts=cbt,missnumspec=2,cond=1)

## You are optimizing lambda0 mu0
## You are fixing lambda1 mu1
## Optimizing the likelihood - this may take a while.
## The loglikelihood for the initial parameter values is -281.755.
##
## Maximum likelihood parameter estimates:
```

```
## lambda0: 0.105805, mu0: 0.000061, lambda1: 0.000000,
## mu1: 0.000000:
## Maximum loglikelihood: -276.809218
```

```
bd_res_ddd
```

```
##   lambda0       mu0 lambda1 mu1 loglik df conv
## 1   0.106 0.0000612       0   0   -277  2    0
```

In this function call, we also specified `missnumspec=2` (two missing species) and `cond=1`, the *conditioning level*,[28] of our fitted model.

Wait a second. Our parameters seem nearly identical to what we estimated using *diversitree* and *phytools* earlier in the chapter, but our likelihood is totally different. What's up?

Well, it turns out that in diversification analyses, we can compute the likelihood of the phylogeny *or* the likelihood of the branching times. *Ironically,*[29] by default, `bd_ML` computes the likelihood of the phylogeny, while *diversitree* and *phytools* use the likelihood of the branching times.

Conveniently, however, `bd_ML` has an argument (`btorph`[30]) that can be used to switch between the two different likelihood calculations.

Even though these different likelihood surfaces have the same optima, let's switch to computing the likelihood of the phylogeny so that our numbers match with what we obtained earlier in the chapter.

```
bd_res_ddd<-bd_ML(brts=cbt,missnumspec=2,cond=1,btorph=0)
```

```
## You are optimizing lambda0 mu0
## You are fixing lambda1 mu1
## Optimizing the likelihood - this may take a while.
## The loglikelihood for the initial parameter values is 18.466.
##
## Maximum likelihood parameter estimates:
## lambda0: 0.105805, mu0: 0.000061, lambda1: 0.000000,
## mu1: 0.000000:
## Maximum loglikelihood: 23.411730
```

```
bd_res_ddd
```

```
##   lambda0       mu0 lambda1 mu1 loglik df conv
## 1   0.106 0.0000612       0   0   23.4  2    0
```

OK, that's much better. Now not only do our parameter estimates equate, but our likelihoods do too!

Next, as promised, we'll fit the *DDD* model that should match our *diversitree* variable-speciation and variable-extinction model.

---

[28] `cond=1` conditions on the "stem or crown age and non-extinction of the phylogeny" (Etienne and Haegeman 2021). We don't need to worry too much about that here, but it matches what we did with *diversitree*.

[29] Ironic, in our opinion, because *DDD* takes the branching times as input, not the tree.

[30] `btorph` is short for "branching times or phylogeny," but don't worry—it took us a second to guess that too!

To do that, we'll use the same bd_ML function as before, but we'll update the argument tdmodel[31] to 1, which should correspond to exponentially changing speciation and/or extinction rates through time. We also need to supply initial values for the optimization now, which we'll do using the argument initparsopt.

```
bdvar_res_ddd<-bd_ML(brts=cbt,missnumspec=2,cond=1,
    tdmodel=1,initparsopt=c(0.1,0.1,0.1,0.1),
    idparsopt=1:4,btorph=0)

## You are optimizing lambda0 mu0 lambda1 mu1
## You are fixing nothing
## Optimizing the likelihood - this may take a while.
## The loglikelihood for the initial parameter values is -139.303.
##
## Maximum likelihood parameter estimates:
## lambda0: 0.105982, mu0: 0.000021, lambda1: 0.000000,
## mu1: 0.064131:
## Maximum loglikelihood: 23.412349

bdvar_res_ddd

##    lambda0       mu0      lambda1     mu1 loglik df
## 1    0.106 0.0000211 0.000000438 0.0641   23.4  4
##    conv
## 1    0
```

10.4.2 Fitting a diversity-dependent diversification model

Finally, we'll go ahead and fit our diversity-dependent model. To fit this model, we'll switch to using a different[32] *DDD* function called dd_ML. This function fits a diversity-dependent model by default—so all we have to do is make sure that our conditioning (cond=1) and our likelihood calculation (btorph=0) match the values for the other models that we fit in this section!

```
dd_res_ddd<-dd_ML(brts=cbt,missnumspec=2,cond=1,
    btorph=0)

## You are optimizing lambda mu K
## You are fixing nothing
## Optimizing the likelihood - this may take a while.
## The loglikelihood for the initial parameter values is 22
##
## Maximum likelihood parameter estimates:
## lambda: 0.146211, mu: 0.026237, K: 219.925708
## Maximum loglikelihood: 23.656078
```

---

[31] tdmodel is short for *time-dependence model.*
[32] Although highly similar.

```
dd_res_ddd
```

```
##    lambda     mu   K loglik df conv
## 1  0.146 0.0262 220   23.7  3    0
```

In this model, we've optimized a parameter, K, that might be best thought of as the "carrying capacity" of the diversification process (Etienne et al. 2012, 2016; Etienne and Haegeman 2012). This is the point at which the rate of speciation (λ) is expected to equal the rate of extinction (μ) under our model.

10.4.3 Comparing between alternative models in *DDD*

In our model of best fit, the MLE of K is quite high (220) compared to the observed number of species in our clade. This suggests that the density dependence is probably not an important factor influencing diversification of Cetacea and thus may not fit our data any better than the other models that we have seen.[33]

Nonetheless, we can compare among our three fitted models of this section and see which is best supported by the data.

Let's make a table.

```
lnL<-c(bd_res_ddd$loglik,bdvar_res_ddd$loglik,
    dd_res_ddd$loglik)
k<-c(bd_res_ddd$df,bdvar_res_ddd$df,
    dd_res_ddd$df)
aic<-2*k-2*lnL
data.frame(model=c("birth-death","variable-rate",
    "density-dependent"),logLik=lnL,df=k,AIC=aic)
```

```
##                 model logLik df    AIC
## 1         birth-death   23.4  2 -42.8
## 2       variable-rate   23.4  4 -38.8
## 3 density-dependent   23.7  3 -41.3
```

We see that the density-dependent diversification model is *not* a particularly good fit to our data, although it is a better fit than the variable-rate model we studied earlier. Overall, constant-rate birth-death is still the most reasonable choice.

10.5 Testing for variation in diversification rates among clades

Within the whales and their relatives, and despite the nonresult we've obtained so far, previous work has shown evidence of extreme variation in diversification rates—but *among clades*, rather than through time or as a function of standing diversity.

In fact, Morlon et al. (2011) showed that we can actually infer variation in speciation and extinction through time—but only *after* we account for variation in rates across clades. We'll explore precisely this type of variation now.

---

[33] And, taken literally, means that the world is not yet full of whales.

As this has been a popular topic of research in recent years, there is a wide range of different models for clade- and lineage-specific diversification rate variation. Many of these methods focus on locating diversification rate shifts in a tree without an a priori hypothesis about where on the phylogeny the rate shift or shifts occurred. One important example of this type of approach is the very popular software *BAMM* (Rabosky 2014), which runs outside of R and is thus beyond the scope of this book.

10.5.1 Clade-specific variation in diversification rate: An empirical example comparing whales and dolphins

For a very simple introduction to variation across clades, we'll just consider the case in which we have a pretty good idea about how the speciation rate might vary on our tree.

In the Cetacea, we could hypothesize (following Morlon et al. 2011 and others) that the parvorder Odontoceti (toothed whales—the dolphins and their relatives[34]), due to their many ecological, phenotypic, and behavioral attributes that distinguish them from the rest of Cetacea, may have diversified under a different process than their whale kin (Morlon et al. 2011).

This is a hypothesis that we can test, and we'll do it by going *back* to the *diversitree* package.

To test this hypothesis, we simply need to identify the node that represents the common ancestor of all Odontoceti so that we can pass the label[35] for this node along to the *diversitree* package.

If we refer to figure 10.2,[36] we can see that the node labeled *n42* marks the common ancestor of the Ondoceti. We'll use this as the split point for our clade-specific diversification analysis.

As we've seen in *diversitree* before, once again we create a likelihood function (this time with make.bd.split), and then we proceed to optimize that function using find.mle. We need to provide four different initial values for the optimization for the speciation and extinction rates both *within* Odontoceti and in the rest of the tree.

Let's proceed.

```
## make birth-death split model
lik.split<-make.bd.split(whale.tree,nodes="n42",
    sampling.f=Ntip(whale.tree)/89)
## optimize model using find.mle
split_mle<-find.mle(lik.split,x.init=c(0.1,0.01,0.1,
    0.01))
split_mle

## $par
##    lambda.1        mu.1    lambda.2        mu.2
## 0.08931157 0.00512242 0.11806054 0.00000208
##
## $lnLik
## [1] 26.4
```

---

[34] A handful of species, such as killer whales, which are commonly referred to as whales but are phylogenetically dolphins.

[35] We labeled our nodes when we made figure 10.2 earlier in the chapter!

[36] And recall our cetacean taxonomy.

```
##
## $counts
## [1] 19
##
## $code
## [1] 2
##
## $gradient
## [1]  46.6    3.4 -88.8  41.7
##
## $method
## [1] "nlm"
##
## $func.class
## [1] "bd.split" "bd"        "dtlik"     "function"
##
## attr(,"func")
## Constant rate birth-death (split tree) likelihood function:
##    * Parameter vector takes 4 elements:
##        - lambda.1, mu.1, lambda.2, mu.2
##    * Function takes arguments (with defaults)
##        - pars: Parameter vector
##        - condition.surv [TRUE]: Condition
##        -                 likelihood on survival?
##        - intermediates [FALSE]: Also return
##        -                 intermediate values?
##    * Phylogeny with 87 tips and 86 nodes
##        - Taxa: Balaena_mysticetus, ...
##    * Reference:
##        - Nee et al. (1994) doi:10.1098/rstb.1994.0068
## R definition:
## function (pars, condition.surv = TRUE, intermediates = FALSE)
## attr(,"class")
## [1] "fit.mle.bd.split" "fit.mle"
```

When we inspect our fitted model, we see that it indicates a *higher* rate of speciation, λ, combined with a *lower* rate of extinction, μ, within Odontoceti when compared to the rest of the cetacean phylogeny. Neat!

Last of all, we can compare this model to our various others[37] using AIC.

```
aic<-AIC(bd_mle,bvar_mle,dvar_mle,bdvar_mle,split_mle)
rownames(aic)<-c("constant-rate","variable-speciation",
    "variable=extinction","variable speciation & extinction",
    "clade-specific variable rate")
print(aic)
```

---

[37] Here we compare only the models that we fit using *diversitree*, but we also could have included the diversity-dependent model from *DDD*.

```
##                                    df   AIC
## constant-rate                       2  -42.9
## variable-speciation                 3  -40.9
## variable=extinction                 3  -40.9
## variable speciation & extinction    4  -38.9
## clade-specific variable rate        4  -44.8
```

In this case, we *do* find that our rate-variable method is a better fit to our data than the other models that we tested! This supports our hypothesis that the dolphins and their relatives have diversified more rapidly than other whales.

The models and methods in this chapter can be combined in various ways. For instance, we might be interested in time-varying rates that vary across clades or density dependence that differs across the tree. Hopefully what we've covered in this chapter is sufficient to give you a foothold into studying these more complicated evolutionary scenarios of variable-rate diversification.

10.6 Practice problems

10.1 Is the constant-rate model the best for the Odontoceti (a subclade of the whale tree)? Or might this clade exhibit interesting dynamics? To find out, prune out this part of the tree, and analyze diversification dynamics. In particular, compare the fit of constant-rate, variable-rate, and diversity-dependent models to this clade.

10.2 Fit all of the models that we describe in this chapter to the anoles, using the tree stored in Anolis.tre that you have encountered in previous chapters. For the clade-specific variable-rate model, you can choose any clade that you fancy.

10.3 You might wonder whether we can tell diversity dependence from time dependence and, if so, how well? Let's explore that a bit. Use the diversity-dependence simulator in *DDD* called dd_sim. Simulate a tree, and then compare the fit of a diversity-dependent model to a simple time-dependent model where speciation rate changes exponentially through time, but extinction is constant. Using AIC, do you prefer the correct model? If so, do you strongly prefer the correct model ($\Delta AIC > 4$) or are they hard to tell apart?

Character-dependent diversification

We know that some parts of the tree of life grow faster than others. In certain cases, we might have an idea of the trait that influences this change in the tempo of diversification. The evolution of flight, for instance, may open up new areas of niche space for clades, permitting rapid diversification in birds, bats, and insects (but see Ikeda et al. 2012). Or perhaps large body size makes a species more vulnerable to extinction compared to its smaller close relatives (Jablonski 2008). In the present chapter, we'll consider this sort of model in which diversification depends on character traits.

So far, we've looked at models for character evolution, both discrete and continuous, as well as models for speciation and extinction rates. Here, for the first time, we'll consider models that include both character evolution *and* diversification.

In this chapter, we'll:

1. Learn how to run the so-called *BiSSE* model (Maddison et al. 2007), which can be used to test for a binary character's effect on speciation and extinction rates.
2. Explore the *MuSSE* model (FitzJohn 2012), a multi-state version of BiSSE for discrete character traits with more than two conditions.
3. Fit some hidden-state (*HiSSE* and *MiSSE*) models (Beaulieu and O'Meara 2016), in which we suppose that diversification is character state dependent, but in which either our observed character has multiple, "hidden" rates of diversification (HiSSE)—or the trait is not observed at all (MiSSE).
4. Finally, test for the effect of a continuous character on the rate of speciation using *QuaSSE* (FitzJohn 2010).

The most basic character-dependent state model is the BiSSE model (Maddison et al. 2007),[1] which allows us to test for the effect of a binary character (a character with two states) on speciation and extinction.

Using this method, we can compare the fit of a null model in which the character evolves under an Mk model (chapter 6) and the tree grows under a constant birth-death process (chapter 9), uninfluenced by our trait, to one in which the rates of new species formation by speciation and loss via extinction depend on the value of our binary phenotypic character (Maddison et al. 2007).

Just as we have done in some previous chapters, to understand the BiSSE model a little bit better, why don't we simulate under the model and visualize our result?

To do this, we can use the package *diversitree* that we've seen in so many previous chapters already (FitzJohn 2012).

```r
library(diversitree)
```

To simulate a BiSSE tree, we'll use the function `tree.bisse`.

`tree.bisse` takes as input a parameter vector containing all the *six*[2] parameters of the BiSSE model. These parameters need to be supplied in the order λ_0, λ_1, μ_0, μ_1, $q_{0,1}$, and $q_{1,0}$. The result can be seen in figure 11.1a.

For comparison, we'll also simulate under a birth-death model of chapter 9, with a binary trait that evolved by the Mk model of chapter 6, but in which there is no dependence of the birth or death rate of our simulation on the discrete character state.

To do this, we can again use the `tree.bisse` function of *diversitree*, but in this case, we'll set $\lambda_0 = \lambda_1$, and so on. The result of this second simulation is shown in figure 11.1b.

```r
## subdivide our plot and set margins
par(mfrow=c(1,2),mar=c(0.1,1.1,2.1,1.1))
## panel a)
## set BiSSE parameter values
pars<-setNames(
    c(0.1,0.3,0.01,0.01,0.03,0.03),
    c("lambda0","lambda1","mu0,","mu1",
    "q01","q10"))
pars

## lambda0 lambda1    mu0,     mu1      q01      q10
##    0.10    0.30    0.01    0.01    0.03    0.03

## simulate BiSSE tree
phy.bisse<-tree.bisse(pars,max.taxa=80,x0=0)
```

---

[1] Short for *binary-state speciation and extinction* model.

[2] The BiSSE model has a maximum of six parameters: one rate of speciation for each binary state, one rate of extinction for each state, and two rates of change between states!

```
h<-history.from.sim.discrete(phy.bisse,0:1)
## plot simulated tree
plot(h,phy.bisse,show.tip.label=FALSE,
        cols=c("black","grey"),cex=0.7)
## add panel label
mtext("(a)",lwd=1,adj=0,font=1)
## panel b)
## set non-BiSSE simulation parameter values
pars<-setNames(
    c(0.2,0.2,0.01,0.01,0.03,0.03),
    c("lambda0","lambda1","mu0","mu1",
    "q01","q10"))
pars
```

```
## lambda0 lambda1     mu0     mu1     q01     q10
##    0.20    0.20    0.01    0.01    0.03    0.03
```

```
## simulate state-independent diversification
phy.nonbisse<-tree.bisse(pars,max.taxa=80,x0=0)
h<-history.from.sim.discrete(phy.nonbisse,0:1)
## plot results
plot(h,phy.nonbisse,show.tip.label=FALSE,
        cols=c("black","gray"),cex=0.7)
## add panel label
mtext("(b)",lwd=1,adj=0,font=1)
```

Comparing panels (a) and (b) of figure 11.1 makes the difference between our simulations quite clear.

In figure 11.1a, clades with the gray state tend to have relatively *short* waiting times between speciation events, which is what we'd expect if having the gray state increased the rate of speciation, as we simulated.

By contrast, in figure 11.1b, our state-independent simulation, there's no particular tendency of black or gray clades to have longer or shorter waiting times between speciation events.

11.2.1 Fitting the BiSSE model to data: An example from coral reef fish

Naturally, the model wouldn't be very useful if we couldn't fit it to our data.

Let's give that a try, once again using the *diversitree* package (FitzJohn 2012).

We're going to compare the fit of a state-dependent BiSSE model to a null model where the states have identical speciation and extinction rates.

To do this, we'll analyze the diversification of a group of fish species called grunts,[3] which can be classified as either reef-dwelling or not (modified from Price et al. 2013).

Our question is: does moving onto reefs affect either the speciation rate or the extinction rate (or both rates) in grunts?

---

[3] A perciform fish family, Haemulidae.

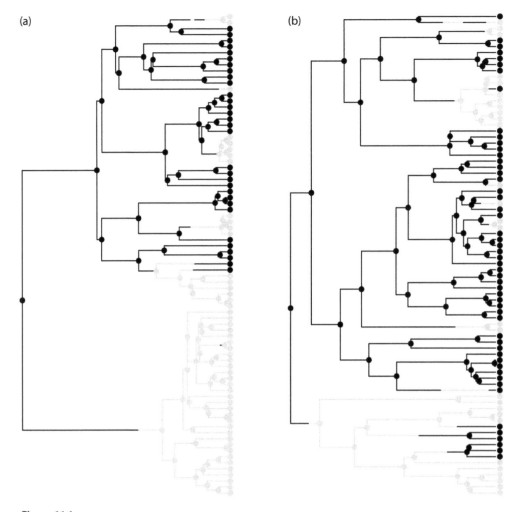

Figure 11.1
(a) Simulated tree under the BiSSE model. The gray state was simulated with a threefold higher rate of speciation than the black state. (b) Simulated tree under a birth-death model and a discrete character under the M*k* model, but in which neither the speciation nor the extinction rate is state dependent.

For this example, we'll use two files, `grunts.phy` and `grunts.csv`, that (just like all the files we use in this volume) can be obtained from the book website.[4]

We've already loaded *diversitree*. Let's also load the *geiger* and *phytools* packages, which we'll use a bit later on.

```
library(geiger)
library(phytools)
```

Now we can read in our tree and read in our data for habitat preference in grunts.

---

[4]http://www.phytools.org/Rbook/.

```
gt<-read.tree("grunts.phy")
print(gt,printlen=2)
```

```
##
## Phylogenetic tree with 50 tips and 49 internal nodes.
##
## Tip labels:
##    Haemulon_scudderi, Haemulon_scudderi_N.sp, ...
##
## Rooted; includes branch lengths.
```

```
gd<-read.csv("grunts.csv",row.names=1,
    stringsAsFactors=TRUE)
head(gd)
```

```
##                          habitat habitat.names
## Pomadasys_panamensis           0       non-reef
## Pomadasys_macracanthus         0       non-reef
## Anisotremus_moricandi          1           reef
## Anisotremus_virginicus         1           reef
## Anisotremus_caesius            1           reef
## Anisotremus_surinamensis       1           reef
##                           trait1    trait2
## Pomadasys_panamensis     -0.10973 -0.221172
## Pomadasys_macracanthus    0.12907 -0.006984
## Anisotremus_moricandi     0.28532  0.071222
## Anisotremus_virginicus    0.60648  0.186049
## Anisotremus_caesius       0.48501  0.076586
## Anisotremus_surinamensis  0.34417  0.192564
##                            trait3
## Pomadasys_panamensis      0.041617
## Pomadasys_macracanthus   -0.043540
## Anisotremus_moricandi     0.154659
## Anisotremus_virginicus    0.233115
## Anisotremus_caesius       0.083559
## Anisotremus_surinamensis  0.202466
```

diversitree requires that our discrete character be coded numerically (0, 1, 2, and so on). Luckily, the authors of this data set already took care of that for us—coding non-reef fish species as 0 and reef species as 1.[5]

Let's plot our tree and data. We could use the plotTree.datamatrix that we learned in chapter 7, but to keep things simple, we just use plotTree in *phytools* combined with the *ape* function tiplabels.

The result can be seen in figure 11.2.

---

[5]If they had not, it wouldn't be hard to recode our factor column, habitat.names, as numerical.

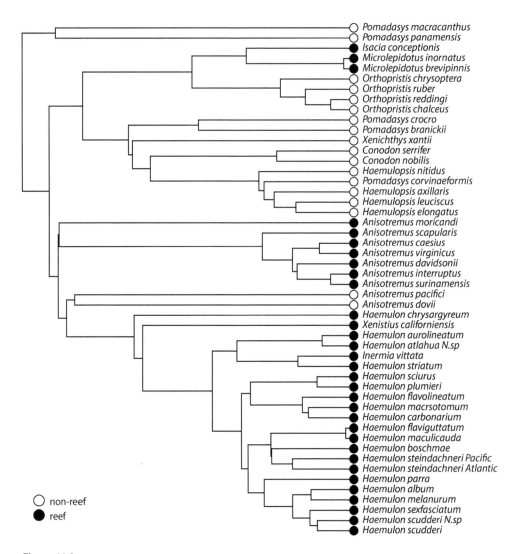

Figure 11.2
Phylogenetic tree of grunts: fish in the family Haemulidae.

```
## extract habitat data
hab<-gd[,1]
## set names
names(hab)<-rownames(gd)
## plot our tree
plotTree(gt,ftype="i",fsize=0.7,
    offset=0.5)
## add tip labels
tiplabels(pie=to.matrix(hab,0:1)[gt$tip.label,],
    piecol=c("white","black"),cex=0.4)
## create legend
```

```
legend("bottomleft",c("non-reef","reef"),
    pch=21,pt.cex=1.6,
    cex=0.8,bty="n",
    pt.bg=c("white","black"))
```

In this code chunk, we first extracted our binary discrete character from the data frame we created previously, then we plotted our tree using plotTree, and finally we added tip labels using tiplabels of *ape*.[6]

At this point, we should be ready to set up and fit our BiSSE model.

Much as we saw in chapters 9 and 10, to fit our model using *diversitree*, we need to first make the likelihood function (using a custom function) and then optimize it (using find.mle).

One difference here is that we'll use the helper function starting.point.bisse to generate reasonable starting values for our optimization.

```
## make BiSSE likelihood function
bisse.model<-make.bisse(gt,hab)
## find reasonable parameter values for
## optimization
p<-starting.point.bisse(gt)
p

## lambda0 lambda1    mu0     mu1    q01    q10
##    48.0    48.0     0.0     0.0    9.6    9.6
```

The particular values generated by starting.point.bisse for λ and μ are just the maximum likelihood estimations (MLEs) for λ and μ that we'd get using the method of Nee, Holmes, et al. (1994) that we learned in chapter 9.

To see that this is true, we can compare our starting point values to the result from fit.bd in *phytools*.

```
fit.bd(gt)

##
## Fitted birth-death model:
##
## ML(b/lambda) = 48
## ML(d/mu)  = 0
## log(L)  = 282.3834
##
## Assumed sampling fraction (rho) = 1
##
## R thinks it has converged.
```

With our BiSSE likelihood function (bisse.model) and our starting point (p), we're now ready to fit our model as follows:

---

[6]The *phytools* function to.matrix converts our discrete character into a matrix of 0s and 1s that we can pass as input to the tiplabels argument pie. We used the same function in chapter 8.

```
## optimize BiSSE model
bisse.mle<-find.mle(bisse.model,p)
bisse.mle
```

```
## $par
##     lambda0    lambda1         mu0         mu1
## 3.6182e+01 6.3121e+01 8.3949e-09 2.2058e-06
##         q01         q10
## 7.6767e+00 9.2447e-07
##
## $lnLik
## [1] 127.98
##
## $counts
## [1] 959
##
## $convergence
## [1] 0
##
## $message
## [1] "success! tolerance satisfied"
##
## $hessian
## NULL
##
## $method
## [1] "subplex"
##
## $func.class
## [1] "bisse"     "dtlik"     "function"
##
## attr(,"func")
## BiSSE likelihood function:
##   * Parameter vector takes 6 elements:
##       - lambda0, lambda1, mu0, mu1, q01, q10
##   * Function takes arguments (with defaults)
##       - pars: Parameter vector
##       - condition.surv [TRUE]: Condition
##                   likelihood on survival?
##       - root [ROOT.OBS]: Type of root
##                   treatment
##       - root.p [NULL]: Vector of root state
##                   probabilities
##       - intermediates [FALSE]: Also return
##                   intermediate values?
##   * Phylogeny with 50 tips and 49 nodes
##       - Taxa: Haemulon_scudderi, ...
##   * References:
```

```
##        - Maddison et al. (2007) doi:10.1080/10635150701607033
##        - FitzJohn et al. (2009) doi:10.1093/sysbio/syp067
## R definition:
## function (pars, condition.surv = TRUE, root = ROOT.OBS,
##     root.p = NULL, intermediates = FALSE)
## attr(,"class")
## [1] "fit.mle.bisse" "fit.mle"
```

In our fitted model, the MLE of λ_1[7] is nearly twice as high as the speciation rate of non-reef fish, λ_0.

To test whether this model significantly better explains our data compared to a null model in which the speciation and extinction rates are constant through the tree, we'll need to fit this model too.

We do that by first making a *constrained*[8] likelihood function and then optimizing it the same way we optimized our BiSSE model. We use the syntax lambda1~lambda0 (and so on) to indicate which parameters should be forced to be equal in our constrained model.

Let's do it.

```
## create constrained null model
bissenull.model<-constrain(bisse.model,
    lambda1~lambda0,mu1~mu0)
## optimize null model
bissenull.mle<-find.mle(bissenull.model,
    p[c(-2,-4)])
```

Instead of printing out the *whole* fitted model object, why don't we just print the fitted model coefficients and likelihood. To do that, we'll use the generic methods coef and logLik.

```
coef(bissenull.mle)

##     lambda0        mu0        q01        q10
## 4.8000e+01 1.7735e-07 7.7556e+00 5.6045e-08

logLik(bissenull.mle)

## 'log Lik.' 126.25 (df=4)
```

The astute reader may notice that the estimated speciation and extinction rates of this model match our birth-death model from fit.bd, but the likelihood does not. This is because our likelihood in the constrained model also includes the probability of the binary character data!

diversitree makes it quite easy to compare models. In fact, the anova method for the *diversitree* object classes we've been working with gives us a likelihood ratio test, in which we use the argument null to specify which of our two models corresponds to our H_0 in the test.

In our case, this should be bissenull.mle. Let's try it.

---

[7]Remember, this is the speciation rate of grunts that live in coral reefs.

[8]We call it *constrained* because we're forcing certain parameters of the model to be equal to each other.

```
## run likelihood-ratio test
bisseAnova<-anova(bisse.mle,
    null=bissenull.mle)
bisseAnova
```

```
##      Df lnLik  AIC ChiSq Pr(>|Chi|)
## full  6   128 -244
## null  4   126 -244  3.47       0.18
```

In addition, we can also pull out the Akaike information criterion (AIC) values from this table and compute Akaike weights—a measure of the weight of evidence in favor of each of the two models.

```
aicw(setNames(bisseAnova$AIC,
    rownames(bisseAnova)))
```

```
##          fit    delta       w
## full -243.96  0.53342 0.43371
## null -244.49  0.00000 0.56629
```

Our coefficients suggest that both speciation (λ) and extinction (μ) rates are *higher* in reef than in non-reef grunts (lambda1 and mu1 are both larger than lambda0 and mu0, respectively). Nonetheless, the likelihood for the full BiSSE model is not sufficiently improved compared to the null, and so the *P*-value for our likelihood ratio test is nonsignificant.

Likewise, the AIC weight of the BiSSE model is smaller than the null (although not by very much), indicating relatively little weight of evidence in support of the state-dependent model.

Overall, we find weak support for a state-dependent diversification in these data.

One thing that's worth noting here is that we did not place any constraints on the character transitions in either model in the previous example. Because of this, both the null and the BiSSE model are asymmetric models in which $q_{0,1}$ and $q_{1,0}$ can differ.

diversitree gives us a tremendous amount of flexibility in this regard. For instance, it's equally straightforward to fit models in which extinction rates (μ) are forced to be the same while λ differs, λ is forced to be the same but μ differs, or any combination of these three different constraints!

11.2.2 Analyzing a BiSSE model using Bayesian MCMC

As we mentioned in the previous chapter, *diversitree* makes it very easy to decide to be Bayesian instead of frequentist whenever it suits us.

Let's reanalyze our BiSSE model from the previous section, but this time using Bayesian MCMC.

To do this, we'll use the same model object we created before; however, we'll need to set *prior probability distributions* for each of our model parameters.

For simplicity,[9] we'll use the same prior for all model parameters.

---

[9] And no other reason, really. In practice, it makes sense to use Bayesian inference if we have a good reason to incorporate prior information, via the prior probability distributions, about our model. In this case, we don't have such information, but we'll use Bayesian inference anyway.

```
prior<-make.prior.exponential(1/(2*0.4))
prior
```

```
## function (pars)
## sum(dexp(pars, r, log = TRUE))
## <bytecode: 0x000000003b2751e8>
## <environment: 0x000000003b2755d8>
```

When we print it out, we can see that our prior is an R probability density function. That makes sense.

Now we're just about ready to start our MCMC. Here we're going to cheat a bit and initiate our MCMC at the MLE for the model parameters,[10] which cuts down on the time needed for the chain to converge to the posterior distribution.[11]

```
## run Bayesian MCMC
bisse.mcmc<-mcmc(bisse.model,bisse.mle$par,
    nsteps=1000,prior=prior,w=0.1,
    print.every=100)
```

```
## 100: {10.906, 13.894, 0.160, 0.306, 3.617, 0.697} -> 60.69
## 200: {5.292, 16.962, 2.367, 0.237, 0.236, 1.606} -> 57.36
## 300: {9.487, 17.981, 0.158, 0.014, 0.072, 3.868} -> 62.67
## 400: {13.348, 13.907, 0.253, 0.349, 2.482, 0.714} -> 60.63
## 500: {8.381, 21.621, 1.166, 0.359, 0.258, 2.346} -> 60.45
## 600: {8.248, 20.313, 2.158, 1.078, 0.329, 5.228} -> 57.12
## 700: {4.226, 20.329, 0.847, 2.185, 1.010, 3.361} -> 55.69
## 800: {9.947, 17.157, 0.443, 0.655, 3.686, 1.952} -> 58.00
## 900: {10.240, 26.414, 0.167, 1.729, 0.307, 4.685} -> 57.12
## 1000: {12.583, 14.347, 0.670, 0.517, 1.813, 0.168} -> 60.87
```

This analysis produces a posterior sample for each of the six parameters in our model.

Since we're mostly interested in the speciation and extinction rates, we can use the profiles.plot function to visualize the posterior distribution of these two different parameters in our model (figure 11.3).

```
## subdivide plot and set margins
par(mfrow=c(1,2),mar=c(5.1,4.1,3.1,2.1))
## set colors for plotting
col<-setNames(c("blue", "red"),
    c("non-reef","reef"))
## create graph of posterior sample for lambda
```

---

[10] We did this in chapter 10 too!

[11] For computational reasons, we're only running the MCMC for a meager 1,000 generations. This is probably not enough to get a good sample from the posterior distribution. Readers replicating this approach with their own data should likely set nsteps to be 10, 100, or even 1,000 times higher than this value, depending on the complexity of their model!

```
profiles.plot(bisse.mcmc[,c("lambda0","lambda1")],
    col.line=col,las=1,bty="n",
    xlab=expression(lambda),cex.axis=0.7)
## add legend & panel label
legend("topright",names(col),pch=15,col=col,
    pt.cex=1.5,bty="n",cex=0.7)
mtext("a)",line=0.5,adj=0)
## create graph of posterior sample for mu
profiles.plot(bisse.mcmc[,c("mu0","mu1")],
    col.line=col,,las=1,bty="n",
    xlab=expression(mu),cex.axis=0.7)
## add legend & panel label
legend("topright",names(col),pch=15,col=col,
    pt.cex=1.5,bty="n",cex=0.7)
mtext("b)",line=0.5,adj=0)
```

One way to measure support for a difference in diversification rate between the two states is by calculating the proportion of our sampled values in which $\lambda_1 > \lambda_0$ (or, likewise, the fraction in which $\mu_1 > \mu_0$).

This quantity should give us the *posterior probability* that $\lambda_1 > \lambda_0$ and thus is another measure of evidence in support of the hypothesis that the speciation rate is higher in Haemulidae found on reefs than in their non-reef cousins.

Let's compute these two values as follows:

```
## lambda
sum(bisse.mcmc$lambda1>bisse.mcmc$lambda0)/
    length(bisse.mcmc$lambda1)

## [1] 0.965

## mu
sum(bisse.mcmc$mu1>bisse.mcmc$mu0)/
    length(bisse.mcmc$mu1)

## [1] 0.515
```

This simple calculation tells us that the posterior probability that the diversification rate is higher in reef than non-reef fishes is high—quite close to 1.0.[12]

This result seems to slightly disagree with our maximum likelihood (ML) analysis, in that one could conclude from it that there is a high posterior probability that habitat state 1 (reef) has a *higher* diversification rate than state 0 (non-reef).

Really, though, we should consider either $\lambda_1 > \lambda_0$ *or* $\lambda_0 > \lambda_1$ as interesting, while our calculation only considered one of these two possibilities, making it effectively analogous to a one-tailed P-value.

---

[12] The posterior probability that μ_1 is greater than μ_0, on the other hand, is little better than flipping a coin!

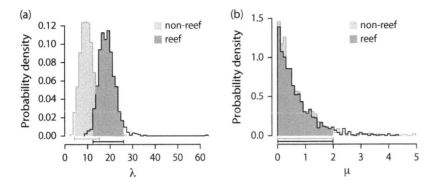

Figure 11.3
(a) Posterior distribution for the reef and non-reef speciation rates in Haemulidae fishes. (b) Posterior distribution for the extinction rates.

If we'd only hypothesized that reefs increase grunt speciation rates, we might be convinced! As it is, we can again consider this as quite weak evidence for an effect of character state on diversification rates, and there is not much reason to prefer the BiSSE model over the simpler null hypothesis.

11.3 Multi-state speciation and extinction (MuSSE) model

BiSSE was the first method that allowed us to test a hypothesis of state-dependent diversification, but since then, a number of other useful models have also been developed.

The simplest of all of these is called *MuSSE*, short for multi-state speciation and extinction (FitzJohn 2012).

This model is suitable for cases in which our discrete character has more than two states, and we're interested in the effect of these different character values on speciation, extinction, or both.

The basic idea underlying the MuSSE model is the same as BiSSE (Maddison et al. 2007); however, for multi-state character data, there can be *a lot* of parameters to estimate, as we'll be adding up to two diversification parameters (λ and μ) for each level of the trait, not to mention the trait transition rates between states!

11.3.1 An empirical example of the MuSSE method using terapontid fishes

Let's try out the MuSSE method using a new data set for Terapontidae, a family of Indo-Pacific fishes, also known as "grunters,"[13] that can be carnivorous (state 1 in our data set), omnivorous (state 2), or herbivorous (state 3) (Davis et al. 2016).

Our question is thus: does this dietary state affect speciation or extinction rates of the grunters?

For this section of the chapter, you will need two files (`terapontidae.phy` and `terapontidae.csv`), both of which can be downloaded from the book website.

We can start by reading in our data from file.

---

[13] Really, ichthyologists?

```
## read tree
tt<-read.tree("terapontidae.phy")
print(tt,printlen=2)

##
## Phylogenetic tree with 38 tips and 37 internal nodes.
##
## Tip labels:
##     Terapon_jarbua, Pelsartia_humeralis, ...
##
## Rooted; includes branch lengths.

## read data
td<-read.csv("terapontidae.csv",row.names=1)
head(td)

##                            Diet    Animal     Plant
## Amniataba_affinis            2   0.28185  -0.28185
## Amniataba_caudavittatus      1   2.58669  -2.58669
## Amniataba_percoides          2  -0.23305   0.23305
## Bidyanus_bidyanus            3  -2.19722   2.19722
## Bidyanus_welchi              2   0.84730  -0.84730
## Hannia_greenwayi_1           2   1.19705  -1.19705
##                                  PC1        PC2
## Amniataba_affinis             0.29135  -0.215270
## Amniataba_caudavittatus      -0.63360   0.154685
## Amniataba_percoides           0.19121   0.079852
## Bidyanus_bidyanus             0.48236  -0.783466
## Bidyanus_welchi               0.28914  -0.778327
## Hannia_greenwayi_1           -0.18609  -0.299196

## extract diet as a new vector
diet<-td[,1]
names(diet)<-rownames(td)
```

Let's plot our character data on the tree like we did in figure 11.2. The result can be seen in figure 11.4.

```
## plot phylogeny
plotTree(tt,ftype="i",lwd=1,fsize=0.7,offset=0.5)
## compute binary matrix for trait
pp<-to.matrix(diet,1:3)[tt$tip.label,]
## rename columns to match our traits
colnames(pp)<-c("carnivory","omnivory","herbivory")
## add tip labels
tiplabels(pie=pp,piecol=c("black","gray","white"),
    cex=0.5)
## create legend
```

```
legend("topleft",colnames(pp),pch=21,
    pt.bg=c("black","gray","white"),pt.cex=1.6,
    bty="n",cex=0.9)
```

Our next step, just as we did for BiSSE, is to build our likelihood function.

Here, however, a new wrinkle appears. Much as we did for birth-death models,[14] we have to specify an estimate of the fraction of the total species in our clade that are represented in our input phylogeny.

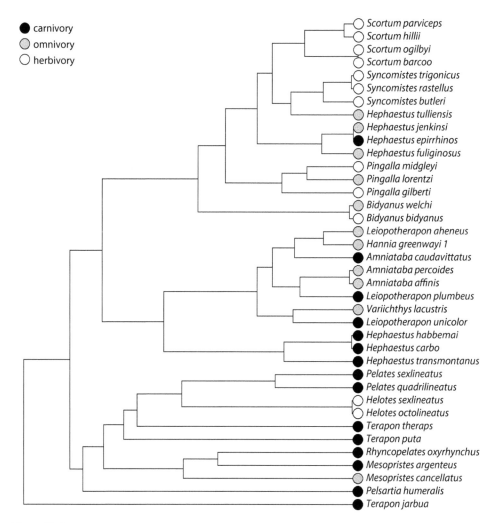

Figure 11.4
Phylogenetic tree of Tetrapontidae with a three-state dietary character mapped onto the tips.

---

[14] And the BiSSE model, although we didn't have to worry about it in the grunt example as our tree was fully sampled.

This can be hard enough under normal circumstances, especially for groups whose diversity is poorly known; however, with state-dependent speciation and extinction (SSE) models, it's even worse because in that case, we have to estimate separate sampling fractions for each level of our discrete trait!

Intuitively, this should make sense.

If our reconstructed phylogeny contained (for instance) relatively few carnivorous fish species, this could be because carnivory was associated with a lower speciation rate, a higher extinction, or both ... *or* because our sample of species happened (by random chance) to contain few carnivorous fishes relative to their representation in the fully sampled tree.

In the case of tetrapontid fishes, we think that our reconstructed tree contains 58 percent of carnivorous species, 78 percent of omnivorous species, and 85 percent of the herbivores. Note that these are not the fractions of terminal taxa in our tree that have each state.[15] These quantities are the fractions of fish species in each state from the *full* phylogeny that we included in our particular, empirical tree!

```
rho<-setNames(c(0.58,0.78,0.85),
    1:3)
rho

##    1    2    3
## 0.58 0.78 0.85
```

Now we can make our likelihood function, which we'll do as follows using the function make.musse.

```
general.musse.model<-make.musse(tree=tt,
    states=diet,k=3,sampling.f=rho)
```

Next, before we optimize our model, we again might like to pick some reasonable starting values for our optimization. We'll do this using the *diversitree* helper function starting.point.musse.[16]

```
p<-starting.point.musse(tt,k=3)
p

##  lambda1  lambda2  lambda3      mu1      mu2
## 0.087259 0.087259 0.087259 0.010139 0.010139
##      mu3      q12      q13      q21      q23
## 0.010139 0.015424 0.015424 0.015424 0.015424
##      q31      q32
## 0.015424 0.015424
```

---

[15]Which is why our percentages don't add to 100!

[16]Technically, and just as before, we could have just picked random starting parameter values—but as our models get more and more complicated, it becomes increasingly useful to choose values that are at least on the same order of magnitude as our ML solution!

Woah! We can already see that our model has a lot of parameters: twelve in total. This is perhaps too many to be reasonably estimated from our relatively modest-sized tree of figure 11.4.

As such, what we're going to do is first try to *simplify* our model in a few different biologically sensible ways and then proceed to generate some different models that focus in on our questions of scientific interest.[17]

There is a bit of an art to this.

One piece of advice that we'd give is that it's best if the decisions on which models to include are based on biology, rather than trolling through an endless space of models to find the ones that fit best.[18]

The first simplification that we propose is to constrain our character evolution model to be *ordered*.

We should remember this model from chapter 6: it just means that only some types of transitions between character states are permitted, while others are not. In our tetrapontid data, we'll suppose that only transitions from *carnivory* → *omnivory* (and back again) and only transitions from *omnivory* → *herbivory* (and back again) are allowed, while other types of evolutionary changes are excluded from occurring.

To create this model likelihood function, we'll use the *diversitree* `constrain` function that we learned about in the previous section. Here we use the syntax q13~0 to indicate the types of changes that are disallowed.[19]

```
ordered.musse.model<-constrain(general.musse.model,
    q13~0,q31~0)
```

This constrained model now has a total of ten parameters,[20] which we can list using the function `argnames` as follows.

```
argnames(ordered.musse.model)

##   [1] "lambda1"  "lambda2"  "lambda3"  "mu1"
##   [5] "mu2"      "mu3"      "q12"      "q21"
##   [9] "q23"      "q32"
```

Our new model is a bit simpler, although it still has a lot of parameters. We'll fit it to our data but also consider further simplification.

Since we're mostly interested in the pattern of species diversification, we can *also* try an even less parameterized trait evolution model in which we set all transition rates[21] equal to each other.

We can actually do this by *constraining* our constrained model! Isn't that convenient? We just run the `constrain` function again and use the syntax q23~q12 to set any parameter value equal to any other, as we wish.

---

[17] In fact, this is what we'd recommend when working with complicated statistical methods generally!

[18] Likewise, *trolling through an endless space of models* has a name—it's called data-dredging, and it will often have the effect of increasing our experiment-wise type I error rate if we don't take careful measures to correct for the number of different analyses we've undertaken in our study!

[19] We're fixing their rates at zero—which is the same thing.

[20] Still a lot!

[21] But still sticking with the ordered scenario.

```
orderedER.musse.model<-constrain(ordered.musse.model,
    q23~q12,q32~q12,q21~q12)
```

This new model now has a total of seven parameters: three speciation rates, three extinction rates, and just one character transition rate.

```
argnames(orderedER.musse.model)

## [1] "lambda1" "lambda2" "lambda3" "mu1"
## [5] "mu2"     "mu3"     "q12"
```

We're going to want to compare each of these models to ones in which the speciation and extinction rates *do not* vary as a function of the character trait.

To do that, we just need to further constrain our two constrained models by setting speciation (lambda3~lambda1 and so on) and extinction (mu3~mu1 and so on) equal one to the other.

```
## create null models
ordered.musse.null<-constrain(ordered.musse.model,
    lambda3~lambda1,lambda2~lambda1,
    mu3~mu1,mu2~mu1)
orderedER.musse.null<-constrain(orderedER.musse.model,
    lambda3~lambda1,lambda2~lambda1,
    mu3~mu1,mu2~mu1)
```

The ordered.musse.null model should now have six parameters,[22] while the orderedER.musse.null model should have only three.[23] We can verify this using argnames.

```
argnames(ordered.musse.null)

## [1] "lambda1" "mu1"     "q12"     "q21"
## [5] "q23"     "q32"
```

```
argnames(orderedER.musse.null)

## [1] "lambda1" "mu1"     "q12"
```

We've built four models, but we haven't yet fit them to our data and tree. Let's do that now using find.mle.

We're going to have to modify our starting.point.musse vector, p, each time to match the specific parameterization of our model. The cool thing is, we can just do p[argnames(model)] to pull out the right arguments for the model called model.

Let's see.

---

[22] A single speciation rate (λ), an extinction rate (μ), and four character transition rates.

[23] A speciation rate, an extinction rate, and a single character transition rate.

```
## fit ordered MuSSE model
ordered.musse.mle<-find.mle(ordered.musse.model,
    x.init=p[argnames(ordered.musse.model)])
## fit ordered ER MuSSE model
orderedER.musse.mle<-find.mle(orderedER.musse.model,
    x.init=p[argnames(orderedER.musse.model)])
## fit ordered null model
ordered.musse.null.mle<-find.mle(ordered.musse.null,
    x.init=p[argnames(ordered.musse.null)])
## fit ordered ER null model
orderedER.musse.null.mle<-find.mle(orderedER.musse.null,
    x.init=p[argnames(orderedER.musse.null)])
```

We won't look at the results of our fitted models just yet. Instead, let's start by compiling them into a single table to see which model is best supported by our data.

```
musseAnova<-anova(orderedER.musse.null.mle,
    ER.null=ordered.musse.null.mle,
    ER.MuSSE=orderedER.musse.mle,
    MuSSE=ordered.musse.mle)
musseAnova

##           Df lnLik AIC ChiSq Pr(>|Chi|)
## minimal    3 -159 325
## ER.null    6 -157 326  4.98      0.173
## ER.MuSSE   7 -155 325  8.18      0.085 .
## MuSSE     10 -152 324 14.81      0.038 *
## ---
## Signif. codes:
## 0 '***' 0.001 '**' 0.01 '*' 0.05 '.' 0.1 ' ' 1
```

This shows us that we *can* reject the null model (labeled `minimal` in our table) in favor of the full, non-equal-rate MuSSE model. Let's look at the model coefficients of just this fitted model: `ordered.musse.mle`. We'll do that using the generic function `coef` as follows.

```
coef(ordered.musse.mle)

##     lambda1    lambda2    lambda3        mu1
## 3.6075e-02 1.4837e-01 2.1926e-01 3.9355e-02
##         mu2        mu3        q12        q21
## 4.0606e-07 1.3025e-01 9.7172e-08 1.1173e-01
##         q23        q32
## 2.6165e-02 5.1182e-02
```

This shows us that in the best-fitting model, the speciation rate, λ, increases monotonically from carnivore $\to$ omnivore $\to$ herbivore. Likewise, the extinction rate, μ, is higher in herbivore lineages than for any other state in the tree. That suggests that herbivorous tetrapontids turn over (speciate more quickly, but also more frequently go extinct) more than lineages with other dietary regimes (Davis et al. 2016).

Although the differences in AIC among models are quite small,[24] this result nonetheless suggests that there may be significant heterogeneity in diversification rates in Tetrapontidae and that it could be linked to diet.

11.4 Hidden-state speciation and extinction (HiSSE) model

One issue with both BiSSE and MuSSE methods[25] is not the models that they consider—it's those that they don't (Rabosky and Goldberg 2015)!

In particular, the typical approach to applying SSE models,[26] illustrated above, involves first fitting a character-dependent model, then fitting a constant-rate birth-death model (that is, a character-*independent* diversification model), and then comparing the two. The constant-rate birth-death character-independent model is treated as a null hypothesis, and normal statistical analysis[27] could then proceed accordingly.

What if, however, the null hypothesis is indeed false—but not because of state-dependent diversification? In other words, what if speciation and extinction rates are *not* constant, as assumed by the null, but also *do not* vary as a function of our character, as modeled in the alternative (Rabosky and Goldberg 2015)?

Perhaps, instead, the rate of diversification or extinction varies based on some other, unmeasured characteristic or feature of our diversifying lineages.

In that case, the null hypothesis (constant-rate diversification) may fit the data quite poorly. The alternative (our SSE model) is probably also poor but could end up the best of two wrong models and, as such, would misleadingly cause us to *prefer* the character-dependent model. In this case, we'd be choosing the "least bad" of two bad models.[28]

In this section, we'll discuss a potential solution to this problem, which is called HiSSE—the hidden-state speciation and extinction model (Beaulieu and O'Meara 2016).

The HiSSE model allows us to consider the possibility that speciation and/or extinction rates vary, but not in such a way that follows the specific value of measured trait. Instead, our model posits, the diversification rate changes as a function of a hidden character trait (Beaulieu and O'Meara 2016). Now that we've added a hidden trait, we can *also* consider the possibility that speciation or extinction varies as a function of both our measured *and* our unobserved trait!

As you might imagine, with all these possibilities, things can get complicated fast!

We'll try to keep our models as simple as possible here but encourage you to more fully explore the large space of possible HiSSE models in your own research.

11.4.1 Fitting the HiSSE model to data: An empirical example using Haemulidae

Just to see how the HiSSE model works, we'll go back to our phylogeny and data for grunts[29] (Haemulidae) from earlier in the chapter (Price et al. 2013).

---

[24] Which is kind of unsurprising, given the relatively modest size of our data set.

[25] As they are typically used.

[26] BiSSE, MuSSE, and the like.

[27] Such as the oft-maligned null hypothesis testing.

[28] This problem is by no means unique to SSE models. In fact, it likewise afflicts virtually all statistical methods that involve null hypothesis testing. This is why it's always useful to keep in mind that when we reject a null hypothesis, that's all we've done! Doing so doesn't necessarily mean that our alternative hypothesis is correct.

[29] Note that these are grunts not grunters!

The HiSSE model is implemented in[30] the *hisse* R package by Beaulieu and O'Meara (2016).

Unlike the other packages and functions that we've used so far, *hisse* performs ML optimization diversification after first reparameterizing the model to be in terms of *extinction fraction*, ϵ (defined as μ/λ), and *species turnover*[31] ($\lambda + \mu$), rather than λ and μ directly.[32]

Since extinction fraction and turnover are both linear transformations of λ and μ, there should be no effect on inference.

Furthermore, it's fairly straightforward (as we'll see) to back-transform our model parameterization to λ and μ in the function output! Although there's no right or wrong parameterization of birth-death models, we'll sometimes back-transform to λ and μ, just so that our results are more comparable to what we saw with the same data set earlier in the chapter.

The first step in our HiSSE analysis is to load the package that we're going to use: *hisse*.[33]

```
library(hisse)
```

Just like the *corHMM* package[34] from chapter 7, *hisse* requires that we first compile our data into a data frame with two columns.[35]

```
## create input data frame for hisse
hd<-data.frame(Genus.species=rownames(gd),
    x=gd[,"habitat"])
head(hd)

##                  Genus.species x
## 1         Pomadasys_panamensis 0
## 2      Pomadasys_macracanthus 0
## 3        Anisotremus_moricandi 1
## 4       Anisotremus_virginicus 1
## 5         Anisotremus_caesius 1
## 6     Anisotremus_surinamensis 1
```

Even though the models are closely related, the way we set up R model fitting in *hisse* is quite different from what we saw for *diversitree* in chapters 9 and 10, as well as earlier in the current chapter.

First, we need to create a design matrix for our transition matrix between states of our discrete character, including with hidden states. This is *highly* similar to what we did with discrete character models, absent the consideration of speciation and extinction, in chapters 6 and 7. As such, we expect that the concept of the design matrix should be quite familiar by now.

---

[30] Somewhat predictably.

[31] This is called *turnover* because if speciation (λ) and extinction (μ) are both high, for a given value of net diversification, then on average, lineages won't last very long before they either go extinct (or pseudo-extinct) and are replaced (i.e., turned over) by new ones. Note that—just to make things confusing—exactly what is called "extinction fraction" by *hisse* is sometimes referred to as "turnover" elsewhere in the literature! Isn't that crazy?

[32] A discussion of why this is done is beyond the scope of this chapter; however, suffice it to say that this can make solving the model much easier!

[33] If you haven't used *hisse* before, then you'll have to install it from CRAN using *install.packages*.

[34] Which is also by Beaulieu and O'Meara (Beaulieu et al. 2020).

[35] As with *corHMM*, we could've also just formatted our input file this way and then read it into the correct format directly!

To build our design matrix, we'll use the *hisse* function `TransMatMakerHiSSE`.[36]

```
## create HiSSE design matrix
rates.hisse<-TransMatMakerHiSSE(hidden.traits=1)
rates.hisse

##        (0A) (1A) (0B) (1B)
## (0A)    NA    2    5   NA
## (1A)     1   NA   NA    5
## (0B)     5   NA   NA    4
## (1B)    NA    5    3   NA
```

If this matrix seems suspiciously familiar, it's probably because it's virtually identical to the design matrices we used for our fitted hidden-rates models obtained using the *corHMM* package in chapter 7!

To properly learn *hisse*, we're going to fit a series of models now, including some that we learned earlier in the chapter.

The first of these is the BiSSE model of Maddison et al. (2007).

This model should have the same fit as we found for the equivalent model earlier in this chapter, but if we fit it using the *hisse* machinery, it'll be easier to compare the fit and estimated parameters of the model to those of our subsequent HiSSE models.

The way we fit a BiSSE model using *hisse* is by first building a trait transition design matrix in which we set `hidden.states` to be `FALSE` and then by setting `turnover=c(1,2)` and `eps=c(1,2)` to indicate that we want to estimate separate values of turnover and extinction fraction for each of the two different *observed* values of our discrete state. This corresponds exactly to the BiSSE model. We can then go right ahead and fit the model to our data.

```
## create hisse design matrix for BiSSE model
rates.bisse<-TransMatMakerHiSSE(hidden.traits=0)
## fit BiSSE model using hisse
bisse.hmle<-hisse(gt,hd,turnover=c(1,2),
    eps=c(1,2),hidden.states=FALSE,
    trans.rate=rates.bisse)

## Initializing...
## Finished. Beginning simulated annealing...
## Finished. Refining using subplex routine...
## Finished. Summarizing results...
```

If we inspect this fitted model, we should see right away that the model *log-likelihood* matches almost exactly the result that we obtained from *diversitree* earlier in the chapter. That's a really good start.

```
bisse.hmle
```

---

[36] Although we could've made this matrix ourselves—and if we need to change it, we can.

```
##
## Fit
##              lnL             AIC             AICc
##           127.98          -243.96          -242.01
##           n.taxa n.hidden.states
##            50.00             1.00
##
## Model parameters:
##
## turnover0A turnover1A      eps0A        eps1A
## 3.6182e+01 6.3121e+01 2.0612e-09 2.0612e-09
##      q0A1A      q1A0A
## 7.6767e+00 2.0612e-09
```

We can also back-transform our estimated values of ϵ and turnover to λ and μ. Using a little bit of algebra, we write a simple custom function to do just that.

```
## custom function to back-transform turnover and
## extinction-fraction to lambda & mu
repar.bd<-function(object,k=2){
    pars<-object$solution
    tt<-pars[grep("turnover",names(pars))][1:k]
    ee<-pars[grep("eps",names(pars))][1:k]
    lambda<-tt/(1+ee)
    mu<-tt-lambda
    nn<-sapply(strsplit(names(tt),"turnover"),
        function(x) x[2])
    matrix(c(lambda,mu),k,2,dimnames=list(nn,
        c("lambda","mu")))
}
repar.bd(bisse.hmle)
```

```
##    lambda         mu
## 0A 36.182 7.4576e-08
## 1A 63.121 1.3010e-07
```

We'll learn more about writing custom R functions in chapter 13, but see if you can figure out how this function works!

The values for λ and μ that we obtain from repar.bd match up[37] to what we obtained for the same model using *diversitree* earlier in the chapter.

While we're at it, why don't we fit the "normal" BiSSE null model—in other words, constant-rate birth-death.

We do *that* by using the same trait transition design matrix as for our BiSSE model but then by setting both turnover and eps to c(1,1): in other words, one category of rate for each parameter!

---

[37] With very small differences due to numerical estimation error.

We're going to call this the CID model, which stands for character-independent diversification.

```
## fit CID model using hisse
cid.mle<-hisse(gt,hd,turnover=c(1,1),
    eps=c(1,1),hidden.states=FALSE,
    trans.rate=rates.bisse)

## Initializing...
## Finished. Beginning simulated annealing...
## Finished. Refining using subplex routine...
## Finished. Summarizing results...

cid.mle

##
## Fit
##             lnL           AIC           AICc
##          126.25       -244.49        -243.61
##           n.taxa n.hidden.states
##            50.00            1.00
##
## Model parameters:
##
## turnover0A turnover1A       eps0A        eps1A
## 4.8000e+01 4.8000e+01 2.0612e-09 2.0612e-09
##       q0A1A        q1A0A
## 7.7553e+00 2.0612e-09
```

This fitted model likewise matches closely what we obtained for our null model using *diversitree* earlier in the chapter. The parameter estimates should also match what we'd get from the fit.bd in the *phytools* that we learned in chapter 9, although the likelihood is different.[38]

Let's check using our repar.bd custom function.

```
## fit birth-death model using phytools
fit.bd(gt)

##
## Fitted birth-death model:
##
## ML(b/lambda) = 48
## ML(d/mu) = 0
## log(L) = 282.3834
##
```

---

[38]This is because the likelihood of our character-independent model here also takes into account the probability of the trait data, as we mentioned before, even though the trait doesn't affect diversification. If it didn't, the log-likelihoods of our different models would not be comparable.

```
## Assumed sampling fraction (rho) = 1
##
## R thinks it has converged.
```

```
## reparameterize CID model in terms of lambda
## and mu
repar.bd(cid.mle,1)
```

```
##      lambda           mu
## 0A       48  9.8935e-08
```

Now that we have these preliminaries out of the way, let's get to fitting some HiSSE models!

The first model that we can try to fit is one in which there is *no* influence of our measured trait on diversification but in which there *is* an influence of a hidden character. Beaulieu and O'Meara (2016) refer to this as the CID-2 model, because it is a character-independent, two-rate model.

To simplify, we'll first assume that the rate of transition between our different levels of the hidden character is equal. This is not strictly necessary but will have the effect of reducing the parameterization of our fitted model. To do this, we'll first copy our rates.hisse design matrix into a new object and then set all permissible transitions to the same rate.

```
## create CID-2 design matrix
rates.cid2<-rates.hisse
rates.cid2[!is.na(rates.cid2)]<-1
rates.cid2
```

```
##         (0A) (1A) (0B) (1B)
## (0A)     NA    1    1   NA
## (1A)      1   NA   NA    1
## (0B)      1   NA   NA    1
## (1B)     NA    1    1   NA
```

Next, we fit our CID-2, hidden-rate model by setting both turnover and eps to c(1,1,2,2).

The order of these rate category vectors has to correspond to the row order of the rates.cid2 design matrix we just created.

As such, we should be able to see that we're assigning discrete character states (0A) and (1A) to one rate category but (0B) and (1B) to another.

The way we know this is a hidden-rate-only model is because our states have been grouped solely by the hidden character (A vs. B) and not by our observed binary trait.

Does this make sense?

```
## fit CID-2 model using hisse
cid2.mle<-hisse(gt,hd,f=c(1,1),
    turnover=c(1,1,2,2),eps=c(1,1,2,2),
    hidden.states=TRUE,trans.rate=rates.cid2)
```

```
## Initializing...
## Finished. Beginning simulated annealing...
## Finished. Refining using subplex routine...
## Finished. Summarizing results...
```

```
cid2.mle
```

```
##
## Fit
##             lnL            AIC            AICc
##          126.64         -243.27         -241.91
##          n.taxa n.hidden.states
##           50.00           2.00
##
## Model parameters:
##
## turnover0A turnover1A       eps0A       eps1A
## 5.6747e+01 5.6747e+01 2.0612e-09 2.0612e-09
##       q0A1A       q1A0A       q0A0B       q1A1B
## 7.1348e+00 7.1348e+00 7.1348e+00 7.1348e+00
## turnover0B turnover1B       eps0B       eps1B
## 2.0629e-09 2.0629e-09 6.4955e-07 6.4955e-07
##       q0B1B       q1B0B       q0B0A       q1B1A
## 7.1348e+00 7.1348e+00 7.1348e+00 7.1348e+00
```

```
## reparameterize to lambda & mu
repar.bd(cid2.mle,k=4)
```

```
##        lambda        mu
## 0A 5.6747e+01 1.1696e-07
## 1A 5.6747e+01 1.1696e-07
## 0B 2.0629e-09 1.3399e-15
## 1B 2.0629e-09 1.3399e-15
```

Inspecting our fitted model, we should see that our rates of speciation (λ) and extinction (μ) differ *only* as a function of our hidden character (which assumes state A or B) and not as a function of the observed discrete state.

hisse also has some functionality to *visualize* phylogenetic rate variation under our hidden-rate model.

To do that, we'll start by using the *hisse* function MarginReconHiSSE to obtain a marginal reconstruction of the observed and hidden states on the phylogeny and then the separate plotting function plot.hisse.states to graph them on the tree.

```
## obtain marginal reconstructions under CID-2 model
cid2.recon<-MarginReconHiSSE(phy=gt,data=hd,
    f=cid2.mle$f,pars=cid2.mle$solution,
    hidden.states=2)
```

```
## Calculating marginal probabilities for 49 internal nodes...
## Finished. Calculating marginal probabilities for 50 tips...
## Done.
```

```
## create a plot of the rates on the tree
cid2.map<-plot.hisse.states(cid2.recon,
    rate.param="speciation",
    show.tip.label=TRUE,type="phylogram",
    fsize=0.6,legend.position=c(0,0.3,0,0.3))
```

plot.hisse.states of figure 11.5 is not a particularly flexible plotting method, but it does return a list of two objects to the user that consists of a reconstruction of the observed

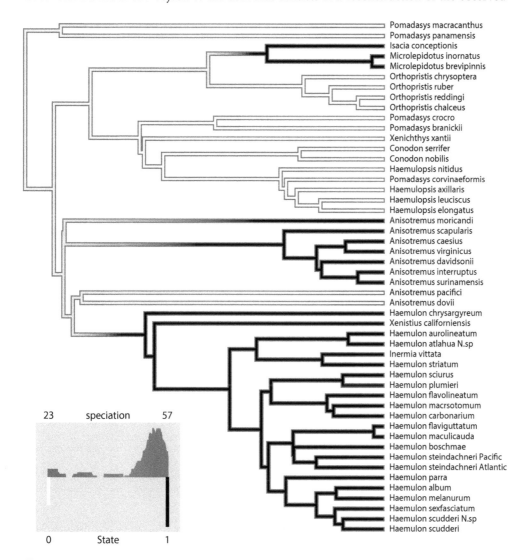

Figure 11.5
HiSSE reconstruction mapped onto the tree. The outline shows the posterior probability of being in either the high (red) or low (blue) diversification rate.

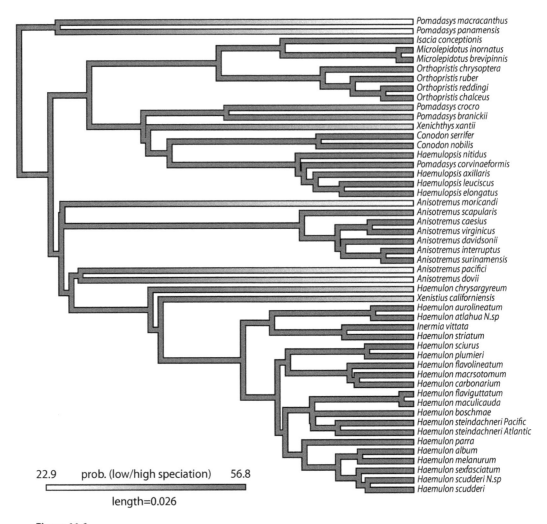

Pomadasys macracanthus
Pomadasys panamensis
Isacia conceptionis
Microlepidotus inornatus
Microlepidotus brevipinnis
Orthopristis chrysoptera
Orthopristis ruber
Orthopristis reddingi
Orthopristis chalceus
Pomadasys crocro
Pomadasys branickii
Xenichthys xantii
Conodon serrifer
Conodon nobilis
Haemulopsis nitidus
Pomadasys corvinaeformis
Haemulopsis axillaris
Haemulopsis leuciscus
Haemulopsis elongatus
Anisotremus moricandi
Anisotremus scapularis
Anisotremus caesius
Anisotremus virginicus
Anisotremus davidsonii
Anisotremus interruptus
Anisotremus surinamensis
Anisotremus pacifici
Anisotremus dovii
Haemulon chrysargyreum
Xenistius californiensis
Haemulon aurolineatum
Haemulon atlahua N.sp
Inermia vittata
Haemulon striatum
Haemulon sciurus
Haemulon plumieri
Haemulon flavolineatum
Haemulon macrsotomum
Haemulon carbonarium
Haemulon flaviguttatum
Haemulon maculicauda
Haemulon boschmae
Haemulon steindachneri Pacific
Haemulon steindachneri Atlantic
Haemulon parra
Haemulon album
Haemulon melanurum
Haemulon sexfasciatum
Haemulon scudderi N.sp
Haemulon scudderi

22.9   prob. (low/high speciation)   56.8

length=0.026

Figure 11.6
Reconstruction of the diversification rate mapped onto the tree from our HiSSE CID-2 model.

discrete character (which we're not that interested in) and a reconstruction of the hidden character.

We can then regraph these trees, as we see fit, using a generic plot method from *phytools*.

```
## create CID-2 plot using phytools
plot(setMap(cid2.map$rate.tree,c("white","red")),
        fsize=c(0.5,0.8),leg.txt="prob. (low/high speciation)",
        dig=2)
```

When we see this alone, it's evident that the low speciation rate tends to be reconstructed in parts of the tree with long branches—and high rates in regions of the tree with short branches (figure 11.6).

This is perfectly sensible because, just as we saw in figure 11.1, a high rate of speciation should create shorter distances between nodes!

One word of caution, though, in interpreting this plot.

At first glance, it appears as if the speciation rate is changing *continuously* over the edges of the tree. In fact, our model was for only *two* speciation rates—a low rate and a high rate. Our reconstruction thus actually represents the continuous *probability* of being in either the low rate (in our case, 22.9) or the high rate (56.8).

hisse also lets us fit a four-level hidden-rate model, independent of the value of any discrete character state. This should allow us to capture an even higher complexity of heterogeneity in diversification across our phylogeny.

To do this, we will need to build a *new* transition matrix between states and then assign turnover and eps the value c(1,1,2,2,3,3,4,4). We do this because we want each of the levels of our hidden state to have the same rate, independent of the corresponding value of the *observed* binary character, right?

```
## create design matrix for CID-4 model
rates.cid4<-TransMatMakerHiSSE(hidden.traits=3)
rates.cid4[!is.na(rates.cid4)]<-1
rates.cid4

##      (0A) (1A) (0B) (1B) (0C) (1C) (0D) (1D)
## (0A)  NA    1    1   NA    1   NA    1   NA
## (1A)   1   NA   NA    1   NA    1   NA    1
## (0B)   1   NA   NA    1    1   NA    1   NA
## (1B)  NA    1    1   NA   NA    1   NA    1
## (0C)   1   NA    1   NA   NA    1    1   NA
## (1C)  NA    1   NA    1    1   NA   NA    1
## (0D)   1   NA    1   NA    1   NA   NA    1
## (1D)  NA    1   NA    1   NA    1    1   NA

## fit CID-4 model
cid4.mle<-hisse(gt,hd,f=c(1,1),
    turnover=c(1,1,2,2,3,3,4,4),
    eps=c(1,1,2,2,3,3,4,4),
    hidden.states=TRUE,
    trans.rate=rates.cid4)

## Initializing...
## Finished. Beginning simulated annealing...
## Finished. Refining using subplex routine...
## Finished. Summarizing results...

cid4.mle

##
## Fit
##           lnL              AIC            AICc
##        129.16          -240.32         -235.82
##        n.taxa n.hidden.states
##         50.00            4.00
##
```

```
## Model parameters:
##
## turnover0A turnover1A        eps0A        eps1A
## 2.0612e-09 2.0612e-09 4.0620e-03 4.0620e-03
## turnover0B turnover1B        eps0B        eps1B
## 2.0612e-09 2.0612e-09 2.2981e-03 2.2981e-03
## turnover0C turnover1C        eps0C        eps1C
## 6.6636e+01 6.6636e+01 2.0612e-09 2.0612e-09
## turnover0D turnover1D        eps0D        eps1D
## 2.0612e-09 2.0612e-09 1.3433e-04 1.3433e-04

## reparameterize model to lambda & mu
repar.bd(cid4.mle,8)

##         lambda         mu
## 0A 2.0528e-09 8.3386e-12
## 1A 2.0528e-09 8.3386e-12
## 0B 2.0564e-09 4.7258e-12
## 1B 2.0564e-09 4.7258e-12
## 0C 6.6636e+01 1.3735e-07
## 1C 6.6636e+01 1.3735e-07
## 0D 2.0609e-09 2.7684e-13
## 1D 2.0609e-09 2.7684e-13
```

We won't plot this model, but we could if we wanted to, just as for the CID-2 model, using
plot.hisse.states.

Finally, the last HiSSE model that we'll fit is the "full" model in which both our discrete
state *and* a four-level hidden state can influence the rates of speciation and extinction in our
phylogeny. Hopefully, it's evident how this model has all previous models (including the BiSSE
model) as a special case.

```
## fit full HiSSE model
hisse.mle<-hisse(gt,hd,f=c(1,1),
    hidden.states=TRUE,
    turnover=c(1,2,3,4,5,6,7,8),
    eps=c(1,2,3,4,5,6,7,8),
    trans.rate=rates.cid4)

## Initializing...
## Finished. Beginning simulated annealing...
## Finished. Refining using subplex routine...
## Finished. Summarizing results...

hisse.mle

##
## Fit
##              lnL          AIC          AICc
```

```
##           129.73           -225.46            -206.33
##         n.taxa n.hidden.states
##          50.00              4.00
##
## Model parameters:
##
## turnover0A turnover1A         eps0A         eps1A
## 2.0612e-09 2.0612e-09 3.0000e+00 3.0000e+00
## turnover0B turnover1B         eps0B         eps1B
## 5.4198e+01 7.6318e+01 1.7120e-08 2.0612e-09
## turnover0C turnover1C         eps0C         eps1C
## 2.0612e-09 2.0612e-09 5.1871e-04 3.0000e+00
## turnover0D turnover1D         eps0D         eps1D
## 2.0619e-09 2.0612e-09 3.0000e+00 3.0000e+00
```

```
repar.bd(hisse.mle,8)
```

```
##          lambda         mu
## 0A 5.1529e-10 1.5459e-09
## 1A 5.1529e-10 1.5459e-09
## 0B 5.4198e+01 9.2786e-07
## 1B 7.6318e+01 1.5730e-07
## 0C 2.0601e-09 1.0686e-12
## 1C 5.1529e-10 1.5459e-09
## 0D 5.1547e-10 1.5464e-09
## 1D 5.1529e-10 1.5459e-09
```

To compare the results of all our different models, we'll assemble them into a single table.

Since *hisse* doesn't have one, for fun, we thought we'd show how to write a simple S3 logLik method[39] for the object classes produced by the various *hisse* functions that we used: "hisse.fit".

```
## our logLik methods
logLik.hisse.fit<-function(x,...){
    lik<-x$loglik
    attr(lik,"df")<-(x$AIC+2*lik)/2
    lik
}
```

This just lets us pull out the log-likelihoods, number of estimated parameters, and AIC values more efficiently from each model so that we can build our results table without having to iterate column by column through the table.[40] Let's do it.

---

[39] This is easier than you probably imagined, right?

[40] Note that as soon as a properly designed logLik method exists for a particular object class, there's no need to write a separate AIC method. The generic AIC method works just fine!

```
## print a table of results
data.frame(
    model=c("CID","BiSSE","HiSSE CID-2",
        "HiSSE CID-4","HiSSE full"),
    logL=sapply(list(cid.mle,bisse.hmle,
        cid2.mle,cid4.mle,hisse.mle),
        logLik),
    k=sapply(list(cid.mle,bisse.hmle,
        cid2.mle,cid4.mle,hisse.mle),
        function(x) attr(logLik(x),"df")),
    AIC=aic<-sapply(list(cid.mle,bisse.hmle,
        cid2.mle,cid4.mle,hisse.mle),
        AIC),
    Akaike.weight=unclass(aic.w(aic))
)
```

```
##            model    logL   k     AIC Akaike.weight
## 1            CID 126.25   4 -244.49   0.411099019
## 2          BiSSE 127.98   6 -243.96   0.314858306
## 3    HiSSE CID-2 126.64   5 -243.27   0.223005048
## 4    HiSSE CID-4 129.16   9 -240.32   0.051007411
## 5    HiSSE full 129.73  17 -225.46   0.000030214
```

We see here essentially no support for the full HiSSE model.

Most relevant, though, is that the model that receives the most support is the simplest, character-independent model that we started with! The rest of the model support is spread across the BiSSE and CID-2 models, hinting (perhaps) at the *possibility* of some heterogeneity in diversification. The main effect, then, of considering a hidden-state character was decreasing the support for models where our measured character had some influence on diversification.

11.4.2 HiSSE models without a discrete trait

In all of the models that we fit in the previous subsection, we supplied a discrete character trait *even if* that trait was not hypothesized to affect the rate of diversification. This was necessary because the total likelihood that we computed for each model was the likelihood of our diversification model parameters—*and* the trait!

Under many circumstances, however, we might be interested in fitting a hidden-state diversification model but without any discrete trait. This can be done using the *hisse* package too.

For this analysis, let's use another tree from the book website, pipefishes.phy, which is a phylogeny of pipefishes, seahorses, and seadragons[41] modified from Hamilton et al. (2017).

```
## read pipefish tree from file
pipefish.tree<-read.tree(file="pipefishes.phy")
```

---

[41] The percomorph fish family Syngnathidae.

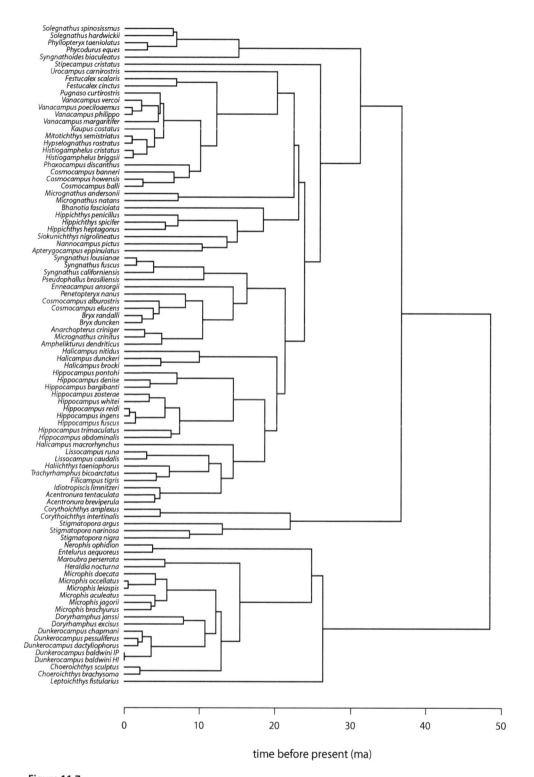

Figure 11.7

Phylogeny of pipefishes, seahorses, and their kin from Hamilton et al. (2017).

We can start by plotting our phylogeny, but this time, why don't we plot it pointing upward and with a vertical time-before-present axis?

In R, the easiest way to do that is by plotting the tree *downward*—but then flipping the direction of our *y*-axis in the plot! Here is what that looks like (figure 11.7).

```
plotTree(pipefish.tree,ftype="i",fsize=0.5,
    lwd=1,mar=c(1.1,4.1,0.1,0.1),
    ylim=c(49,-15),
    direction="downwards")
title(ylab="time before present (ma)")
axis(2,at=seq(0,50,by=10),las=1)
```

To fit our model now, we'll use a different *hisse* package function called MiSSE[42] as follows.

First, we can specify the sampling fraction, the argument f, in this case, to be Ntip(pipefish.tree)/169 because we think that there are about 169 species in total in the syngnathid fish family.

Next, we just go ahead and fit our model!

```
## set sampling fraction
rho<-Ntip(pipefish.tree)/169
## fit MiSSE model
misse2.mle<-MiSSE(pipefish.tree,f=rho,
    turnover=c(1,2),eps=c(1,2))

## Initializing...
## Finished. Beginning simulated annealing...
## Finished. Refining using subplex routine...
## Finished. Summarizing results...

misse2.mle

##
## Fit
##              lnL            AIC            AICc
##           297.90         605.81          606.51
##           n.taxa n.hidden.states
##            92.00            2.00
##
## Model parameters:
##
## turnover0A        eps0A  turnover0B          eps0B
## 2.1141e-01   1.5997e-01  6.5141e-02     2.0612e-09
##          q0
## 3.0016e-02

repar.bd(misse2.mle,2)
```

---

[42] We're not sure what MiSSE stands for, but it might be "missing" state-dependent speciation and extinction.

```
##      lambda           mu
## 0A 0.182257 2.9156e-02
## 0B 0.065141 1.3426e-10
```

Let's graph the modeled rate heterogeneity on the phylogeny using `MarginReconMiSSE`, which works pretty much the same way as `MarginReconHiSSE` from the last section (figure 11.8).

```
## conduct marginal reconstruction
misse2.recon<-MarginReconMiSSE(phy=pipefish.tree,
    f=Ntip(pipefish.tree)/169,
    pars=misse2.mle$solution,
    hidden.states=2)

## Calculating marginal probabilities for 91 internal nodes...
## Finished. Calculating marginal probabilities for 92 tips...
## Done.

## graph MiSSE model on the tree
misse2.map<-plot.misse.states(misse2.recon,
    rate.param="speciation",edge.width=3,
    show.tip.label=TRUE,type="phylogram",
    fsize=0.5,legend.position=c(0,0.3,0.85,1),
    rate.colors=c("yellow","darkblue"))
```

If we want to compare this to a homogeneous birth-death model, we can do that using MiSSE too.

We might fit this simpler birth-death model as follows.

```
## fit birth-death model using MiSSE
misse1.mle<-MiSSE(pipefish.tree,
    f=Ntip(pipefish.tree)/169,
    turnover=1,eps=1)

## Initializing...
## Finished. Beginning simulated annealing...
## Finished. Refining using subplex routine...
## Finished. Summarizing results...

misse1.mle

##
## Fit
##              lnL            AIC            AICc
##          -299.49         602.97          603.11
##            n.taxa n.hidden.states
##             92.00            1.00
##
```

```
## Model parameters:
##
## turnover0A        eps0A
##    0.21661      0.33410
```

repar.bd(misse1.mle,1)

```
##        lambda         mu
## 0A 0.16236 0.054247
```

Finally, let's compare the two models, just as we did with our grunter HiSSE analysis from the previous section.

```
## copy logLik method for different object
## class: "misse.fit"
logLik.misse.fit<-logLik.hisse.fit
## compile our results
data.frame(
    model=c("MiSSE-1","MiSSE-2"),
    logL=sapply(list(misse1.mle,misse2.mle),
        logLik),
    k=sapply(list(misse1.mle,misse2.mle),
        function(x) attr(logLik(x),"df")),
    AIC=aic<-sapply(list(misse1.mle,misse2.mle),
        AIC),
    Akaike.weight=unclass(aic.w(aic))
)
```

```
##      model     logL k     AIC Akaike.weight
## 1 MiSSE-1 -299.49 2 602.97       0.80492
## 2 MiSSE-2 -297.90 5 605.81       0.19508
```

This result shows that there is relatively little support for the multi-rate MiSSE model compared to our homogeneous-rate birth-death model!

11.5 Quantitative-trait speciation and extinction (QuaSSE) model

To end the chapter, we also wanted to briefly introduce one more SSE model called QuaSSE, the quantitative-state speciation and extinction model (FitzJohn 2010).

In principle, this model is very similar to BiSSE and MuSSE but applies to the scenario in which we hypothesize that a *continuously valued* (i.e., quantitative) character, rather than discrete trait, influences the diversification rates in our clade.

One complication that naturally arises when we consider quantitative traits is that the speciation or extinction rates could be related to a continuously valued character trait in more than one way (FitzJohn 2010).

For instance, let's think about how speciation could vary as a function of overall body size.

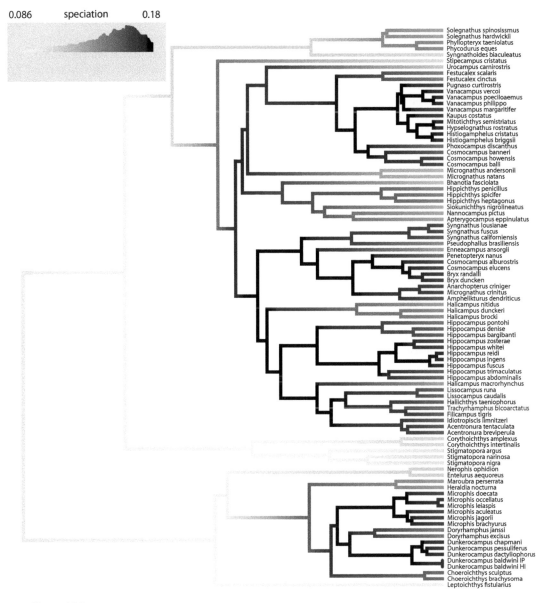

0.086 speciation **0.18**

Figure 11.8
Marginal reconstruction of the rate of speciation on the tree of pipefishes and their kin.

The relationship between the two could be flat (in other words, no relationship)—but it could also be linearly or curvilinearly increasing or decreasing.[43]

Likewise, there's no theoretic requirement that λ and μ both share the same type of relationship with our quantitative trait. For example, μ might decrease for larger values of the quantitative trait while λ increases, or vice versa.

As such, when we fit a QuaSSE model to data, we must also decide a priori a functional form for the relationship between diversification rate and the trait!

---

[43] The speciation rate could even go up and then down, or down and then up!

11.5.1 Fitting the QuaSSE model: An empirical example using scale insects

To see how the QuaSSE method works, we'll use a data set and phylogeny of scale insects (superfamily Coccoidea) from Hardy et al. (2016). Hardy et al. (2016) analyzed a data set containing over 400 species. To make the exercise a little easier to reproduce, we decided to subsample their data set here so that it included only 115 of the species in that study. As such, our results may differ from theirs!

The files we'll use are `Coccoidea_phylogeny.tre` and `Coccoidea_hosts.csv`. As always, these files can be downloaded from the book website.

Hardy et al. (2016) hypothesized that host diet specialization might affect the rate of diversification in their group. They measured diet specialization as the number of host plant families, but since these range in value from 1 to 109,[44] Hardy et al. (2016) decided it made more sense to treat this value as a continuous trait.[45]

We can start by reading in the tree and data from file in a typical way.

```
## read tree adn data from file
scale_insect.tre<-read.tree(file="Coccoidea_phylogeny.tre")
scale_insect.data<-read.csv(file="Coccoidea_hosts.csv",
    row.names=1)
```

As we've done in prior chapters, let's use `name.check` from the *geiger* package to help identify any inconsistencies between our phylogeny and our trait data.

```
## check tree and data to ensure matching
chk<-name.check(scale_insect.tre,scale_insect.data)
summary(chk)

## 357 taxa are present in the tree but not the data:
##      Asterolecaniidae_Asterodiaspis_quercicola,
##      Asterolecaniidae_Planchonia_stentae,
##      Beesoniidae_Beesonia_napiformis,
##      Beesoniidae_Gallacoccus_heckrothi,
##      Coccidae_Ceroplastes_ceriferus,
##      Coccidae_Ceroplastes_floridensis,
##      ....
##
## To see complete list of mis-matched taxa, print object.
```

This shows us that there are a number of species present in the phylogeny that are absent from our trait data. We can prune these taxa from our phylogeny before continuing.

```
## prune mismatched taxa from the tree
scale_insect.pruned<-drop.tip(scale_insect.tre,
    chk$tree_not_data)
```

---

[44] One to 81 in our subsampled data.

[45] We agree that this is pretty sensible.

Next, we can pull out the number of host families for each species and log-transform the result.

We needn't have done this, but log-transformation has the effect of making equivalent proportional changes equal. For instance, on a log-scale, a change from one to two host families is equal in magnitude to an evolutionary change from ten to twenty families (and so on), which kind of makes sense.

This is our quantitative character.

```
ln.hosts<-setNames(log(scale_insect.data$host.families),
    rownames(scale_insect.data))
```

For fun, let's use the contMap function of *phytools* to project our continuous trait onto the tree (figure 11.9). contMap uses the techniques of ancestral character estimation that we learned about in chapter 8. It does not take into account variation in λ and μ that is correlated with our phenotypic trait.

```
## visualize a continuous character map of host plant
## number on scale insect tree
host.map<-contMap(scale_insect.pruned,ln.hosts,
    plot=FALSE)
host.map<-setMap(host.map,c("yellow","darkblue"))
plot(host.map,lwd=c(2,5),outline=FALSE,ftype="off",
    leg.txt="ln(host families)",legend=60)
```

Now we're going to fit our models.

Loosely following Hardy et al. (2016), we'll fit a total of three different QuaSSE models. First, we can fit a model in which extinction (μ) is constant, and speciation (λ) changes as a function of the number of host families. Next, we'll fit a model in which λ is fixed and μ changes. Finally, we'll fit a model in which both λ and μ are allowed to vary as a function of the trait. As our null, we'll add a constant λ and μ model as well.

In each case, we'll use a *linear* relationship—although, as we mentioned before, we could have used other functional forms.

Our QuaSSE model needs us to indicate our sampling fraction, ρ, just as did other diversification methods in chapters 9 and 10. By some accounts, the Coccoidea has about 8,000 species, so let's use that as our denominator of ρ.[46]

```
rho<-Ntip(scale_insect.pruned)/8000
```

The QuaSSE model can be a bit of a pain to fit, so it's important that we start our optimization with reasonable starting values of our model parameters. In fact, we'd go as far as to say that if we don't start from reasonable values, there's not much chance of converging on the correct ML solution!

We could try to use the *diversitree* function starting.point.quasse to get starting values for our QuaSSE optimization. For many SSE optimizations, this tactic should work reasonably well.

---

[46]Ntip(scale_insect.pruned) gives us the number of taxa in our tree.

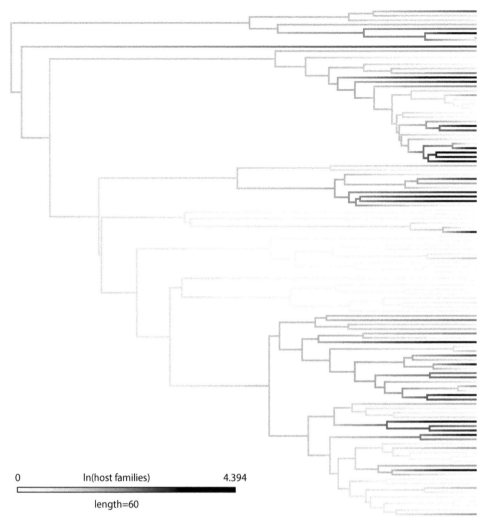

Figure 11.9
Number of host families (on a log-scale) mapped onto the phylogeny of scale insects using contMap.

Unfortunately, starting.point.quasse[47] isn't set up to take into account the sampling fraction, ρ, which we know is quite low for our data set.

Instead of starting.point.quasse, then, let's use *phytools* fit.bd to get starting values of λ and μ[48] and fitContinuous from *geiger* to obtain a starting value for the *diffusion* parameter of QuaSSE.[49]

```
## run fit.bd and fitContinuous to get starting
## values for our QuaSSE optimization
```

---

[47] And, likewise, the other starting.point functions offered by the *diversitree* package.
[48] Taking into account ρ.
[49] Which corresponds to the rate of evolution of the character.

```
bd<-fit.bd(scale_insect.pruned,rho=rho)
bm<-fitContinuous(scale_insect.pruned,ln.hosts)
p<-setNames(c(bd$b,bd$d,bm$opt$sigsq),
    c("lambda","mu","diffusion"))
p
```

```
##    lambda       mu diffusion
##   1.08219  1.03185   0.15662
```

As we said before, we're going to need to make a linear function to fit the QuaSSE model we're interested in. Let's do that with make.linear.x.[50]

```
## define range of x
xr<-range(ln.hosts)+c(-1,1)*20*
    p["diffusion"]
## make linear model for QuaSSE
linear.x<-make.linear.x(xr[1],xr[2])
linear.x
```

```
## function (x, c, m)
## {
##     x[x < x0] <- x0
##     x[x > x1] <- x1
##     ans <- m * x + c
##     ans[ans < 0] <- 0
##     ans
## }
## <bytecode: 0x0000000024e10d68>
## <environment: 0x0000000024e120e0>
```

Now, let's build our first likelihood function using make.quasse, with the fixed μ model.[51]

```
## make QuaSSE likelihood function for variable
## lambda and constrain
lik.lambda<-make.quasse(scale_insect.pruned,
    ln.hosts,lambda=linear.x,mu=constant.x,
    sampling.f=rho,states.sd=0.1)
lik.lambda<-constrain(lik.lambda,drift~0)
```

---

[50]The function make.linear.x builds a new function that we'll proceed to pass to our QuaSSE model-fitting function. We need to give make.linear.x the lower and upper limits of our continuous trait. It will often make sense to define these values via an expansion of the range of our trait, as we do here.

[51]In addition to the arguments we've explained already, we're also going to set states.sd, which is the standard error of the trait means for our quantitative character, to be the constant value of 0.1. This is arbitrary but corresponds to about 10 percent standard error on the log-scale. We can set it to be a smaller value, but it cannot be zero.

We constrained one parameter, drift~0. The drift parameter in make.quasse allows the mean of our character evolution process to change through time. We're not going to worry about that here.

All right, now let's try to optimize it using find.mle. To do that, we first need to update our starting values vector (p) so that its parameterization corresponds to the model we want to fit. Another word of caution—fitting this model may take some time!

```
## subsample starting parameter values to match
## the model we're fitting
pp<-setNames(c(p["lambda"],0,p["mu"],
    p["diffusion"]),argnames(lik.lambda))
pp
```

```
##       l.c       l.m       m.c diffusion
##   1.08219   0.00000   1.03185   0.15662
```

```
## fit our first QuaSSE model
lambda.mle<-find.mle(lik.lambda,x.init=pp,
    control=list(parscale=0.1),lower=rep(0,4))
lambda.mle
```

```
## $par
##       l.c       l.m       m.c diffusion
##   0.419538  0.021227  0.340787  0.113193
##
## $lnLik
## [1] -272.8
##
## $counts
## [1] 804
##
## $convergence
## [1] 0
##
## $message
## [1] "success! tolerance satisfied"
##
## $hessian
## NULL
##
## $method
## [1] "subplex"
##
## $par.full
##       l.c       l.m       m.c     drift diffusion
##   0.419538  0.021227  0.340787  0.000000  0.113193
##
## $func.class
```

```
## [1] "constrained" "quasse"      "dtlik"        "function"
##
## attr(,"func")
## QuaSSE likelihood function:
##    * Parameter vector takes 4 elements:
##       - l.c, l.m, m.c, diffusion
##    * Function constrained (original took 5 elements):
##       - drift ~ 0
##    * Function takes arguments (with defaults)
##       - pars: Parameter vector
##       - ...: Additional arguments to underlying function
##       - pars.only [FALSE]: Return full parameter vector?
##    * Phylogeny with 115 tips and 114 nodes
##       - Taxa: Diaspididae_Chionaspis_americanas, ...
##    * Reference:
##       - FitzJohn (2010) doi:10.1093/sysbio/syq053
## R definition:
## function (pars, ..., pars.only = FALSE)
## attr(,"class")
## [1] "fit.mle.quasse" "fit.mle"
```

Next, we're going to build our fixed λ model. We'll do that much the same way—this time just setting mu equal to linear.x in make.quasse instead of lambda as follows.

```
## make QuaSSE likelihood function for variable
## mu and constrain
lik.mu<-make.quasse(scale_insect.pruned,
    ln.hosts,lambda=constant.x,mu=linear.x,
    sampling.f=rho,states.sd=0.1)
lik.mu<-constrain(lik.mu,drift~0)
## fit variable mu model
pp<-setNames(c(p[c("lambda","mu")],0,
    p["diffusion"]),argnames(lik.mu))
mu.mle<-find.mle(lik.mu,x.init=pp,
    control=list(parscale=0.1),lower=rep(0,4))
```

Instead of printing the whole model, once again, let's just print the coefficients and the log-likelihood.

```
coef(mu.mle)

##      l.c      m.c      m.m diffusion
##  0.644297 0.541901 0.014627 0.119508

logLik(mu.mle)

## 'log Lik.' -276.05 (df=4)
```

Third, we'll fit a "full" model in which both λ *and* μ are free to vary as (potentially different) linear functions of our quantitative trait: number of host plant families. This model is the hardest to fit, so we're going to give it all the help we can by starting it off with the MLEs from each of the two previous fitted models.

```
## create full likelihood function
lik.full<-make.quasse(scale_insect.pruned,
    ln.hosts,lambda=linear.x,mu=linear.x,
    sampling.f=rho,states.sd=0.1)
lik.full<-constrain(lik.full,drift~0)
## fit full QuaSSE model
pp<-setNames(c(lambda.mle$par[1:2],mu.mle$par[2:3],
    p["diffusion"]),argnames(lik.full))
pp
```

```
##      l.c        l.m        m.c        m.m diffusion
##   0.419538   0.021227   0.541901   0.014627   0.156616
```

```
full.mle<-find.mle(lik.full,x.init=pp,
    control=list(parscale=0.1),lower=rep(0,5))
## print model coefficients and log-likelihood
coef(full.mle)
```

```
##      l.c        l.m        m.c        m.m diffusion
##   0.47957    0.15310    0.40390    0.14222    0.14299
```

```
logLik(full.mle)
```

```
## 'log Lik.' -268.72 (df=5)
```

Last of all, we can fit our "null" model in which the continuous character evolves, and our phylogeny adds and loses lineages by speciation and extinction, but there's no correlation between our trait and the rates of diversification.

Under this model, we should expect to find more or less the *same* MLEs of the speciation and extinction rates, although the likelihood is different because it takes into account the evolution of our trait.

```
## likelihood function for character-
## independent model
lik.cid<-make.quasse(scale_insect.pruned,
    ln.hosts,lambda=constant.x,mu=constant.x,
    sampling.f=rho,states.sd=0.1)
lik.cid<-constrain(lik.cid,drift~0)
argnames(lik.cid)
```

```
## [1] "l.c"        "m.c"        "diffusion"
```

This model has only three parameters: the speciation rate, the extinction rate, and the diffusion rate (i.e., the rate of trait evolution under our model).

Let's fit it.

```
## fit CID QuaSSE model and print coefficients
cid.mle<-find.mle(lik.cid,x.init=p,
    control=list(parscale=0.1),lower=rep(0,3))
coef(cid.mle)

##     lambda          mu diffusion
##    1.09563     1.04557   0.11318

logLik(cid.mle)

## 'log Lik.' -280.69 (df=3)
```

As we did for our other SSE models, we can finish by comparing them using the `anova` generic function.

```
anova(cid.mle,variable.lambda=lambda.mle,
    variable.mu=mu.mle,full.model=full.mle)

##                   Df lnLik AIC ChiSq Pr(>|Chi|)
## minimal            3  -281 567
## variable.lambda    4  -273 554 15.78    7.1e-05 ***
## variable.mu        4  -276 560  9.27     0.0023 **
## full.model         5  -269 547 23.94    6.3e-06 ***
## ---
## Signif. codes:  0 '***' 0.001 '**' 0.01 '*' 0.05 '.' 0.1 ' ' 1
```

This tells us that we can *reject* constant-rate speciation and extinction in favor of any of the three variable-rate models and that our best model is in fact our variable λ and variable μ model.

To get a sense of what these different models really *mean* in terms of how λ and μ might vary with our quantitative trait, let's plot them.

Remember, each model shows how scale insect speciation (λ), extinction (μ), or both[52] varies as a function of the number of host families, on a logarithmic scale.

```
## subdivide plotting area
par(mfrow=c(2,2))
## a) plot constant rate (CID) QuaSSE model
plot(NULL,xlim=range(ln.hosts),ylim=c(0,1.5),bty="n",
    xlab="log(host families)",
    ylab=expression(paste(lambda," or ",mu)))
mtext("(a)",line=1,adj=0)
clip(min(ln.hosts),max(ln.hosts),0,1.5)
abline(h=cid.mle$par["lambda"],lwd=2)
abline(h=cid.mle$par["mu"],lwd=2,col="gray")
## b) plot variable lambda QuaSSE model
```

---

[52]Or neither—in the case of our character-independent model.

```
plot(NULL,xlim=range(ln.hosts),ylim=c(0,1.5),bty="n",
    xlab="log(host families)",
    ylab=expression(paste(lambda," or ",mu)))
mtext("(b)",line=1,adj=0)
clip(min(ln.hosts),max(ln.hosts),0,1.5)
abline(a=coef(lambda.mle)["l.c"],
    b=coef(lambda.mle)["l.m"],lwd=2)
abline(h=coef(lambda.mle)["m.c"],lwd=2,
    col="gray")
legend(x=3,y=1.5,
    c(expression(lambda),expression(mu)),
    lwd=2,col=c("black","gray"),bty="n")
## c) plot variable mu QuaSSE model
plot(NULL,xlim=range(ln.hosts),ylim=c(0,1.5),bty="n",
    xlab="log(host families)",
    ylab=expression(paste(lambda," or ",mu)))
mtext("(c)",line=1,adj=0)
clip(min(ln.hosts),max(ln.hosts),0,1.5)
abline(h=coef(mu.mle)["l.c"],lwd=2)
abline(a=coef(mu.mle)["m.c"],
    b=coef(mu.mle)["m.m"],lwd=2,col="gray")
## d) plot variable lambda and mu QuaSSE model
plot(NULL,xlim=range(ln.hosts),ylim=c(0,1.5),bty="n",
    xlab="log(host families)",
    ylab=expression(paste(lambda," or ",mu)))
mtext("(d)",line=1,adj=0)
clip(min(ln.hosts),max(ln.hosts),0,1.5)
abline(a=coef(full.mle)["l.c"],
    b=coef(full.mle)["l.m"],lwd=2)
abline(a=coef(full.mle)["m.c"],
    b=coef(full.mle)["m.m"],lwd=2,col="gray")
```

What we see from figure 11.10 is that depending on our QuaSSE model assumptions (constant λ, constant μ, or variable λ and μ), we might make very different inferences about how the rate of diversification is affected by our quantitative trait.

If we decide to accept the model with the lowest AIC (variable λ and μ), then we might say that our data support a conclusion that both speciation and extinction rates of scale insects rise with an increase in the number of host families—although λ at a faster rate than μ.[53]

11.6 Practice problems

11.1 Using the data set from chapter 6 on squamate toes, test for an effect of having limbs (and toes) on diversification rates. First, change the multistate character to a binary character contrasting species with no legs (and no toes) and species with legs (and one

---

[53] This result is qualitatively similar to what was shown by Hardy et al. (2016), although they analyzed a much larger data set, used Bayesian MCMC, and did not log-transform the number of host families!

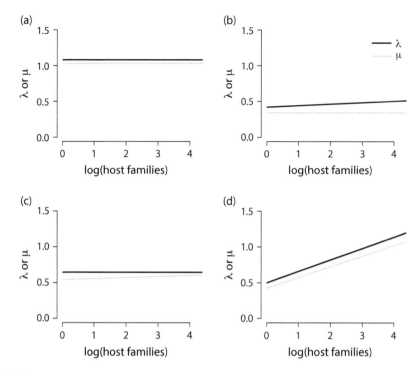

Figure 11.10
Fitted ML QuaSSE models. Models are (a) constant speciation and extinction, (b) variable speciation, (c) variable extinction, and (d) variable speciation and extinction.

or more toes). Then do a BiSSE analysis on this character. You can assume that sampling is random with respect to the two traits[54] and that there are about 10,900 species of squamates. Don't forget to prune out *Sphenodon punctatus* (it's not a squamate). Compare the models and parameters using both AIC and Bayesian analyses. What do you conclude?

11.2 Carry out a QuaSSE analysis using the primate eye size data from chapter 3. Use a sampling fraction of $90/449 = 0.20$[55], and compare the fits of the same set of models used in this chapter.

11.3 In our QuaSSE example for scale insects, we only considered constant and linear models. *diversitree* can do much more than that! Try fitting a sigmoid function (`sigmoid.t`) and an exponential function (`exp.t`). As a small warning, the documentation for this function is a bit lacking, but we think you can figure it out!

---

[54] In other words, that both limbed and limbless lizards have the same sampling fraction.

[55] The number 449 is the average of the two figures, 376 and 522, that we gave you for practice problem 9.3 from chapter 9!

Biogeography and phylogenetic community ecology

12

Species are distributed across the earth in patterns that still beguile us.

How, for instance, did the ancestors of the Fijian iguana get to that archipelago, when the nearest related species live thousands of miles away (Keogh et al. 2008)? Why do whole suites of distantly related taxa, from trees to fungi to mammals, show similar patterns of deep divergence across continents (Cracraft 1988)? And how do the rich communities that can be found in particular places form over deep time (Webb et al. 2002)?

These questions can be addressed with techniques from biogeography and community ecology.

In this chapter, and in contrast to prior sections of the book in which we focused on the physical traits of organisms, we'll be modeling the macroevolutionary processes that determine the places where species are found as well as the evolutionary structure of ecological communities.

This is a vast field. Indeed, a few years ago, a whole book was published (Swenson 2014) that is focused entirely on the topic of phylogenetic and functional community ecology in R.[1]

Nonetheless, the models and methods for phylogenetic biogeography and community ecology are closely related to the trait and diversification models that we've studied in previous chapters. As such, we thought it would be of value to overview these important topics as well.

In this chapter, we will thus:

1. Learn how to reconstruct species' ranges through time using ancestral area reconstruction under maximum likelihood.
2. Examine how to compare the fit of a set of biogeographic models to data on the phylogenetic relatedness and the spatial arrangement of species.
3. Finally, see how to test a limited set of alternative hypotheses for community structure using phylogenetic comparative data.

---

[1] In fact, that's the title of the book!

In chapter 8, we studied the endeavor of ancestral character estimation for discrete and continuously valued phenotypic traits. We learned that this can be done accurately when important model assumptions are met and examined conditions in which ancestral state estimation tends to be biased.

Contemporary geographic distribution is an important attribute of the extended phenotype of living organisms—and, indeed, much like their physical features, the geographic distributions of species are expected to change through time and among lineages.

Just as we can use phylogenetic methods to estimate the ancestral features of hypothetical ancestral taxa in our tree, we can similarly reconstruct the ancestral ranges of ancestral species.

Estimating ancestral areas is *similar* to reconstructing the ancestral physical features of organisms but includes some important differences.

For instance, if we're reconstructing the history of a discrete physical feature that is *only* represented in a monomorphic condition (e.g., *blue*, *red*, or *green*) in the extant lineages of our tree, we usually assume that hypothetical ancestral taxa each had *one* of the three states, not a combination of two or three.[2]

This assumption makes little sense when we consider geographic areas. To the contrary, it's normally pretty reasonable to assume that colonization of area *B* from area *A* should include[3] a period of time in which the colonizing lineage is found in both areas.

In addition, in the character models of this book, we typically assume that nothing particularly interesting goes on with trait evolution at the moment of speciation.[4]

If speciation tends to occur by geographic isolation, then this assumption likewise makes very little sense. To the contrary, under a scenario of allopatric speciation, we would expect many more shifts in geographic area to be associated with speciation events in the tree than not!

As such, ancestral area reconstruction is most often done using models that can take these different nuances into consideration. The most important such model is one called the *dispersal-extinction-cladogenesis* model, or *DEC* (Ree et al. 2005; Ree and Smith 2008).

12.2.1 The DEC model

The DEC model is named for the three processes that we tend to think are likely to dominate the evolution of species distributions over macroevolutionary time. Species can move from one area to another via *dispersal* (D). Species can go (locally) *extinct* from an area (E). Finally, species can undergo *cladogenesis*,[5] splitting from one lineage into two (C) (Ree et al. 2005).

In a subsequent section, we'll add a fourth process to this set of three that is called *jump-dispersal speciation* (Matzke 2012, 2014).

---

[2] We learned about an exception to this, the polymorphic trait model, in chapter 7. Indeed, this model is closely related to the biogeographic evolution models that we'll consider in the present section of the book!

[3] At least transiently.

[4] There are exceptions to this generality. The most prominent among these are variants of Eldredge and Gould's (1972) paleontological model of punctuated equilibrium, in which a disproportionate amount of evolution tends to accrue with speciation—intervened by long periods of stasis. The earliest application of this model to phylogenetic data (to our knowledge) comes from Bokma (2002, 2008), and more recently, Goldberg and Igic (2012) created ClaSSE, a BiSSE extension to test for cladogenetic change.

[5] That is to say, *speciation*.

To run a DEC ancestral area reconstruction, we need a phylogenetic tree with branch lengths, probably one that is ultrametric,[6] such that the branch lengths of the tree are either in units of time or are *proportional* to time.

We also need species distribution data for all the taxa at the tips of our phylogeny.

DEC takes discrete areas as input values (Ree et al. 2005; Ree and Smith 2008), so we always need to start by dividing the various places in which our organisms live into a relatively small set of discretely valued areas and then classify each taxon in our tree as being present or absent from each area. This presence/absence information will form the input data of our phylogenetic biogeographic analysis.

Unless our species are uniformly distributed across a set of evenly spaced and similarly sized islands in the Caribbean, there is more art than science to this categorization step.

We need to specify the right number of areas to capture the process of biogeographical interest, but not too many such that our model ends up with absurd numbers of parameters to estimate—compared to the number of taxa we have in our tree!

A reasonable rule of thumb might be that if our areas are such that none or very few of the taxa of our tree are found in more than one area, our areas may be too *large* and too *few*. By contrast, if lots of the species in our phylogeny are found in numerous areas, then our areas may be too *small* and too *many*.

12.2.2 Fitting a DEC model to data: An empirical example using fungus-growing ants

To see how a DEC model is fit to data, we'll use an empirical example: fungus-farming ants in the tribe Attini from Branstetter et al. (2017).

The files we'll use for this analysis are `attine-tree-pruned.tre` and `attine-distribution-data.txt`, both of which are available through the book website.[7]

As usual, our first step will be to load the packages that we'll be using in this R session.

For us, these will be *ape* and *phytools*, plus a new package that we haven't see before called *BioGeoBEARS* (Matzke 2013).

Unlike most of the other packages of the book,[8] however, *BioGeoBEARS* is not presently available from CRAN. As such, the easiest way to install *BioGeoBEARS* is directly from the package authors' development GitHub repository, which can be done using the package *devtools* (Wickham et al. 2020).[9]

As soon as we have *BioGeoBEARS* installed, we can load it in the normal way.

We'll also load the *phytools* package, which we intend to use in a moment as well.

```
## load packages
library(phytools)
library(BioGeoBEARS)
```

Next, we can proceed to read in our phylogenetic tree from file using `read.tree` from *ape*.[10]

---

[6]Unless your analysis includes extinct lineages!

[7]http://www.phytools.org/Rbook/.

[8]With the exception of *bayou*, from chapter 5.

[9]Installing *BioGeoBEARS* using *devtools* is easy. First, get *devtools* from CRAN, and then simply execute the command `devtools::install_github("nmatzke/BioGeoBEARS")` from the R command prompt. That's all there is to it!

[10]*ape* is automatically loaded with *phytools*, as we've seen in previous chapters.

```
## read tree from file
ant.tree<-read.tree("attine-tree-pruned.tre")
print(ant.tree,printlen=2)

##
## Phylogenetic tree with 84 tips and 83 internal nodes.
##
## Tip labels:
##   Acanthognathus_ocellatus_1492, Daceton_armigerum_1448, ...
##
## Rooted; includes branch lengths.
```

One peculiarity of the *BioGeoBEARS* package is that instead of taking its input in the typical way of other R packages,[11] *BioGeoBEARS* instead requires specially formatted input data files.[12] As such, we need our *geographic data* to be in a standardized input file format.

To see what this looks like, let's just print the first handful of lines of our input data file `attine-distribution-data.txt`.

```
writeLines(readLines("attine-distribution-data.txt",10))

84  5 (A B C D E)
Acanthognathus_ocellatus_1492     01100
Acromyrmex_echinatior_Genome      01000
Acromyrmex_balzani_1487 00100
Acromyrmex_heyeri_1477    00100
Acromyrmex_lundi_1476     00100
Acromyrmex_octospinosus_1475      01100
Acromyrmex_striatus_1490      00100
Acromyrmex_versicolor_1474    10000
Apterostigma_auriculatum_1578     01100
```

This file format is essentially an adaptation of the well-known PHYLIP data file format[13] (Felsenstein 1993).

The first line of the input file contains two integer numbers: the number of taxa in the input data and tree (in our case, 84) and the number of characters—in our case, regions (5). This latter value is then followed by a list of the names of the five regions separated by spaces inside a pair of round parentheses: (A B C D E).

All subsequent lines of the input file first contain the taxon label and then a string of five 0s and 1s depending on whether each taxon is present (1) or absent (0) from each of our five regions.

---

[11] That is, in the form of standard R objects.

[12] This is much more similar to what we'd expect for a stand-alone software, rather than a package designed to run in R!

[13] http://rosalind.info/glossary/phylip-format/.

Even though our *BioGeoBEARS* function will use this file[14] as input, we can nonetheless read in the file to visualize the distribution of the species of our phylogeny across the five biogeographic regions of our study.

To do this, we'll use a *BioGeoBEARS* function[15] to read in our data and then plot the presence/absence of each taxon from each region using `plotTree.datamatrix` from *phytools*.

In our input data, we coded the five regions as A through E. These letters correspond to *Nearctic*, *Middle America*, *South America*, *Afrotropics*, and *Australasia*: the five general areas of the globe in which attine ants are found.

We can use the names of these areas in our plot.[16]

```
## read data from file
ant.data<-getranges_from_LagrangePHYLIP(lgdata_fn=
    "attine-distribution-data.txt")
## extract a presence/absence matrix
tmp<-ant.data@df
tmp[,1:5]<-lapply(tmp[,1:5],factor)
## re-name the columns of the matrix to correspond
## to our geographical areas
colnames(tmp)<-c("Nearctic","Middle America",
    "South America","Afrotropics","Australasia")
## set the colors we'll use for plotting
colors<-setNames(replicate(ncol(tmp),
    setNames(c("white","darkgray"),0:1),
    simplify=FALSE),colnames(tmp))
## graph presence/absence using plotTree.datamatrix
object<-plotTree.datamatrix(ant.tree,tmp,fsize=0.5,
    yexp=1.1,header=TRUE,xexp=1.25,colors=colors)
## add a legend
legend("topleft",c("species absent","species present"),
    pch=22,pt.bg=c("white","darkgray"),pt.cex=1.3,
    cex=0.8,bty="n")
```

Our phylogeny in figure 12.1 contains eighty-four different species of fungus-farming attine ants.

Here is where we get to the part in which the *BioGeoBEARS* package behaves unlike every other R package that we've used in the book.

As we mentioned already, so far throughout this book, every function of every package that we've used has operated within the typical R paradigm in which we use a function or method to read in our data from file. This, in turn, created one or multiple objects in memory within our

---

[14]Not an R object created from it.

[15]Called `getranges_from_LagrangePHYLIP`.

[16]Astute readers might notice that to access elements of the object created by `getranges_from_LagrangePHYLIP`, we use @ rather than the more familiar $ character. This is because *BioGeoBEARS* uses what is called the *S4 object system*—instead of the S3 system that is used by all the other R phylogenetic packages that we have seen in the book.

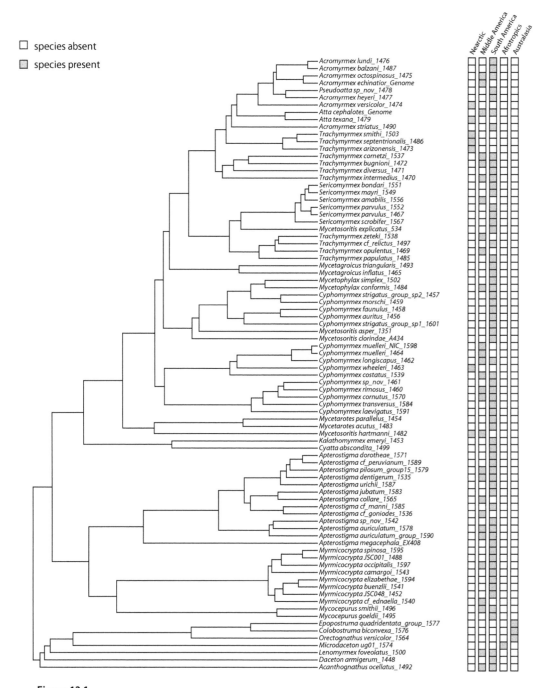

Figure 12.1
Geographic area plot of Attini ants.

R session. We then proceeded to pass these objects on to other functions that run our analysis and returned a result.

BioGeoBEARS, for better or for worse, simply doesn't work this way.

Instead, as we've said, *BioGeoBEARS* is going to take *files* as input and then run the operations that it needs to extract the data it needs from these files internally and then produce a result.

Before we do that, we need to define the conditions of our *BioGeoBEARS* analysis.[17]

One important decision that *BioGeoBEARS* requires us to make before fitting the DEC model is to set the *maximum* number of areas that a single species is allowed to occupy. Logically, this can be as large as the total number of different areas in our analysis—which is five, in our case, remember—but it can also be smaller.

It makes sense to choose a value that is consistent with the size and geographic distribution of our areas, as well as with the biology[18] of the living organisms in our study.

For instance, since in our analysis the areas are continents or subcontinents, and because our organisms are ants, we decided to set the maximum number of areas that a species could occupy to two.[19]

```
max_range_size<-2
```

Our next step is to specify the control conditions of our *BioGeoBEARS* DEC model run.

We do this by creating a single, long object that is just a list of all the different pieces of information that the *BioGeoBEARS* model optimizer will need to fit the model. We'll do this list using the function define_BioGeoBEARS_run.

The way define_BioGeoBEARS_run works is essentially by creating a list of run parameters under the default setting, which we could proceed to update one-by-one just as we'd normally update elements in a regular list in R.

Alternatively, we can supply our desired run option values as function arguments[20] in define_BioGeoBEARS_run and the list will be automatically populated with the values we supplied.

```
## create run object
bgb_run<-define_BioGeoBEARS_run(
    num_cores_to_use=1,
    max_range_size=max_range_size,
    trfn="./attine-tree-pruned.tre",
    return_condlikes_table=TRUE)
## update definition of list element geogfn
bgb_run$geogfn<-"./attine-distribution-data.txt"
```

---

[17] As we'll see in a second, these conditions *include* the file names of our input files but also various control parameters for our analysis.

[18] Especially the dispersal ability.

[19] Some invasive ant species—like the fire ant (*Solenopsis invicta*) and the pavement ant (*Tetramorium caespitum*)—may be found in more than two continents; however, this is only due to anthropogenic activities, rather than the "natural" biogeographic processes of dispersal, extinction, and speciation, and is not true, so far as we know, of any of the fungus-eating ants of our data set.

[20] Most of them, at least. This didn't seem to work for geogfn: the location of our distribution data!

BioGeoBEARS includes a "checking" function, check_BioGeoBEARS_run, which we can now run on our run object to ensure that it's complete. If the function returns TRUE, we know that we should be good to go!

```
check_BioGeoBEARS_run(bgb_run)

## [1] TRUE
```

Now we can optimize our DEC model using maximum likelihood.

```
DEC.fit<-bears_optim_run(bgb_run)
```

bears_optim_run prints out a lot of different information as it fits our model.[21]

The model object, in our case called DEC.fit, is *very* long, and *BioGeoBEARS* doesn't include the typical print or summary methods for the type of object that is produced by bears_optim_run.

Nonetheless, we can see the optimized model by just printing out the fitted model object component called optim_result.

```
DEC.fit$optim_result

##                p1         p2     value fevals gevals
## bobyqa 0.0062679 0.0017895 -143.03         38     NA
##           niter convcode  kkt1 kkt2 xtime
## bobyqa       NA        0 FALSE TRUE  0.65
```

Although they're not well identified in our output here, the parameter p1 is the MLE *dispersal* rate between areas. The parameter p2 is the estimated local *extinction* rate within each area.

Let's plot the ancestral areas that are estimated by bears_optim_run. Our object contains the marginal likelihoods of each geographic region or combination of regions at each node of the tree.

We'll visualize this using the function plot_BioGeoBEARS_results, which (by default) shows the area with the *highest* marginal likelihoods.[22]

```
## subdivide our plotting area using layout
layout(matrix(1:2,1,2),widths=c(0.2,0.8))
## set plotting parameters
par(mar=c(4.1,0.1,3.1,0.1),cex=0.8)
## plot legend and tree
plot_BioGeoBEARS_results(DEC.fit,
    analysis_titletxt="DEC model",
    plotlegend=TRUE,tipcex=0.4,statecex=0.4)
```

---

[21] Which we've mostly suppressed here.

[22] We could also show the scaled marginal likelihoods for each state at each node using pie charts at the nodes—as we did for marginal ancestral character reconstruction in chapter 8—by setting the argument plotwhat="pies".

```
##      LnL nparams        d         e
## 1 -143.03       2 0.0062679 0.0017895
```

In the code chunk above, we actually created two plots: one of the legend and the second of our tree and reconstructed states. To graph them on the same plotting device, we set `layout(matrix(1:2,1,2),widths=c(0.2,0.8))`, which divides the device into two columns—the first encompassing 20 percent of the width of the device and the second comprising the remaining 80 percent (figure 12.2).

12.2.3 The DEC+J model: Adding an additional dispersal mode

Now, let's fit a different model: DEC+J (Matzke 2012, 2014).

In addition to dispersal, extinction, and cladogenesis, this model also includes an additional process called *jump speciation* (J), in which a lineage disperses to a new location and simultaneously speciates.

This model has led to some controversy in the literature. Although we're going to largely ignore this here, we refer interested readers to Ree and Sanmartin (2018) and Klaus and Matzke (2020) for more information.

Once we have a DEC+J fit, we'll then be able to compare the reconstructions as well as evaluate the relative fit of this model, compared to our previous DEC model. Let's see.

To set up this model, we'll need to identify starting values for our maximum likelihood optimization.

We've decided to use the MLEs of the dispersal and extinction rates from our DEC model and to set the starting value of our jump speciation rate as 0.0001.[23]

```
## set starting values for optimization
dstart<-DEC.fit$outputs@params_table["d","est"]
estart<-DEC.fit$outputs@params_table["e","est"]
jstart<-0.0001
```

Now let's update the *BioGeoBEARS* run object, `bgb_run`, that we created for our DEC analysis in the previous section.

```
## update run object with new starting parameter values
bgb_run$BioGeoBEARS_model_object@params_table["d",
    "init"]<-dstart
bgb_run$BioGeoBEARS_model_object@params_table["d",
    "est"]<-dstart
bgb_run$BioGeoBEARS_model_object@params_table["e",
    "init"]<-estart
bgb_run$BioGeoBEARS_model_object@params_table["e",
    "est"]<-estart
```

---

[23] This lattermost quantity doesn't have any particular significance—except for being > 0 and small. In a real analysis, as with all maximum likelihood optimization, it is probably wise to try different various different starting values and see if you obtain the same result!

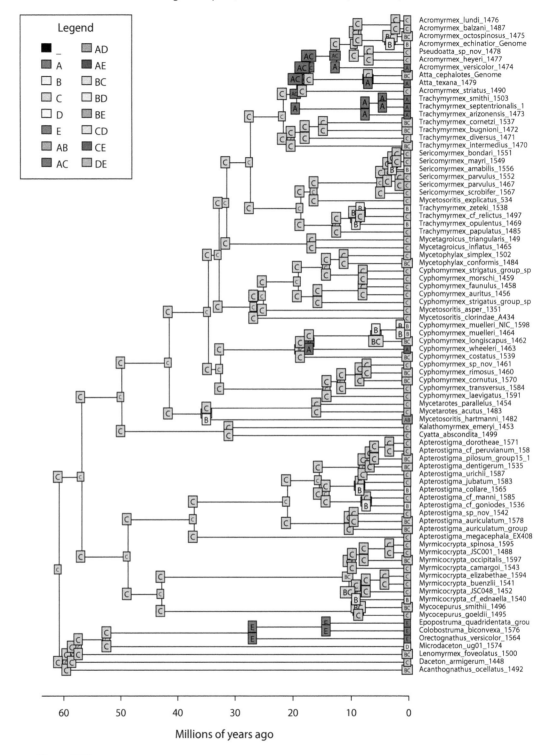

DEC model
ancstates: global optim, 2 areas max. d=0.0063; e=0.0018; LnL=−143.03

Legend

- ■ _
- ▨ A
- □ B
- ▨ C
- □ D
- ▨ E
- ▨ AB
- ▨ AC
- ▨ AD
- ▨ AE
- ▨ BC
- ▨ BD
- ▨ BE
- □ CD
- ▨ CE
- ▨ DE

Millions of years ago

Figure 12.2
Reconstructed ancestral areas for Attini ants under the DEC model: (A) Nearctic, (B) Middle America, (C) South America, (D) Afrotropics, and (E) Australasia.

Next, we need to set the jump speciation parameter, j, to be a free parameter in the model. It was fixed at zero before.

Once we've done that, we can also assign it an initial value for optimization as we did for the other parameters of the model.

```
## update jump speciation parameter to be estimated
bgb_run$BioGeoBEARS_model_object@params_table["j",
    "type"]<-"free"
## set initial value for optimization
bgb_run$BioGeoBEARS_model_object@params_table["j",
    "init"]<-jstart
bgb_run$BioGeoBEARS_model_object@params_table["j",
    "est"]<-jstart
```

Last, let's check to make sure that our *BioGeoBEARS* run is correctly configured using the same check_BioGeoBEARS_run function that we used in the previous section.

```
check_BioGeoBEARS_run(bgb_run)

## [1] TRUE
```

Perfect!

Now, let's optimize this object using bears_optim_run. Just as before, this analysis will print out a lot of information to our display buffer as it runs.

We've suppressed this here, but we can find the fitted model details in optim_result of our fitted model object.

```
DEC_J.fit<-bears_optim_run(bgb_run)
DEC_J.fit$optim_result

##                  p1      p2       p3    value fevals
## bobyqa  0.0040463  1e-12  0.021333 -136.03     61
##          gevals niter convcode kkt1 kkt2 xtime
## bobyqa      NA    NA         0 FALSE TRUE  1.02
```

We should see three different parameters, now, instead of the two we estimated before.

These are p1, which corresponds to our dispersal rate; p2, which is our estimated rate of local extinction; and p3, which is our jump speciation rate.

Just as we did for the DEC model in the previous section, we can plot our DEC+J jump-dispersal speciation results in a similar way using plot_BioGeoBEARS_recon as follows.

```
## subdivide plot device
layout(matrix(1:2,1,2),widths=c(0.2,0.8))
## set plotting parameters
par(mar=c(4.1,0.1,3.1,0.1),cex=0.8)
```

```
## plot legend and tree
plot_BioGeoBEARS_results(DEC_J.fit,
    analysis_titletxt="DEC+J model",
    plotlegend=TRUE,tipcex=0.4,statecex=0.4)

##
## NOTE: multiple states tied
##
## Note: in get_ML_probs(), picking the first state in the tie;
##       use unlist_TF=FALSE to see all states.

##       LnL nparams        d      e        j
## 1 -136.03       3 0.0040463 1e-12 0.021333
```

Overall, the reconstructions in figures 12.2 and 12.3 are quite similar.

For example, both agree that the most likely ancestral area for Attini ants is C, which we can see from figures 12.1 and 12.2 corresponds with South America.

On the other hand, some of the details of the reconstructions of the two figures differ. For example, several nodes in the DEC+J model are reconstructed as being in state C (South America; figure 12.2), while in the DEC model, the most likely state for the same nodes is AC (North *and* South America; figure 12.2).

Let's now compare the overall fit of the two models.

We'll do this by pulling out the log-likelihood of each one and then using a custom *BioGeoBEARS* function to build a table of model fits.

```
## obtain log-likelihoods from each model
logL.DEC<-get_LnL_from_BioGeoBEARS_results_object(DEC.fit)
logL.DECJ<-get_LnL_from_BioGeoBEARS_results_object(DEC_J.fit)
## assemble the results into a summary table
AIC.table<-AICstats_2models(logL.DECJ,logL.DEC,numparams1=3,
    numparams2=2)
print(AIC.table)

##    alt null  LnLalt LnLnull DFalt DFnull DF
## 1           -136.03 -143.03     3      2  1
##    Dstatistic        pval        test       tail
## 1     14.005 0.00018232 chi-squared one-tailed
##      AIC1   AIC2  AICwt1    AICwt2
## 1 278.06 290.06 0.99753 0.0024664
##    AICweight_ratio_model1 AICweight_ratio_model2
## 1                  404.45              0.0024725
```

This table is a bit difficult to wade through, but if we look carefully, we can see that our likelihood ratio test statistic (Dstatistic) is around 14, so compared to a χ^2 distribution with 1 degree of freedom,[24] we strongly reject the null model (DEC) in favor of the alternative model.

---

[24]Because there is only one more parameter in the DEC+J model compared to the DEC model.

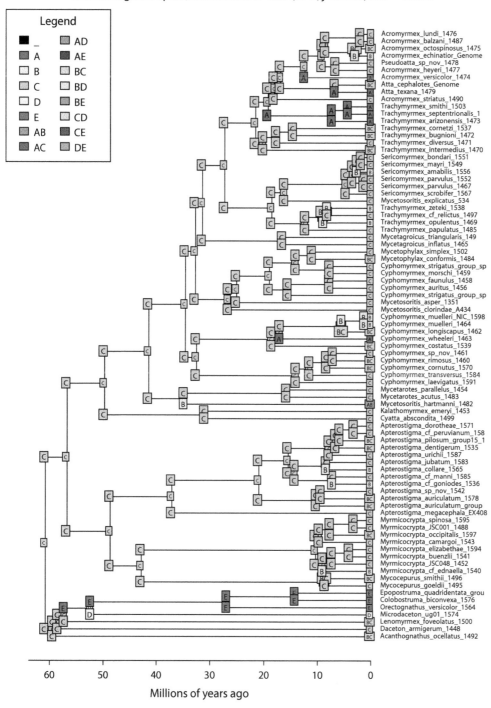

DEC+J model
ancstates: global optim, 2 areas max. d=0.004; e=0; j=0.0213; LnL=−136.03

Figure 12.3
Reconstructed ancestral areas for Attini ants under the DEC+J model. Areas are as in figure 12.2.

Akaike information criterion (AIC) values paint a similar picture, with almost all Akaike model weight falling to the DEC+J jump-dispersal speciation model.[25]

There are several other models implemented in the *BioGeoBEARS* package in addition to the DEC and the DEC+J models. We focused on these two merely because they are the two most popular for reconstructing ancestral areas and comparing the fit of alternative biogeographic scenarios using phylogenies in R!

12.3 Phylogenetic community ecology

Phylogenetic community ecology is a rapidly growing field of research in which scientists have begun leveraging phylogenies and the methods of phylogenetic comparative biology to better understand the process and pattern of ecological community assembly and maintenance (reviewed in Webb et al. 2002).

Most analyses in phylogenetic community ecology center on how the species in a local community are arranged or distributed on a phylogenetic tree.

The phylogenetic tree in phylogenetic community ecology typically contains all of the taxa in a regional *species pool*. These are assumed to be the set of species that could conceivably co-occur in a local community—say, all of the tree species found in the entire region that surrounds (and includes) the local community.

Inferences in phylogenetic community ecology tend to be focused on the phylogenetic distribution of the species in local ecological communities on the regional species tree.

For instance, if species in local communities tend to be more *clustered* on the regional species tree,[26] this pattern is often taken as evidence that local communities are assembled via a process of *habitat filtering*—in which a key trait or set of traits, shared only by close relatives on the phylogeny, is required for a species to persist locally (Webb et al. 2002).

By contrast, if local communities consist of species that tend to be distributed widely[27] across the phylogeny, or *overdispersed*, this pattern is frequently interpreted as indicating that local community assembly is driven by competition in which closely related (and thus ecologically similar) species cannot easily coexist (Webb et al. 2002).

12.3.1 Analyzing phylogenetic communities: An empirical example using plants of the San Juan Islands

To see how an analysis of phylogenetic community structure might typically proceed, we'll use some data from an excellent study by Marx et al. (2016), in which the authors obtained community-level presence/absence information from eighty different plant communities sampled across the San Juan Islands of Washington state in the United States.

For this analysis, we'll use the R package *picante* (Kembel et al. 2010), which can be installed from CRAN.

```
library(picante)
```

Our data files for this exercise are SJtree.phy, a phylogenetic tree containing the species in the regional pool, and SJ_ComMatrix.csv, a large matrix containing presence and absence data for the species in the tree from each of eighty plant communities.

---

[25] This is precisely the behavior predicted by Ree and Sanmartin (2018) in their note about the DEC+J model, so we recommend reading that article if you find the same result in your own data!

[26] Than, for example, you would expect by chance.

[27] Once again, compared to a null model.

Let's start by reading in our data from file.

```
## read in tree from file
sj.tree<-read.tree("SJtree.phy")
print(sj.tree,printlen=2)
```

```
##
## Phylogenetic tree with 366 tips and 365 internal nodes.
##
## Tip labels:
##   Achillea_millefolium, Anthemis_arvensis, ...
##
## Rooted; includes branch lengths.
```

```
## read in data matrix from file
sj.data<-read.csv("SJ_ComMatrix.csv",row.names=1)
```

Our data consist of presence (1) and absence (0) information for a large number of different plant communities.

We can visualize these data in a highly similar way to figure 12.1 using plotTree.datamatrix; however, we have a larger tree, as well as more columns in our data matrix, so that presents us with some difficulties.

First, due to the large number of rows (360) and columns (80) in our data matrix, the black lines between our rows and columns would ordinarily be too thick (relative to the spaces between them) to make our visualization readable. We'll erase them using par.[28]

Since our lines are now erased, it makes sense for us to use a very light absence color (rather than white) so that we can see a clearer outline of our presence/absence grid. We decided to set the color to be semitransparent using the *phytools* function make.transparent.[29]

Finally, we'll *expand* our x dimension, so that we have more space to see our plotted grid.

```
## set colors
colors<-setNames(replicate(ncol(sj.data),
    setNames(c(make.transparent(palette()[4],0.15),
    palette()[2]),0:1),
    simplify=FALSE),colnames(sj.data))
## make foreground color transparent
par(fg="transparent")
## plot tree and data matrix
plotTree.datamatrix(sj.tree,sj.data,yexp=1,header=TRUE,
    xexp=1.6,fsize=0,colors=colors)
## reset foreground color to black
par(fg="black")
## add legend
legend(x=0,y=40,c("absent","present"),pch=15,
    col=colors[[1]],bty="n",pt.cex=1.5)
```

---

[28] We do this by setting par(fg="transparent"), which changes the default foreground color in our plot from black to invisible.

[29] We could have also used the base R function rgb to make transparent colors.

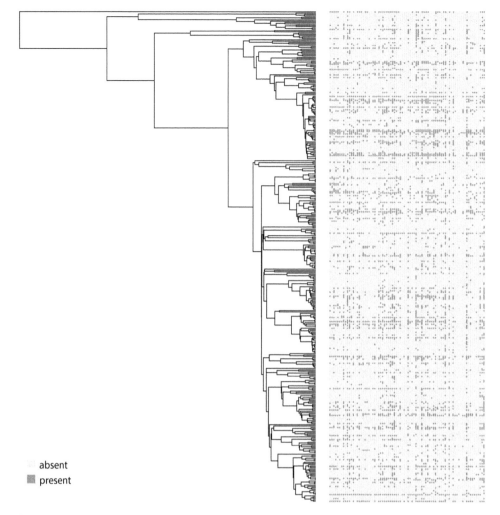

absent
present

Figure 12.4
Presence/absence data for eighty different plant communities from the San Juan Islands archipelago. Data and phylogenetic tree from Marx et al. (2016).

The result can be seen in figure 12.4.

For our next step, let's calculate one simple measure of community diversity for each of our eighty different communities: phylogenetic diversity, or PD.

PD measures the sum of the phylogenetic branch length spanned by a community[30] and is commonly used in ecology and conservation biology for a variety of different purposes.

The logic underlying measurement of the diversity of a community in this way is philosophically similar to that of measuring functional diversity, in which we might consider not only the taxa present in an area but also the variety of functional roles in the ecological community that they play.

In the case of PD, we assume that closely related taxa (by virtue of their shared evolutionary history) are more similar than distantly related ones. By measuring diversity via the sum of the

---

[30]Including, typically, an additional edge from the MRCA of our community phylogeny to the global root of our regional species tree.

spanning edge lengths from the phylogeny of the species found there, we should be able to get a sense of the evolutionary or phylogenetic variety of the species present at each site.

Obviously, phylogenetic diversity will tend to increase as a function of species richness.[31] Later, we'll look at the relationship between PD and local community species richness.

To compute PD for each site, we'll use the *picante* function pd. pd takes a community data matrix as its first argument (samp) and our community phylogeny as its second.

It requires, however, that we supply our input matrix with communities in rows and species in columns—precisely the opposite of how we have it now!

This is easy to remedy using the *transpose* function, t, which we can even apply to our community data matrix within our pd function call as follows.

```
allPd<-pd(t(sj.data),sj.tree)
```

The object that's returned by pd is a data frame with two columns.

The first (labeled PD) contains the phylogenetic diversity of each community. The second (SR) contains integer values indicating the species richnesses. Let's look at the first few rows of our data frame to get an idea.

```
head(allPd)
```

```
##                          PD SR
## Aleck_Rock           5840.5 76
## Barren_Island        3973.2 45
## Battleship_Island    5922.6 76
## Blind_Island         5167.0 60
## Boulder_Island       6083.1 82
## Broken_Point_Island  3365.7 46
```

We can see here that our PD values are in the thousands, while our species richness values (not all shown) vary between 1 and 169.

With PD, we *always* need to keep in mind that the specific values that we get depend on the units of the edge lengths of our tree. Here, since our tree is in millions of years, our PD measure is also in millions of years!

As we mentioned before, PD tends to vary as a function of species richness. Since we have already computed both values for our San Juan Islands plant communities, let's graph them and see (figure 12.5).

```
## set margins
par(mar=c(5.1,4.1,1.1,2.1))
## plot PD as a function of species richness
plot(allPd[,2:1],bty="n",pch=21,bg="gray",cex=1.5,
    las=1,cex.axis=0.7,cex.lab=0.9,
    xlab="Species richness",
    ylab="Phylogenetic diversity")
```

---

[31] The raw number of species found at a place.

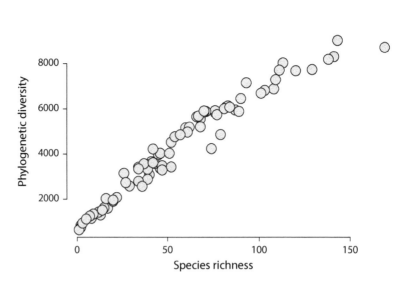

Figure 12.5

Phylogenetic diversity as a function of plant community species richness across eighty plant communities in the San Juan Islands.

What we see here is very interesting and typifies the relationship between species richness and PD: phylogenetic diversity increases with species richness—but at a decreasing rate.

To really understand this pattern, we'll need to compare the phylogenetic relationships among the species in our communities to a null model.

The null model that we'll use is one in which communities are assembled by random selection from the regional species tree, without regard to the structure of the phylogeny.

This is quite an easy process to simulate for any particular assemblage. For example, let's just take the first community, `"Aleck_Rock"`, from our San Juan Islands data set. To make a null distribution of PD for this community, we just need to randomly resample the 0s and 1s from the corresponding column[32] a large number of times, each time recalculating PD.

Let's try it.[33]

```
## create a null distribution of species
## assemblages for the site 'Aleck_Rock'
null.Aleck_Rock<-cbind(sj.data[,"Aleck_Rock"],
    sapply(1:999,function(i,x) sample(x),
    x=sj.data[,"Aleck_Rock"]))
rownames(null.Aleck_Rock)<-rownames(sj.data)
```

We won't look at this null data set (it's too big to print!); however, it should consist of a data frame with 366 rows, corresponding to the 366 species in our tree, and 1,000 columns.

The first column is our original presence/absence data for the site denominated `"Aleck_Rock"`, and the remaining 999 columns are randomized communities with the same number of 1s and 0s. Make sense?

---

[32] Or row, in our transposed data frame.

[33] We can leave our observed community as one of the assemblages. This is typical of permutation tests, of which this is a type.

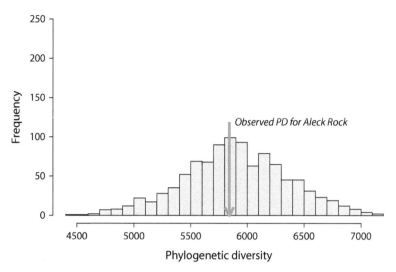

Figure 12.6
Phylogenetic diversity of the plant community on Aleck Rock compared to null assemblages generated by randomly resampling taxa from the regional species tree.

Now, let's compute PD for each one of these. This is easy using pd of *picante* as follows.

```
null.pd<-pd(t(null.Aleck_Rock),sj.tree)
```

If we look at just the top part of this data frame, we should see (1) that the first assemblage has the same PD as did our site (Aleck Rock) in the original data set and (2) that every community of our 999 subsequent null assemblages has the same species richness (seventy-six) as our original assemblage.

Let's plot our null assemblage and overlay our observed value of PD for Aleck Rock (figure 12.6).

```
## set plotting parameters
par(mar=c(5.1,4.1,1.1,2.1))
## create a histogram of the null distribution of
## PD for Aleck Rock
h<-hist(null.pd$PD,breaks=20,
    xlab="Phylogenetic diversity",
    main="",las=1,cex.axis=0.7,cex.lab=0.9,
    ylim=c(0,250))
## add an arrow & text showing the observed PD
arrows(allPd["Aleck_Rock",1],1.2*max(h$counts),y1=0,
    lwd=3,col=palette()[4],length=0.1,lend=1)
text(x=allPd["Aleck_Rock",1],y=1.2*max(h$counts),
    "Observed PD for Aleck Rock",pos=4,
    cex=0.8,font=3)
```

This shows us that the observed PD for Aleck Rock is almost exactly what we would expect by chance if we assembled the plant community there by randomly picking seventy-six species from the regional species tree!

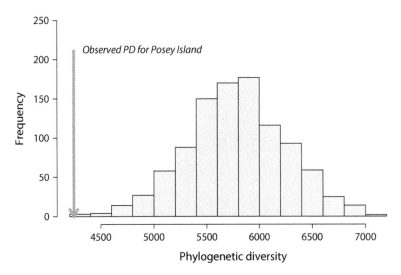

Figure 12.7
Phylogenetic diversity of the plant community on Posey Island compared to null assemblages generated by randomly resampling taxa from the regional species tree.

This is just one of our eighty communities. We could undertake the same analysis on a different community, this time Posey Island ("Posey_Island" in our data frame).

Posey Island has a plant species assemblage of very similar size (seventy-four vs. seventy-six) compared to Aleck Rock.

```
## generate 999 random species assemblages for the
## site 'Posey_Island'
null.Posey_Island<-cbind(sj.data[,"Posey_Island"],
    sapply(1:999,function(i,x) sample(x),
    x=sj.data[,"Posey_Island"]))
rownames(null.Posey_Island)<-rownames(sj.data)
## compute PD for all null assemblages
null.pd<-pd(t(null.Posey_Island),sj.tree)
## plot null distribution and observed PD for
## Posey Island
par(mar=c(5.1,4.1,1.1,2.1))
h<-hist(null.pd$PD,breaks=20,
    xlab="Phylogenetic diversity",
    main="",las=1,cex.axis=0.7,cex.lab=0.9,
    ylim=c(0,250))
arrows(allPd["Posey_Island",1],1.2*max(h$counts),y1=0,
    lwd=3,col=palette()[4],length=0.1,lend=1)
text(x=allPd["Posey_Island",1],y=1.2*max(h$counts),
    "Observed PD for Posey Island",pos=4,
    cex=0.8,font=3)
```

In this case, there we find much *less* phylogenetic diversity in the local community compared to that expected by chance (figure 12.7).

We haven't calculated a *P*-value yet, but comparing our observed value of PD to the null distribution that we obtained by randomly assembling null communities from the regional species pool, it's evident that Posey Island's PD is significantly low in value.

Luckily, when we have a large number of communities, such as our eighty plant assemblages from the San Juan Islands, we don't need to compute null distribution for each community one by one.

Instead, we can use the *picante* function `ses.pd`. `ses.pd` computes PD for each of our communities, and it also simulates a null distribution and computes *p*-values.

Let's see.

```
pd.test<-ses.pd(t(sj.data),sj.tree,null.model="richness")
head(pd.test)
```

```
##                       ntaxa pd.obs pd.rand.mean
## Aleck_Rock               76 5840.5       5878.3
## Barren_Island            45 3973.2       4301.7
## Battleship_Island        76 5922.6       5903.1
## Blind_Island             60 5167.0       5087.3
## Boulder_Island           82 6083.1       6139.8
## Broken_Point_Island      46 3365.7       4350.2
##                       pd.rand.sd pd.obs.rank
## Aleck_Rock                445.29         468
## Barren_Island             423.23         211
## Battleship_Island         463.38         517
## Blind_Island              439.43         575
## Boulder_Island            451.92         457
## Broken_Point_Island       414.59           6
##                         pd.obs.z pd.obs.p runs
## Aleck_Rock             -0.084884    0.468  999
## Barren_Island          -0.776258    0.211  999
## Battleship_Island       0.042014    0.517  999
## Blind_Island            0.181257    0.575  999
## Boulder_Island         -0.125534    0.457  999
## Broken_Point_Island    -2.374567    0.006  999
```

Here we set our null model to be `"richness"`, which corresponds to the species resampling procedure that we did above.[34]

If we cross-check our two different communities (Aleck Rock and Posey Island), we should see that the result from `ses.pd` corresponds closely with what we already found.[35]

---

[34] There are a number of other different null models also implemented in `ses.pd`. To learn more, we refer readers to the user manual of *picante*.

[35] Keep in mind that as this is a random permutation method, your own results may differ slightly.

```
pd.test[c("Aleck_Rock","Posey_Island"),]

##              ntaxa pd.obs pd.rand.mean pd.rand.sd
## Aleck_Rock      76 5840.5       5878.3     445.29
## Posey_Island    74 4240.1       5781.7     444.64
##           pd.obs.rank  pd.obs.z pd.obs.p runs
## Aleck_Rock        468 -0.084884    0.468  999
## Posey_Island        1 -3.467005    0.001  999
```

In practice, phylogenetic community ecologists now more often use measures of community phylogenetic diversity *other* than the simple PD that we have been working with so far.

Two of the most popular of these are the mean pairwise phylogenetic distance between all species in a community (normally denominated MPD) and the mean nearest taxon phylogenetic distance (MNTD). In general, MPD will typically be correlated with traditional PD. MNTD, on the other hand, is a different way to look at the assembly of ecological communities from a phylogenetic perspective because it considers only the minimum evolutionary distance between each taxon and its closest relative within the community.

To conduct an analysis using MPD or MNTD, we must first compute a matrix called the *patristic distance matrix*.

This is simply the $N \times N$ matrix (for N species in our tree) containing the sum of the branch lengths separating every pair of species on our tree.[36] To compute this matrix, we'll use the function cophenetic from the *ape* package that we've used in every chapter of the book so far.

```
phydistmat<-cophenetic(sj.tree)
```

With this matrix in hand, we're already ready to measure the MPD or MNTD of our communities.[37]

To compute MPD, we will use the function ses.mpd.[38] ses.mpd takes the presence/absence data frame as its first input, but the second input is the patristic distance matrix from cophenetic instead of the phylogeny. We'll use the same null model (null.model="richness") as we did for PD above.

```
clustResult_mpd<-ses.mpd(t(sj.data),phydistmat,
    null.model="richness")
head(clustResult_mpd)

##                   ntaxa mpd.obs mpd.rand.mean
## Aleck_Rock           76  355.87        345.82
## Barren_Island        45  350.57        345.51
## Battleship_Island    76  336.13        346.30
## Blind_Island         60  355.18        346.51
```

---

[36]The patristic distance matrix contains all of the information of our tree, so we haven't thrown out any knowledge about phylogenetic relations with this step.

[37]Both of these different metrics use the patristic distance matrix instead of the tree itself.

[38]Much like our ses.pd function earlier in the chapter.

```
## Boulder_Island              82   363.30          345.40
## Broken_Point_Island         46   279.82          345.23
##                       mpd.rand.sd mpd.obs.rank
## Aleck_Rock                   26.399          666
## Barren_Island                36.916          558
## Battleship_Island            27.050          368
## Blind_Island                 32.363          633
## Boulder_Island               25.351          754
## Broken_Point_Island          36.576           29
##                     mpd.obs.z mpd.obs.p runs
## Aleck_Rock            0.38092     0.666  999
## Barren_Island         0.13692     0.558  999
## Battleship_Island    -0.37583     0.368  999
## Blind_Island          0.26811     0.633  999
## Boulder_Island        0.70599     0.754  999
## Broken_Point_Island  -1.78828     0.029  999
```

Like `ses.pd` above, `ses.mpd` reports *P*-values (in the column `"mpd.obs.p"`); however these are one-tailed. That means that any *P*-value below 0.05[39] corresponds to communities that are significantly phylogenetically clustered compared to the null, whereas *P*-
values in excess of 0.95[40] are communities that are significantly overdispersed compared to
the null.[41]

Let's separate out communities with significantly *low* values of MPD.

```
smallmpd<-which(clustResult_mpd$mpd.obs.p < 0.05)
clustResult_mpd[smallmpd,]
```

```
##                               ntaxa mpd.obs
## Broken_Point_Island              46  279.82
## Crab_Island                      13  233.30
## East_Sucia_8_Island               8  233.98
## Pole_Island                      39  285.56
## Posey_Island                     74  290.07
## Ripple_Island                    52  286.56
## Unnamed__1_near_Long_Island      47  284.23
##                          mpd.rand.mean
## Broken_Point_Island             345.23
## Crab_Island                     349.32
## East_Sucia_8_Island             354.72
```

---

[39] Or, perhaps, 0.025—if we are really doing a two-tailed analysis.

[40] Or 0.975.

[41] Of course, if our communities were randomly assembled from the regional pool, then we would expect
about 5 percent of them to be significantly clustered or overdispersed, at an $\alpha = 0.05$ level. As such, we should
probably pay more attention to the excess number of clustered or overdispersed communities, or the relationship between phylogenetic diversity and other ecological factors, rather than simply whether or not some
communities have significantly nonrandom diversity!

```
## Pole_Island                             348.17
## Posey_Island                            346.04
## Ripple_Island                           345.61
## Unnamed__1_near_Long_Island             344.60
##                               mpd.rand.sd
## Broken_Point_Island                36.576
## Crab_Island                        73.805
## East_Sucia_8_Island                98.624
## Pole_Island                        41.223
## Posey_Island                       27.868
## Ripple_Island                      34.802
## Unnamed__1_near_Long_Island        35.613
##                               mpd.obs.rank
## Broken_Point_Island                   29
## Crab_Island                            6
## East_Sucia_8_Island                   39
## Pole_Island                           49
## Posey_Island                          16
## Ripple_Island                         40
## Unnamed__1_near_Long_Island           42
##                               mpd.obs.z mpd.obs.p
## Broken_Point_Island             -1.7883     0.029
## Crab_Island                     -1.5720     0.006
## East_Sucia_8_Island             -1.2242     0.039
## Pole_Island                     -1.5189     0.049
## Posey_Island                    -2.0083     0.016
## Ripple_Island                   -1.6967     0.040
## Unnamed__1_near_Long_Island     -1.6953     0.042
##                               runs
## Broken_Point_Island            999
## Crab_Island                    999
## East_Sucia_8_Island            999
## Pole_Island                    999
## Posey_Island                   999
## Ripple_Island                  999
## Unnamed__1_near_Long_Island    999
```

We see that a total of six out of our eighty communities have significantly lower values of MPD in a one-tailed test than we expected by chance.

This is barely more than the expected number of significantly clustered phylogenetic clustered communities (four) if species assembly occurred by random chance in the San Juan Islands archipelago.

Overall, we find relatively little evidence for phylogenetic clustering or overdispersion when we measure phylogenetic diversity using traditional PD or MPD. Analysis of MNTD can be conducted in a similar way using *picante*—we just substitute the function ses.mntd. We'll leave that for you to do on your own as practice.

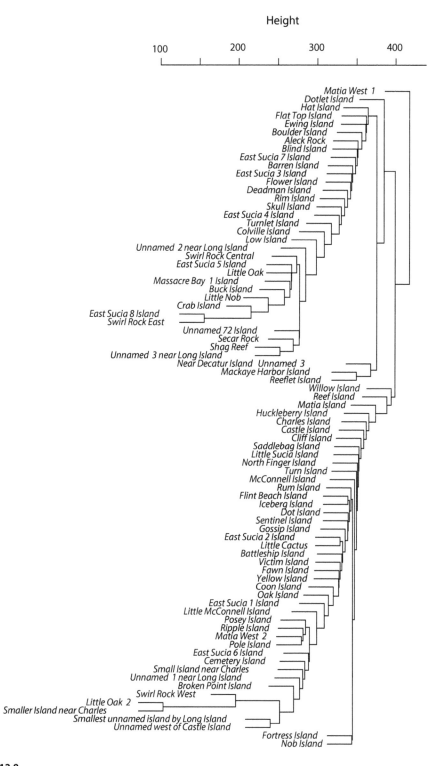

Figure 12.8
Dendogram showing the phylogenetic similarity of eighty plant communities from the San Juan Islands.

Last, it's not too difficult to compare phylogenetic similarity and dissimilarity *between* different communities.[42]

To do this, we'll use the *picante* function `comdist`. `comdist` computes the distance between communities by calculating the mean pairwise phylogenetic distance between each assemblage. It takes our presence/absence matrix as input, along with our patristic distances.

```
D.comdist<-comdist(t(sj.data),phydistmat)
```

`D.comdist` is a distance matrix, in which the pairwise elements are the distances between the species assemblages of difference communities—while taking phylogenetic relationships into account. It's a big matrix—too big for us to print. However, we can visualize the similarity and dissimilarity of the different communities in the set using a dendrogram—a tree-like clustering visualization. Communities that cluster together are more similar in the phylogenetic diversity composition compared to communities farther apart in the dendrogram.

```
## create a community dendrogram
community_dendro<-hclust(D.comdist)
## remove underscore character from site names
community_dendro$labels<-gsub("_"," ",
    community_dendro$labels)
## graph community dendrogram
plot(community_dendro,font=3,cex=0.5,main="",xlab="")
```

Here we see a dendrogram that shows which communities are most similar, as measured by the phylogenetic distances between the species that make them up (figure 12.8). This can be a useful way to visualize broad patterns of diversity across a set of communities.

12.4 Practice problems

12.1 Explore the effects of the `max_range_size` argument on the behavior of ancestral area reconstruction in *BioGeoBEARS*. One place to start is to set this parameter to 1, rerun the analyses, and compare the results—and then repeat this with `max_range_size <- 3`. How does this analysis affect your conclusions about attine ant distributions?

12.2 Pretend that the attine areas are communities, and carry out an analysis of phylogenetic community structure on these areas. Note: If you find significantly lower PD than expected at this scale, is there an alternative explanation for this pattern other than habitat filtering? If so, what is it?

12.3 Rerun the phylogenetic community structure analysis for San Juan Island plants using mean nearest taxon distance (MNTD). How do your results compare to PD and MPD? Make a table summarizing your results across all communities.

---

[42] Computing phylogenetic diversity for each community is analogous to measuring α-diversity. By contrast, if we ask how species composition changes *between* communities while taking phylogeny into account, we are measuring something more akin to phylogenetic β-diversity. For a more complete discussion of this, see Swenson (2014).

Plotting phylogenies and comparative data 13

One of the greatest advantages of undertaking statistical analysis in the R computational environment is the powerful toolset that it offers in data visualization. This statement applies as much to the graphing phylogenies and phylogenetic data as it does to data of any type.

In fact, throughout this book, we've already demonstrated numerous useful techniques for plotting phylogenies, phylogenetic models, and phylogenetic comparative data in R.

For instance, in chapter 1, we saw how to plot phylogenies in various styles using the "phylo" object class plot method (plot.phylo) from the *ape* package, as well as using plotTree from *phytools*. We also learned how to project a phylogeny in a two-dimensional phenotype space, which we did using the *phytools* function phylomorphospace.

In chapter 4, we learned how to visualize a Brownian motion simulation of evolution on the tree (using simBMphylo) and how to plot likelihood surfaces and null distributions for the popular λ and K measures of phylogenetic signal, respectively.

In chapter 5, we saw how to graph more than one phylogeny in a multipanel figure,[1] how to plot a phylogeny with mapped regimes using the *phytools* plot.simmap method, and how to graph a barplot next to a tree using plotTree.barplot.

In chapter 6, we learned how to plot a fitted Mk discrete character evolution model, and in chapter 7, we saw how to visualize multiple discrete characters next to the tips of a plotted phylogeny (using the *phytools* function plotTree.datamatrix), how to plot a fitted Pagel (1994) model, how to visualize the results of hidden-rates analysis (using *phytools* as well as plotMKmodel in the *corHMM* package), how to plot a polymorphic character evolution model, and how to graph a simulation of the threshold trait evolution model.

In chapter 8, we saw how to visualize the observed and estimated ancestral states for a phylogeny using a color gradient (with the function contMap), how to project a tree in a space defined by time (on the horizontal axis) and phenotype (using the function phenogram),

---

[1] Using the *base* R function par.

how to show estimated ancestral values for a discrete trait at nodes using pie graphs, and how to graphically summarize the results of stochastic character mapping.

In chapter 9, we learned how to create a lineage-through-time plot, how to plot large phylogenies using custom axes and settings, and how to graph the likelihood surface or a posterior distribution from a model-based analysis of speciation and extinction rates on the tree.

In chapter 10, we saw how to add special node labels to a tree, and we learned how to add various graph elements, such as vertical dashed lines, to a phylogeny that was already plotted, and in chapter 11, we learned how to graph the results from different state-dependent diversification models.

In some cases, our graphs from previous chapters took advantage of completely "canned"[2] plotting routines.

In other situations, however, we took existing plotting functions and used our knowledge of R graphics to modify the graph, add elements to it, or combine multiple types of visualizations.

As we've covered so many different plotting functions already in R, in this chapter, we *will not* survey canned phylogeny plotting routines.[3] Instead, we *will*:

1. Develop a stronger understanding of the R plotting device so that we can begin to write our own scripts for graphing phylogenies or phylogenetic data in R.
2. See a general algorithm for drawing trees so that we can write our own custom plotting functions in the R language.

Rather than monotonously re-reviewing plotting methods that we've already learned in earlier chapters, then, the purpose of this chapter is to help *empower* you to customize your own plotted phylogenies in R and maybe even develop new methods of your own!

13.2 Phylogenies in the R plotting environment

Throughout the book, we've used many different plotting methods to visualize phylogenies and the results of phylogenetic comparative analyses in R.

In some cases, this was via entirely prepackaged routines in which we simply used an existing R phylogenetic plotting method to create our visualization—in some cases updating default argument values, in others not.

Frequently, however, we used the wide range of R plotting routines to modify or add features and plot elements to our plotted phylogeny or phylogenetic model.

In this part of the chapter, we'll demonstrate how to do the same with your own graphed trees and models in R. It's also worth mentioning that there is a different way to manipulate tree plots using the package *ggtree*, which we will not cover here (for details and examples, see Yu 2020; Yu et al. 2017, 2018).

13.2.1 Phylogenies are plotted in the standard R graphical device

The first principle to remember when plotting phylogenies in R is that *phylogenies are plotted using standard R graphical devices.*[4]

---

[2] "A program which has been written to solve a particular problem, is available to users of a computer system, and is usually fixed in form and capable of little or no modification" (Avallone et al. 2003), from The Free Dictionary.

[3] Although many others exist that we have not covered, they tend to function in similar ways to the methods we've shown so they shouldn't be too hard to learn!

[4] The only exception to this is the very small number of methods that use the three-dimensional visualization library *rgl* (Murdoch and Adler 2021). Although these tools are pretty neat, they're out of scope for this chapter.

This means that any function that can be used to modify plotting parameters or add plot features and elements to a standard plot in R should work equally well for a plotted phylogeny.

To see what we mean, why don't we start with the very simple example that we saw at the beginning of the book—a tree of vertebrate species.

Here, we've updated the species names to Latin binomials, and we added edge lengths in millions of years.[5]

```
## load the phytools package
library(phytools)
## read vertebrate tree from Newick string
vert.tree<-read.tree(text=
    "(Carcharodon_carcharias:473,(Carassius_auratus:435,
    (Latimeria_chalumnae:413,(((Homo_sapiens:73.8,
    Lemur_catta:73.8):22.2,(Myotis_lucifugus:79,
    (Sus_scrofa:62,(Megaptera_novaeangliae:56,
    Bos_taurus:56):6):17):17):216,(Iguana_iguana:280,
    Turdus_migratorius:280):32):101):22):38);")
```

Let's plot this tree using the standard R graphical device.

Although we might've done this just using a simple call of *ape*'s generic `plot` method for this object class, what we'll do instead is first create a plot area and set the parameters of that plotting area.

We can start this using a call of the function `par`.

`par`, as we've seen in prior chapters, is a very handy base R function to set the plotting parameters of our active R graphical device. In this case, we're going to use it to set the margins of the plot, which we'll assign the value `c(1.1,1.1,4.1,4.1)`.

The way to read this is bottom, left, top, and right margin widths: 1.1 lines, 1.1 lines, 4.1 lines, and 4.1 lines.

```
par(mar=c(1.1,1.1,4.1,4.1))
```

Next, we'll create a new plot. We're not going to actually *plot* anything yet[6]—we're just going to set up our plotting area.

In this step, we can assign the limits of the *x*- and *y*-axes,[7] which we'll do using the arguments `xlim` and `ylim`, respectively.

We're also going to turn off the R default of plotting a box around our plotting area,[8] and we'll set it so that our plot is created without axes.[9]

```
plot(NA,xlim=c(0,693),ylim=c(1,Ntip(vert.tree)),
    bty="n",axes=FALSE,xlab="",ylab="")
```

---

[5] The latter we obtained from the super cool website http://www.timetree.org/ by Hedges et al. (2006).
[6] This is accomplished by setting x=NA in the function call, as we've done.
[7] In the units of our plot.
[8] Using the argument bty="n".
[9] axes=FALSE.

Now that our plot is created, let's add axes—but let's put them *above* and *to the right* of our plot area, instead of where they are usually plotted.[10]

The way this is accomplished is by using the axis function argument side.[11]

```
axis(4,at=1:Ntip(vert.tree),cex.axis=0.8)
axis(3,cex.axis=0.8)
```

While we are at it, we also can set the font size for the labels of our axis using the argument cex.axis.[12]

Next, let's put some grid lines on our plot using the function grid. This will give us a better sense of how our tree plot is organized in our plotting area.

```
grid()
```

Finally, we can go ahead and plot our tree. We set the plotTree argument add to be TRUE so that our plotted tree is graphed onto our open plot, rather than by opening an entirely new plot!

```
plotTree(vert.tree,ftype="i",fsize=0.7,
    xlim=c(0,650),ylim=c(1,Ntip(vert.tree)+1),
    mar=par()$mar,add=TRUE)
```

We can see several things from the graph of figure 13.1.

First, the vertical positions of the tips of the tree are plotted from 1 through N for N tips.

Strictly speaking, this *need not* always be true—for instance, the plot method developer might've decided to plot the tip heights from 0 through 1 in increments of $1/(N-1)$.

Nonetheless, this is the most common convention for right- and left-facing plotted phylogenies in R.[13]

Second, the x-axis starts at 0 and then runs forward toward the present in the units of the branch lengths of our tree.

Thus, because the edge lengths of our input phylogeny were in units of millions of years, so is the x-axis of our plot!

13.2.2 Adding graphical elements to a plotted tree

To see how we can use our newfound knowledge about plotting a phylogeny in the R graphical device, let's use the relatively simple example of showing geological periods with boxes using the standard geological color scale.[14]

---

[10] That is, below and to the left: the default in R.

[11] Just as with par(mar), the order is bottom, left, above, and then right.

[12] Remember, cex is short for *character expansion factor* in R, although it's often used to expand or shrink lots of different kinds of plot elements!

[13] For up- and down-facing trees, we could just flip this and say that this was the convention for the x-axis of the plot instead.

[14] Ours came from https://timescalefoundation.org/ (Gradstein et al. 2020).

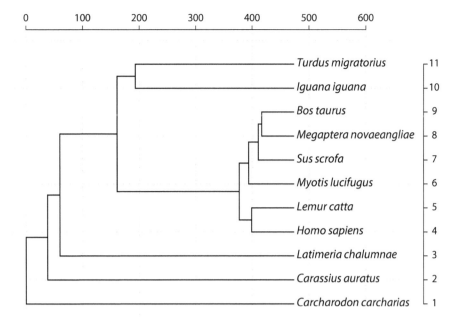

Figure 13.1
Plotted tree of vertebrate species with gridlines and axes.

Before we do that, why don't we remind ourselves about how to plot a tree *backward*—so that the *x*-axis runs from right to left in[15] millions of years before the present.

We first attempted this in chapter 10.

The way we managed this, remember, was by *flipping* the order of the limits of our *x*-axis,[16] so that the bigger value is first and the smaller one second.[17] Then we can just plot our tree in a *rightward* direction and it will appear to face *leftward*!

We'll set the argument value of xlim to c(473,-177) because 473 is the *total depth* of our tree,[18] while −177 is the amount of space we want to leave *right* of the tips to plot our tip labels (figure 13.2).[19]

```
plotTree(vert.tree,ftype="i",fsize=0.7,
    direction="leftwards",xlim=c(473,-177),
    ylim=c(1,Ntip(vert.tree)),mar=c(5.1,1.1,1.1,1.1))
axis(1,at=seq(0,400,by=100),cex.axis=0.8)
mtext("millions of years",side=1,at=200,line=3,
    cex=0.9)
```

We could've used the function title to plot our axis title, but we decided to use mtext instead because it gives us a bit more control of the horizontal position of our label.

---

[15]For our case.

[16]Specified using the argument xlim.

[17]This causes our axis to run "backward," so to speak.

[18]The maximum value of the sum of the edge lengths from the root to any tip.

[19]The number 177 is not a magic number—it just happened to work out for our particular tree!

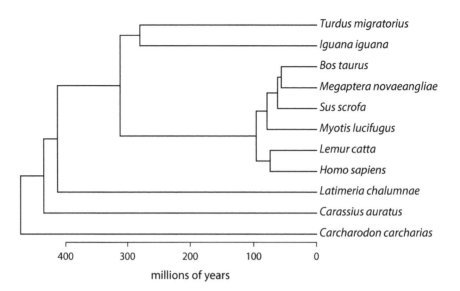

Figure 13.2
A plotted tree with the x-axis flipped.

We think it looks better in the dead middle of our plotted axis, which we set to run from 0 to 400 millions of years before the present day.[20]

So far so good!

Now, let's set our colors for the geological timescale boxes.

To do that, we can use the base R function rgb, which takes as input the level of red, green, and blue[21] of our colors.

Since the root of our vertebrate tree is 473 million years before the present (Hedges et al. 2006), we know that we have to cover three geological eras: the Paleozoic, the Mesozoic, and the Cenozoic.

```
cols<-setNames(c(rgb(153,192,141,0.5,max=255),
    rgb(103,197,202,max=255),
    rgb(242,249,29,max=255)),
    c("Paleozoic","Mesozoic","Cenozoic"))
cols

##    Paleozoic    Mesozoic    Cenozoic
## "#99C08D00"   "#67C5CA"   "#F2F91D"
```

We also need to set a date range for each era. The Paleozoic began 570 million years before the present day, but since our tree doesn't go that far back, why don't we set it to "start" at 500 millions years before present.

---

[20] The default would've been to put the axis title at the midpoint of the x-axis limits of our plot.

[21] Plus an optional α transparency level.

```
eras<-matrix(c(
    500,245,
    245,66.4,
    66.4,0),3,2,byrow=TRUE,
    dimnames=list(names(cols),
    c("start","end")))
eras

##           start   end
## Paleozoic 500.0 245.0
## Mesozoic  245.0  66.4
## Cenozoic   66.4   0.0
```

Just like we did before, we can begin by setting up our plot area. Then we'll proceed to add our various elements to our plot piece by piece.

```
par(mar=c(3.1,2.1,1.1,1.1))
ylim<-c(0.5,1.2*Ntip(vert.tree))
plot(NA,xlim=c(473,-177),ylim=ylim,
    bty="n",axes=FALSE,xlab="",ylab="")
```

This is just want we did before. Remember, we use xlim=c(473,-177) so that our horizontal axis runs from right to left.

Next, we can use polygon to add our geological time periods as follows.

```
for(i in 1:length(cols)){
    polygon(x=c(eras[i,],eras[i,2:1]),
        y=c(rep(ylim[1],2),
        rep(Ntip(vert.tree)+0.5,2)),
        col=make.transparent(cols[i],0.5),
        border="gray")
    text(mean(eras[i,]),Ntip(vert.tree)+0.6,
        names(cols)[i],adj=c(0,0.25),srt=45,
        cex=0.8)
}
```

In this for loop, for every value of i from 1 through the number of rows in our eras matrix, we plot a polygon (using the function polygon).

We're also using the function text to add text elements to our plot. In this case, we'll add the name of each period above the corresponding polygon. As we do so, we can angle them off horizontally (the default) using the text argument srt.

Now let's add a horizontal axis, showing our timescale. We do this using the function axis in the typical way. Even though our axis actually runs from right to left, it's not a problem to indicate our ticks in increasing numerical order using seq.

```
axis(1,at=seq(0,400,by=100),cex.axis=0.8)
```

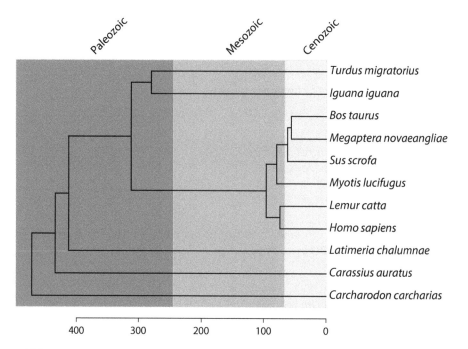

Figure 13.3
Vertebrate phylogeny showing geological eras.

Finally, we'll go ahead and add our phylogeny to the plot.

For this, we will be using the plotTree function of *phytools* because it includes the handy option add=TRUE, which tells R to graph our phylogeny right on top of whatever we had plotted already. The result can be seen in figure 13.3.

```
plotTree(vert.tree,ftype="i",
    fsize=0.7,direction="leftwards",
    xlim=c(473,-177),ylim=ylim,
    mar=par()$mar,add=TRUE)
```

An important thing to keep in mind when we use the add=TRUE option in plotTree (or, likewise, in the S3 plot method for the "simmap" object class, which works the same way) is that if we don't give the function the same values of xlim, ylim, and mar as were used for the rest of the plot, it will reset them to their default values. This could be desirable under some circumstances[22] but is much more likely to mess us up!

Now let's apply what we learned above, but take it to the next level. Instead of plotting geological eras, let's show periods.[23]

Here are the geological timescale colors for the periods covering our vertebrate phylogeny from figures 13.1 through 13.3 (Gradstein et al. 2020).

---

[22] Such as when we're using a different coordinate system for the additional plot elements we need to add to our tree.

[23] For the neontologists like us who need a reminder: periods are nested within eras, so there should be more of them!

```
periods<-matrix(c(508,438,
    438,408,
    408,360,
    360,286,
    286,245,
    245,208,
    208,144,
    144,66.4,
    66.4,23,
    23,2.6,
    2.6,0),11,2,byrow=TRUE,
    dimnames=list(
    c("Ordovician","Silurian",
    "Devonian","Carboniferous","Permian",
    "Triassic","Jurassic","Cretaceous",
    "Paleogene","Neogene","Quaternary"),
    c("start","end")))
cols<-setNames(c(rgb(0,146,112,max=255),
    rgb(179,225,182,max=255),
    rgb(203,140,55,max=255),
    rgb(103,165,153,max=255),
    rgb(240,64,40,max=255),
    rgb(129,43,146,max=255),
    rgb(52,178,201,max=255),
    rgb(127,198,78,max=255),
    rgb(253,154,82,max=255),
    rgb(255,230,25,max=255),
    rgb(249,249,127,max=255)),rownames(periods))
```

Next, we can define our plotting parameters and open the R graphical device for plotting, just as we've done in previous examples.

```
par(mar=c(1.1,4.1,1.1,1.1))
xlim<-c(0.5,1.1*Ntip(vert.tree))
plot(NA,xlim=xlim,ylim=c(508,-110),
    bty="n",axes=FALSE,xlab="",ylab="")
```

In this case, since we intend to plot our tree in an upward direction, we made the *left* margin wider than the others. We also flipped the limits for the x- and y-axes, which makes sense given that our tree is to be plotted facing upward.

It's easy to add our y-axis and axis label,[24] so we'll do that now too:

```
axis(2,at=seq(0,500,by=100),cex.axis=0.7)
mtext("millions of years before present",side=2,
    at=250,las=0,line=3)
```

---

[24]Which here will serve the same function as the x-axis did in our previous plots.

We used mtext instead of title for the same reason as in the prior example.

Next, we can add our various polygons to the tree for each period by using a for loop to iterate over the rows in periods.

```
for(i in 1:nrow(periods)){
    polygon(x=c(rep(xlim[1],2),
        rep(Ntip(vert.tree)+0.5,2)),
        y=c(periods[i,],periods[i,2:1]),
        col=make.transparent(cols[i],0.7),
        border=FALSE)
    text(Ntip(vert.tree)+0.4,
        mean(periods[i,]),
        names(cols)[i],pos=4,
        cex=0.6)
}
```

This time, to facilitate demarcation of our different periods (some of which are similar in color), let's add horizontal dashed lines between each segment. This is easy to do as follows.

```
for(i in 2:nrow(periods))
    segments(x0=xlim[1],y0=periods[i,1],
        x1=Ntip(vert.tree)+0.5,lty="dotted")
```

We're ready to plot our tree, which we can do using plotTree from *phytools* again.

This time, however, to help highlight the plotted phylogeny against its colored background, let's create a "border" effect around our tree by graphing our tree not once but twice.[25]

The first time, we'll set lwd=5 and color="white", and the second time, we can use lwd=1 and the default color ("black").

We'll plot both trees without tip labels by setting ftype="off".

```
plotTree(vert.tree,ftype="off",direction="downwards",
    xlim=xlim,ylim=c(508,-110),lwd=5,color="white",
    mar=par()$mar,add=TRUE)
plotTree(vert.tree,ftype="off",direction="downwards",
    xlim=xlim,ylim=c(508,-110),lwd=1,
    mar=par()$mar,add=TRUE)
```

Finally, to finish up our plot, we can add our tip labels back in.

We're going to print them, though, using the function text as follows.[26]

---

[25]This is a really handy technique that the option add=TRUE in *phytools* facilitates.

[26]Technically, this will only work if our tree is in what *ape* calls "cladewise" order. This will always be the case if the phylogeny is read from file in which it is written as a simple Newick string; however, it is sometimes not the case for NEXUS formatted files. Fortunately, if that's the case, we can simply reorder our tree using reorder.phylo from *ape*.

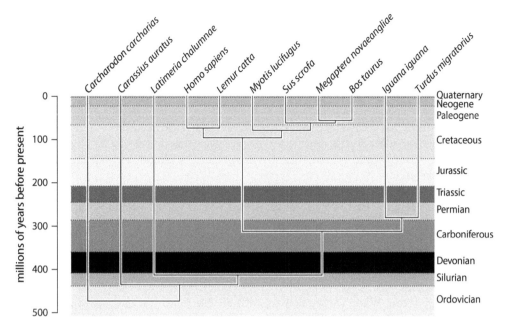

Figure 13.4
Vertebrate phylogeny showing geological periods.

```
for(i in 1:Ntip(vert.tree))
    text(i,-5,sub("_"," ",vert.tree$tip.label[i]),
    srt=45,font=3,cex=0.7,adj=c(0,0.25))
```

Figure 13.4 is still quite simple, but it nonetheless demonstrates a lot of concepts regarding how we can customize plotted phylogenies in R.

We learned the basic coordinate system of a typical phylogeny plot in R, and we saw how easy it can be to superimpose trees onto an existing graph or to add elements of different types to a plotted tree.

What we've learned so far applies primarily to right- or left-facing and up- or down-facing square or slanted phylograms and cladograms. As you've already seen throughout this book, however, this is only a subset of the wide variety of ways in which we can plot phylogenies in R.

In the next section, we'll learn a powerful tool that can be used to help customize phylogenies that have been plotted in a broader range of different styles.

13.2.3 Using `.PlotPhyloEnv` to customize your R phylogenetic plot

An important topic for advanced R users and programmers is that of R environments. In R, an *environment* can be thought of as a container that binds together a set of names or values. An environment, for instance, can consist of all the function names of base R, all the names of functions in the packages that we've loaded in our current R session, or maybe all the data and variables that we read in from file or created during our work in R today.

Most users, and even most R programmers, can largely ignore this concept for the over-whelming majority of our work. When we read in data or when we load a package, R knows where to put the names for that data or the functions of our loaded package.

Likewise, when we need to access elements of a data frame we read from file or use a function in R base, R knows where to look.

As such, many books covering topics in R never mention the concept of environments, and we've largely followed that tradition. Here, though, in this very last chapter, we'll briefly break from it to discuss a particular use for environments that we can exploit to help us make better, customized phylogeny plots in R.

When we're working in R, all the variables and objects we create or read from file are stored in an environment that is called the *global environment* or R_GlobalEnv.

Even though they always exist, other than when accessing function of base R or loaded packages, we rarely have to explicitly access other environments. For plotting phylogenies, however, there's one exception.

Virtually all plotting methods that work with the "phylo" object share one thing in common—that is, they export a special data structure to an environment that is called .PlotPhyloEnv, typically denominated last_plot.phylo. This data structure is essentially just a long list of all the coordinates and plotting parameters of our graphed tree.

Even though it's normally hidden from our view, functions like nodelabels in *ape* use the last_plot.phylo data structure to automatically add elements or modify attributes of a phylogeny plot.

We can use it too. When we see how to, we'll be able to do lots of things with plotted phylogenies in R that would be quite difficult to imagine otherwise.

Let's go back to our vertebrate phylogeny to see a quick example (figure 13.5).

```
plotTree(vert.tree,ftype="i",fsize=0.7,offset=0.5)
nodelabels(bg="white")
tiplabels(bg="white")
```

To access last_plot.phylo data structure from the .PlotPhyloEnv environment, we need to use the base R function get as follows.

```
pp<-get("last_plot.phylo",envir=.PlotPhyloEnv)
```

This is essentially *copying* the data structure from the .PlotPhyloEnv to our working environment, R_GlobalEnv. Let's inspect the object that we've created in so doing.

```
str(pp)

## List of 20
##  $ type            : chr "phylogram"
##  $ use.edge.length : logi TRUE
##  $ node.pos        : num 1
##  $ show.tip.label  : logi TRUE
##  $ show.node.label : logi FALSE
##  $ font            : num 3
##  $ cex             : num 0.7
##  $ adj             : num 0
```

```
##  $ srt             : num 0
##  $ no.margin       : logi FALSE
##  $ label.offset    : num 0.5
##  $ x.lim           : num [1:2]  0 628
##  $ y.lim           : num [1:2]  1 11
##  $ direction       : chr "rightwards"
##  $ tip.color       : chr "black"
##  $ Ntip            : int 11
##  $ Nnode           : int 10
##  $ edge            : int [1:20, 1:2] 12 12 13 13 14 14 15 ...
##  $ xx              : num [1:21] 473 473 473 473 473 473 473 ...
##  $ yy              : num [1:21] 1 2 3 4 5 6 7 8 9 10 ...
```

We can see that this object consists of twenty different elements. Some of them summarize the plotting parameters that we used when we graphed our tree, but others include the coordinate or our plot (specifically pp$xx and pp$yy), as well as attributes of our tree.

It might not be immediately obvious how helpful this information is when we want to customize our tree. What we're going to do now is show two relatively simple examples.

The first involves adding special elements to the tree.

What we'll do to demonstrate this is undertake the task of plotting different color bars on our phylogeny to indicate two different transitions: the evolution of limbs (which occurred on the edge leading to the tetrapods) and the evolution of flight (which occurred twice—on the branch leading to our birds and the branch leading to bats). The result can be seen in figure 13.6.

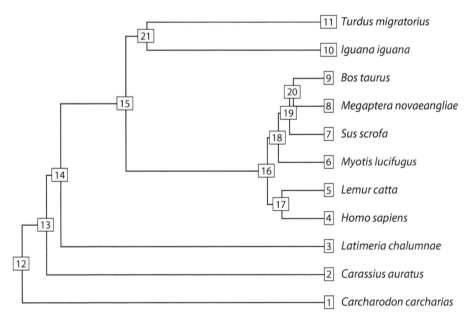

Figure 13.5
Our vertebrate tree again, showing node and tip labels.

```
## plot the tree
plotTree(vert.tree,ftype="i",fsize=0.7)
## get "last_plot.phylo"
pp<-get("last_plot.phylo",envir=.PlotPhyloEnv)
## compute the midpoint x,y coordinates of
## the edge between nodes 14 & 15
xpos<-mean(pp$xx[c(14,15)])+c(-5,5)
ypos<-pp$yy[15]+c(-0.25,0.25)
## add a polygon at that location
polygon(c(xpos,xpos[2:1]),
    y=c(rep(ypos[1],2),rep(ypos[2],2)),
    col="darkgreen")
## compute the midpoint x,y coordinates of
## the edge between nodes 21 & 11
xpos<-mean(pp$xx[c(21,11)])+c(-5,5)
ypos<-pp$yy[11]+c(-0.25,0.25)
## add a polygon at that location
polygon(c(xpos,xpos[2:1]),
    y=c(rep(ypos[1],2),rep(ypos[2],2)),
    col="skyblue")
## compute the midpoint x,y coordinates of
## the edge between nodes 18 & 6
xpos<-mean(pp$xx[c(18,6)])+c(-5,5)
ypos<-pp$yy[6]+c(-0.25,0.25)
## add a polygon at that location
polygon(c(xpos,xpos[2:1]),
    y=c(rep(ypos[1],2),rep(ypos[2],2)),
    col="skyblue")
## add a legend to the plot
legend("topleft",c("limbs","flight"),
    pch=22,pt.bg=c("darkgreen","skyblue"),
    pt.cex=2)
```

In this code chunk, we first plot our tree, just as we've done so many times before. Next, we pull the last_plot.phylo data structure using get. With it, we can now (by referring to figure 13.5), calculate the x and y coordinates of the precise midpoint of each edge. With those values, it becomes very straightforward to add simple polygons to our plot. Finally, we add a legend to our graph to indicate the transitions that correspond with each plotted element.

Lots of phylogeny plotters in addition to plot.phylo and plotTree also export the last_plot.phylo data structure, including some you might not guess.

This includes, for instance, functions for mapping traits onto trees or projecting trees into new spaces, such as the phylomorphospace function that we've seen previously in this book.

In the following code chunk, we'll read some data from the file vert-data.csv, available from the book website,[27] and then plot a phylomorphospace for two traits: log(body size), on the x-axis, and log(clutch or litter size).

---

[27] http://www.phytools.org/Rbook/.

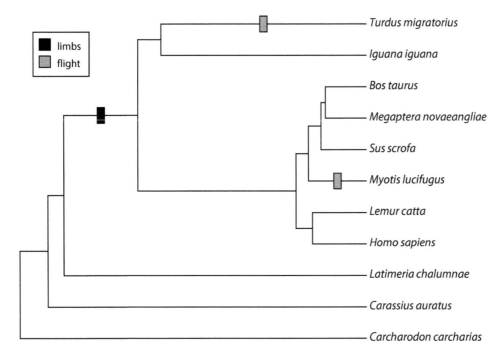

Figure 13.6
Vertebrate tree with changes mapped by extracting `last_plot.phylo` from the `.PlotPhyloEnv` environment.

We'll then proceed to make two modifications to the plot, taking advantage of our `last_plot.phylo` object.

First, we'll recolor the tips of the tree by taxon. This way, we can use a legend to indicate the taxon labels, which is handy because tip labels are often very difficult to decipher in a plotted phylomorphospace.

Then, we'll add arrows to show the ancestor → relationship of each pair of nodes or node and tips. This is also a useful piece of information that would usually be lost in a phylomorphospace plot (figure 13.7).

```
## read data from file
vert.data<-read.csv(file="vert-data.csv",row.names=1)
## load libraries
library(plotrix)
## graph phylomorphospace
phylomorphospace(vert.tree,log(vert.data[,c(1,3)]),
    bty="n",xlab="log(body mass)",
    ylab="log(clutch/litter size)",
    ftype="off",node.size=c(0,0),lwd=2)
## get "last_plot.phylo"
pp<-get("last_plot.phylo",envir=.PlotPhyloEnv)
## color tips by species
n<-Ntip(vert.tree)
```

```
cols<-setNames(gray.colors(n=n,0,1),
    vert.tree$tip.label)
points(pp$xx[1:n],pp$yy[1:n],pch=21,bg=cols,cex=1.6)
## add arrows to indicate direction of each edge
for(i in 1:nrow(pp$edge))
    p2p_arrows(pp$xx[pp$edge[i,1]],pp$yy[pp$edge[i,1]],
        pp$xx[pp$edge[i,2]],pp$yy[pp$edge[i,2]],
        lwd=2,length=0.08)
## add legend
par(font=3)
legend("topright",gsub("_"," ",vert.tree$tip.label),
    cex=0.7,pch=21,pt.bg=cols,pt.cex=1.2,bty="n")
```

Try to figure out the code chunk line by line.[28]

All together, this section is not designed to be in any way comprehensive of the ways the last_plot.phylo can be used to add custom features to your R phylogenetics plot.

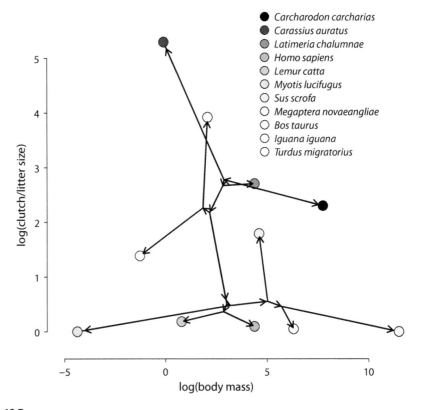

Figure 13.7
Phylomorphospace. Arrows show the ancestor descendant relationship of nodes. Taxa are indicated by color.

---

[28] We included a few comments but left most of our script uncommented on purpose!

Instead, our intent is merely to demonstrate a few of the ways in which it can be used to get you started in exploring custom phylogeny graphing on your own! Hopefully this exploration is fruitful.

13.3 Plotting phylogenies without actually plotting them

Another very useful option that a few phylogeny plotting R functions[29] allow is to *plot* our phylogeny without actually plotting it.

What the heck? Why would we want to do that?

In short, this option[30] opens a graphical device, computes all the coordinates and parameters for our plot, and exports `last_plot.phylo` to the `.PlotPhyloEnv` environment.

This can come in handy in lots of different ways. One simple way is that we can use it, if we want, to add elements to a plotting device before we plot our tree.

For instance, in the following code chunk, we first set `plot=FALSE` and then pull out the coordinates of all the internal nodes to plot them in our open device. Then we can add our tree on *top* of these points—first with thick white edges, then with thin blue edges. This creates the same outline effect that we saw in figure 13.3 earlier in the chapter.

We can see the result in figure 13.8.

```
## assign nicely-spaced branch lengths to our
## vertebrate phylogeny
vt<-compute.brlen(vert.tree)
## open the plotting device and compute
## "last_plot.phylo" without graphing the tree
plotTree(vt,plot=FALSE,ftype="i",offset=0.2,
    mar=rep(2.1,4),type="cladogram",
    nodes="centered",fsize=0.7)
## get "last_plot.phylo"
pp<-get("last_plot.phylo",envir=.PlotPhyloEnv)
## add a grid to the plot
grid()
## add points at all internal nodes
points(pp$xx[1:vert.tree$Nnode+Ntip(vert.tree)],
    pp$yy[1:vert.tree$Nnode+Ntip(vert.tree)],
    pch=16,col=palette()[4],cex=3.5)
## graph our tree on top of these points
plotTree(vt,add=TRUE,ftype="i",offset=0.2,
    color="white",lwd=5,mar=rep(2.1,4),
    type="cladogram",nodes="centered",fsize=0.7)
plotTree(vt,add=TRUE,ftype="i",offset=0.2,
    color=palette()[4],lwd=1,mar=rep(2.1,4),
    type="cladogram",nodes="centered",fsize=0.7)
```

---

[29] Such as `plot.phylo` from *ape* and `plotTree` in *phytools*.
[30] Typically set using the argument `plot=FALSE`.

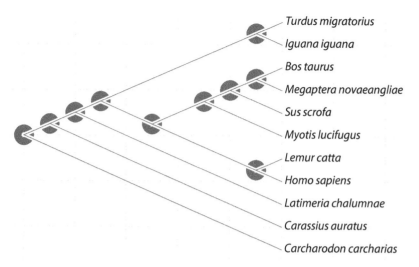

Figure 13.8
Vertebrate phylogeny plotted using the `plot=FALSE` option.

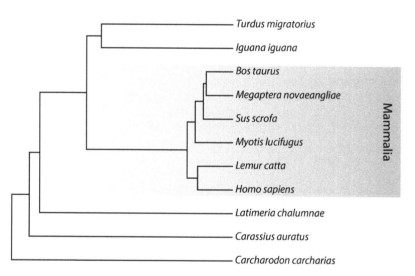

Figure 13.9
Vertebrate phylogeny plotted using the `plot=FALSE` option to show a clade box around mammals.

Likewise, we could similarly take advantage of this feature to add other kinds of background plot elements. For instance, below we show how to add a background clade box around Mammalia.

In this case, what we'll do is, instead of adding a single box, add 100 boxes of different shades on a ramp palette.

In R, the function `colorRampPalette` takes a pair or set of colors as input (we'll use `"white"` and a HEX color for dark gray), and then creates a function that can be used to create a ramp palette between the colors.

```
## open the plotting device and compute
## "last_plot.phylo" without graphing the tree
plotTree(vert.tree,plot=FALSE,mar=rep(2.1,4),ftype="i",
    xlim=c(0,800),fsize=0.7)
## identify the node corresponding the MRCA of mammals
nn<-getMRCA(vert.tree,c("Homo_sapiens","Bos_taurus"))
## get "last_plot.phylo"
pp<-get("last_plot.phylo",envir=.PlotPhyloEnv)
## find x position of the parent node of the mammal MRCA
x0<-pp$xx[phangorn::Ancestors(vert.tree,nn,"parent")]
## define the other edge of the clade box
x1<-par()$usr[2]
## define y limits of clade box
dd<-getDescendants(vert.tree,nn)
y0<-min(pp$yy[dd])-0.25
y1<-max(pp$yy[dd])+0.25
## create the color gradient using colorRampPalette
cols<-colorRampPalette(c("white","#A0A0A0"))(100)
x0<-seq(x0,x1,length.out=101)[1:100]
x1<-seq(x0[1],x1,length.out=101)[2:101]
## plot color gradient
for(i in 1:100){
    polygon(c(x0[i],x1[i],x1[i],x0[i]),
        c(y0,y0,y1,y1),col=cols[i],border=FALSE)
}
## add tree plot
plotTree(vert.tree,mar=rep(2.1,4),add=TRUE,ftype="i",
    xlim=c(0,800),fsize=0.7)
## add clade box label using contrasting color
text(795,mean(c(y0,y1)),"Mammalia",srt=-90,pos=3,
    col="white",font=2,cex=1.2)
```

Line by line here, then, we first plot our tree, without plotting (using `plot=FALSE`, obviously). Next, we get the node number of the common ancestor of all the mammals in our tree. Then, we get the `last_plot.phylo` data structure from our (as yet unplotted) phylogeny. After that, we decide on our left- and rightmost *x*-axis coordinates for the box we'd like to plot. For this graph, we decided to use the height of the immediate *parent* node of the MRCA of Mammalia and the limit of our plot space.[31] We generate our color palette, and then we create 100 different shaded boxes using a `for` loop. Finally, we add our tree to the plot (figure 13.9).

Simple, right?

13.4 Algorithms for drawing trees

For a last exercise of this chapter, we thought we'd try to give a brief overview of a general algorithm for drawing a phylogeny in R.

---

[31] We get this from `par()$usr`.

This algorithm is based on a description by Revell (2014b), but similar computational algorithms are used by all phylogeny plotters written in R or other programming languages.

We'll first give the algorithm for drawing a *slanted phylogram* and then subsequently the algorithm for creating a *traitgram* projection of the tree into a one-dimensional phenotype space.

We're going to first work out the algorithm, and then we'll combine the steps of the algorithm into a function. As such, in this section, we'll also briefly review how to write a custom function in R.

13.4.1 Writing a custom function in R

Writing a function in R is simple and similar to writing a simple program in many other *high-level programming languages*.[32]

In general, the first line of our function should contain the *function definition*: the name of the function and the arguments that it takes as input, if any.

This is followed by a left curly bracket, {.

The code that the function will run on the input arguments or data follows in subsequent lines.

The second to last line of the function contains the value or object to be returned to the user by the function.

Finally, the last line of the function contains a right curly bracket, }.

```
hello<-function(x,y){
    cat("\nHello, World!\n\n")
    x*y
}
```

This function, which is called `hello`, takes two arguments as input: `x` and `y`.

It prints the text `Hello, World!`[33] and then returns the product of `x` and `y` to the user. Let's try it:

```
value<-hello(13,25)

##
## Hello, World!

value

## [1] 325
```

Now that we understand the basic structure of a function in R, let's proceed to write one that is slightly more advanced: taking our phylogeny as input and then computing the mean edge length of all the branches of the tree.

---

[32] A high-level programming language is one that is "strongly abstracted" from the actual operation of the computer.

[33] A *Hello, World!* program is a simple computer program that can be used for almost any programming language and is used to demonstrate the general syntax of the language.

For this function, we'll need to recall what we learned about the structure of the "phylo" object in chapter 1 of this book. Remember, the edge lengths of the tree will be contained in the "phylo" object element edge.length.

We'll also introduce some other error checking, as well as a function, paste, that permits us to combine text and the value of a variable—in our case, the mean edge length of the tree.

Our function will print this value to the screen as well as return it to the user.

```r
meanEdgeLength<-function(phy){
    nm<-deparse(substitute(phy))
    if(!inherits(phy,"phylo")){
        cat(paste("\nInput",nm,
            "is not an object of class \"phylo\".\n"))
        return(NULL)
    } else {
        mel<-mean(phy$edge.length)
        cat(paste("---\nMean edge length of",nm,"is:",
            round(mel,2),"\n---\n"))
        return(mel)
    }
}
```

Now let's try it:

```r
average.branch.length<-meanEdgeLength(vert.tree)

## ---
## Mean edge length of vert.tree is: 137.64
## ---

average.branch.length

## [1] 137.64
```

By contrast, if we try to use it with something that is not an object of class "phylo":

```r
meanEdgeLength(cols)

##
## Input cols is not an object of class "phylo".

## NULL
```

Neat. Our function works exactly as intended!

13.4.2 Algorithm to draw a slanted phylogram

In this part of the chapter, we're going to overview an general algorithm for drawing a right-facing phylogeny in R from a "phylo" object.

Obviously, various functions already exist to do this in *ape*, *phytools*, and other packages.

The purpose of supplying this algorithm is so that you see our general approach to drawing trees—and perhaps apply it to other problems or visualization we haven't even thought of yet!

In our algorithm, we're going to need to do two different traversals of the tree.

The first of these is what is called in phylogenetic analysis[34] a *postorder* traversal (Felsenstein 2004). A postorder traversal goes from the tips of the tree down to the root, passing through every descendant node before its ancestors.

After assigning the vertical position of each tip, in our plotting algorithm, we'll use a postorder traversal to compute the vertical positions of each internal node of the tree.

The second tree traversal will be what is called a *preorder* traversal (Felsenstein 2004).

A preorder traversal goes from the root of the tree toward the tips and must pass through every ancestral node before reaching its descendants.

Fortunately, *ape* contains a very handy `reorder` method for the `"phylo"` object class that sorts the edge order of the tree in such a way as to facilitate each of these two types of traversal.

Our first step is thus to reorder the edges of our phylogeny into what *ape* calls `"cladewise"` order. This reordering has the effect of sorting the edges of the phylogeny such that by passing from the top to the bottom of our `edge` matrix, we'll go through every parental node before reaching its descendants.

```
cw<-reorder(vert.tree,"cladewise")
```

Our next step is to compute *y* positions for each tip in our phylogeny.

As we mentioned earlier in the chapter, we normally space these evenly on the range of 1 through the maximum number of tips in the tree. This needn't be the case—for instance, we could've decided that they should be spaced evenly from 0 to 1 or 0 to 100.

In this step, we'll first create a vector of zeros (called yy, because it will contain the *y* coordinates of our plot) using the function `vector`, and then we'll populate just the first *N* (for our tree of *N* species) with numerical values 1 through *N*.

```
yy<-vector(mode="numeric",length=Ntip(cw)+cw$Nnode)
yy[1:Ntip(cw)]<-1:Ntip(cw)
```

Now we'll reorder the edges of our phylogeny in a `"postorder"` fashion.[35] This reordering puts the edges of the tree in an order such that passing from the top to the bottom of our `edge` matrix now causes us to pass through every daughter edge before reaching its ancestors—the opposite of what we accomplished with `"cladewise"` order.

```
pw<-reorder(cw,"postorder")
```

Next, let's successively iterate over *all* of the internal nodes of the tree and compute their *y* positions as the *average* vertical position of each of their descendants in yy.

---

[34] And, likewise, in graph theory.

[35] We also could've used an *ape*-specific order called `"pruningwise"`. The effect is pretty much the same.

Going through these nodes in "postorder" direction[36] means that we never try to compute the y position of a node before we've set or calculated the y positions of all the descendant nodes. Does this make sense?

```
nn<-unique(pw$edge[,1])
for(i in 1:length(nn)){
    ii<-pw$edge[which(nn[i]==pw$edge[,1]),2]
    yy[nn[i]]<-mean(yy[ii])
}
yy
```

```
## [1]   1.0000  2.0000  3.0000  4.0000  5.0000
## [6]   6.0000  7.0000  8.0000  9.0000 10.0000
## [11] 11.0000  2.3867  3.7734  5.5469  8.0938
## [16]  5.6875  4.5000  6.8750  7.7500  8.5000
## [21] 10.5000
```

At this point, we have all of the vertical coordinates of all our nodes in the tree. Now, we must compute all of the corresponding x coordinates for the same.

To obtain these values, we're going to traverse the tree from the root to the tips because our x coordinates should correspond to the sum of the edge lengths that precede each node in our phylogeny.

An inefficient way to get these would be to go to each node, find all of its ancestors all the way to the root, and then sum their edges.

What we've done by transversing the nodes in a preorder fashion is each time use the x position of the preceding node and sum only the edge length immediately preceding our node of interest.

```
xx<-vector(mode="numeric",length=Ntip(cw)+cw$Nnode)
for(i in 1:nrow(cw$edge))
    xx[cw$edge[i,2]]<-xx[cw$edge[i,1]]+cw$edge.length[i]
xx
```

```
## [1]  473.0 473.0 473.0 473.0 473.0 473.0 473.0
## [8]  473.0 473.0 473.0 473.0   0.0  38.0  60.0
## [15] 161.0 377.0 399.2 394.0 411.0 417.0 193.0
```

At this point, we have the x (in the vector xx) and y (in yy) coordinates of all the internal and tip nodes of the tree.

Let's plot them to see that. The result can be seen in figure 13.10a.

```
plot(xx,yy,bty="n",pch=19,col="black",cex=1.1,
    xlim=c(0,700))
```

Our next step is to draw connecting lines between all the nodes in our tree.

---

[36] That is, using a postorder traversal of the tree.

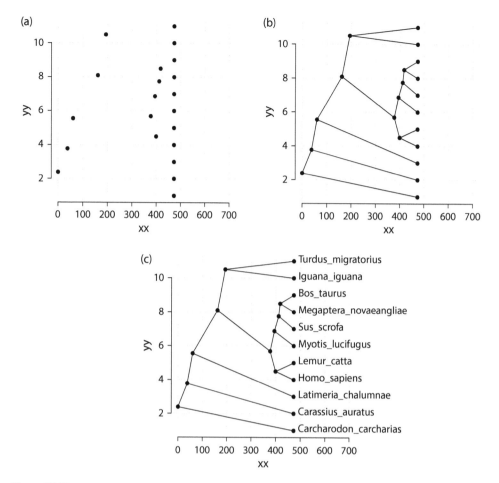

Figure 13.10
Slanted tree algorithm. Each panel, (a), (b), and (c), shows a different step in the algorithm.

We can do this by simply iterating over the edges of the tree (in any order) and drawing a line from the *x*, *y* position of each ancestor to each descendant.

The result of this step can be seen in figure 13.10b.

```
for(i in 1:nrow(cw$edge))
    lines(xx[cw$edge[i,]],yy[cw$edge[i,]])
```

Finally, our last step is to simply add our tip labels to the tree.

We'll do this with the function text, setting pos=4 so that they will be added to the *right* of our tips. The result from this last step can be seen in figure 13.10c.

```
for(i in 1:Ntip(cw))
    text(xx[i],yy[i],cw$tip.label[i],pos=4,cex=0.6)
```

Terrific! Using the general syntax of a function in R that we learned before, let's combine all of these steps into a single R function that simply takes our tree as input.

While we're at it, we can make a few other improvements.

We'll add a check to ensure that the input object is of the correct class.

As we plot our tip labels, we'll also use the base R function `sub` to substitute the underscore character ("_") for a space in each taxon name.

```r
plotSlanted<-function(phy){
    if(!inherits(phy,"phylo")){
        cat(paste("\nInput",nm,
            "is not an object of class \"phylo\".\n"))
        return(NULL)
    } else {
        cw<-reorder(vert.tree,"cladewise")
        yy<-vector(mode="numeric",length=Ntip(cw)+cw$Nnode)
        yy[1:Ntip(cw)]<-1:Ntip(cw)
        pw<-reorder(cw,"postorder")
        nn<-unique(pw$edge[,1])
        for(i in 1:length(nn)){
            ii<-pw$edge[which(nn[i]==pw$edge[,1]),2]
            yy[nn[i]]<-mean(yy[ii])
        }
        xx<-vector(mode="numeric",length=Ntip(cw)+cw$Nnode)
        for(i in 1:nrow(cw$edge))
            xx[cw$edge[i,2]]<-xx[cw$edge[i,1]]+cw$edge.length[i]
        plot(NA,bty="n",axes=FALSE,xlab="",ylab="",
            xlim=c(0,1.4*max(xx)),
            ylim=c(1,Ntip(cw)))
        for(i in 1:nrow(cw$edge))
            lines(xx[cw$edge[i,]],yy[cw$edge[i,]])
        points(xx,yy,pch=21,bg="white",cex=1.1)
        for(i in 1:Ntip(cw))
            text(xx[i],yy[i],sub("_"," ",cw$tip.label[i]),
                pos=4,font=3)
    }
}
```

Now let's try it out on our vertebrate tree. The result is shown in figure 13.11.

```r
par(mar=rep(0.1,4),cex=0.8,lwd=2)
plotSlanted(vert.tree)
```

Excellent. Our function worked perfectly.

13.4.3 Algorithm to project a phylogeny into phenotype space

The last function we'll write is one to project a phylogeny into a one-dimensional phenotypic trait space: the *traitgram* method of Evans et al. (2009). We've used this method in previous parts of the book via the *phytools* function `phenogram`.

Here we'll reuse the `verts-data.csv` file from earlier in the chapter.

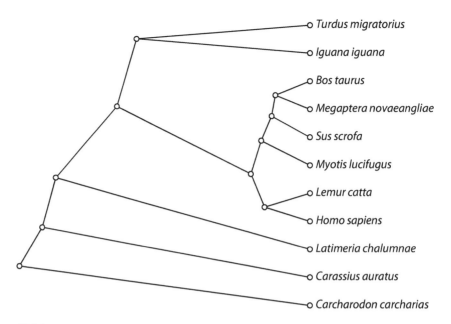

Figure 13.11
Test of the plotSlanted function.

This time, we'll work with the trait body length in meters—on a log-scale.

```
lnBL<-setNames(log(vert.data$Length),
    rownames(vert.data))
lnBL
```

```
## Carcharodon_carcharias         Carassius_auratus
##                1.80829                  -0.96758
##      Latimeria_chalumnae            Iguana_iguana
##                0.69315                   0.69315
##      Turdus_migratorius             Homo_sapiens
##               -2.36446                   0.53063
##             Lemur_catta          Myotis_lucifugus
##               -0.77653                  -2.35388
## Megaptera_novaeangliae                Sus_scrofa
##                2.94444                   0.40547
##               Bos_taurus
##                0.95551
```

To project our tree into a space defined by phenotype (on the horizontal) and time (on the vertical), we first need to compute nodal values for our character.

It usually makes sense for us to use ancestral trait reconstructions (see chapter 8), so we can do that here.

```
anc.states<-fastAnc(vert.tree,lnBL)
```

Now let's combine these values with our observed states at the tips. Together, this will be a vector of all the *x* positions of our nodes.

```
xx<-c(lnBL[vert.tree$tip.label],anc.states)
```

For our next step, we need to compute our vertical positions. This is the height above the root for all the nodes and tips in our phylogeny.

When we worked out the algorithm to plot a tree, we showed you how to do this using a preorder tree traversal.

This time, we'll "cheat"[37] a little bit and use the existing *phytools* function nodeheight[38] to compute the vertical position of all the internal and terminal nodes in our tree.

```
yy<-sapply(1:(Ntip(vert.tree)+vert.tree$Nnode),
    nodeheight,tree=vert.tree)
```

Let's plot these coordinates as our nodes. The result can be seen in figure 13.12a.

```
plot(xx,yy,bty="n",pch=19,col="black",
    ylim=c(0,700))
```

We can then proceed to draw all the edges on our tree, just as we had for a plotted slanted phylogram in the previous section (figure 13.12b).

```
for(i in 1:nrow(vert.tree$edge))
    lines(xx[vert.tree$edge[i,]],
        yy[vert.tree$edge[i,]])
```

Finally, we need to add our tip labels. We'll do that using the function text again—but this time rotating the labels to be vertical by setting the argument srt to 90.

```
for(i in 1:Ntip(vert.tree))
    text(xx[i],1.03*yy[i],vert.tree$tip.label[i],
        adj=c(0,0.25),cex=0.6,srt=90,offset=1)
```

This is pretty good—except that our labels overlap quite a bit with one another in the final figure panel (figure 13.12c).

Let's see if we can improve on this when we combine all of these steps into a new function, as we did for plotSlanted earlier in this section.

---

[37] Actually, this is not cheating at all. In fact, it's a terrific advantage of programming in R that developers can build their packages on top of existing base R and contributed package functionality!

[38] If we know that we're going to need to compute the height of all the nodes in a tree—particularly if that tree is large—it's much more efficient to use a different *phytools* function called nodeHeights. We won't worry about that here, but it doesn't make much difference for such a small phylogeny.

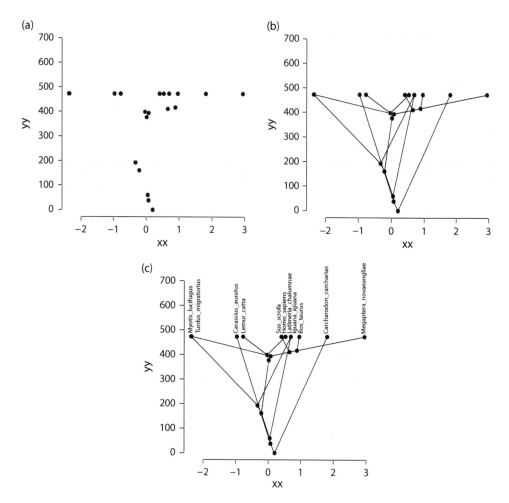

Figure 13.12
Traitgram algorithm. Each panel, (a), (b), and (c), shows a different step of the algorithm.

```
phyloTraitgram<-function(phy,x){
    aa<-fastAnc(phy,x)
    xx<-c(x[phy$tip.label],aa)
    yy<-sapply(1:(Ntip(phy)+phy$Nnode),
        nodeheight,tree=phy)
    plot(NA,bty="n",axes=FALSE,xlim=range(xx),
        ylim=c(0,1.5*max(yy)),
        xlab="trait value",
        ylab="time (millions of years)")
    axis(1)
    ticks<-pretty(yy)
    ticks<-ticks[ticks<=max(yy)]
    axis(2,at=ticks)
    for(i in 1:nrow(phy$edge))
```

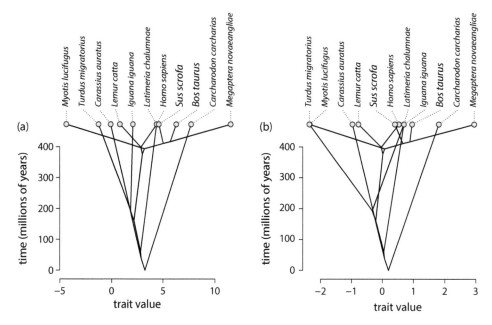

Figure 13.13
Traitgrams for two different characters in vertebrates: (a) body mass in kilograms and (b) body length in meters. Both traits have been transformed to a natural logarithmic scale.

```
            lines(xx[phy$edge[i,]],
                yy[phy$edge[i,]],lwd=2)
    nn<-names(sort(x))
    tips<-setNames(seq(min(x),max(x),length.out=Ntip(phy)),
        nn)[phy$tip.label]
    for(i in 1:Ntip(phy)){
        segments(xx[i],yy[i],tips[i],1.1*yy[i],
            lty="dotted")
        text(tips[i],1.1*yy[i],
            sub("_"," ",phy$tip.label[i]),
            adj=c(0,0.25),cex=0.6,srt=90,
            font=3)
    }
    points(xx[1:Ntip(phy)],yy[1:Ntip(phy)],pch=21,
        bg="gray",cex=1.2)
}
```

Try to see if you can figure out what we did here!
Let's test out our new function:

```
par(mfrow=c(1,2),cex.lab=0.8,cex.axis=0.7)
lnMass<-setNames(log(vert.data[,1]),
    rownames(vert.data))
```

```
phyloTraitgram(vert.tree,lnMass)
mtext("(a)",line=1,adj=0)
phyloTraitgram(vert.tree,lnBL)
mtext("(b)",line=1,adj=0)
```

It works (figure 13.13). Not bad for an afternoon!

13.5 Practice problems

13.1 Make a version of figure 13.4, but for the attine ant phylogenetic tree from chapter 12. For an extra challenge, create your figure so that time goes downward—in other words, with the root of the tree at the top of the page and the tips at the bottom.

13.2 Add the "limbs" and "flight" boxes to the tree that you plotted using plotSlanted. What did you need to change in the code?

13.3 Write your own custom function that plots a phylogenetic tree where the width of the branches is proportional to the number of taxa that descend from that branch. In other words, tip branches have lines that are 1 pixel wide, while a branch that leads to the common ancestor of eight species would be 8 pixels wide. Feel free to start from the code we developed for the function plotSlanted.

References

Aitken, A. C. 1936. On least squares and linear combination of observations. *Proceedings of the Royal Society of Edinburgh* 55:42–48.

Aldrich, J. 1997. R. A. Fisher and the making of maximum likelihood 1912–1922. *Stat. Sci.* 12:162–76.

Avallone, E. A., P. Barry, G. S. Bonn, and W. G. Bowman. (2003). McGraw-Hill Dictionary of Scientific and Technical Terms, 6E. McGraw-Hill.

Beaulieu, J. M., D.-C. Jhwueng, C. Boettiger, and B. C. O'Meara. 2012. Modeling stabilizing selection: Expanding the Ornstein-Uhlenbeck model of adaptive evolution. *Evolution* 66:2369–83.

Beaulieu, J. M., J. C. Oliver, and B. C. O'Meara. 2020. corHMM: Analysis of binary character evolution. R package version 2.3.

Beaulieu, J. M., and B. C. O'Meara. 2020. OUwie: Analysis of evolutionary rates in an OU framework. R package version 2.4.

Beaulieu, J. M., and B. C. O'Meara. 2016. Detecting hidden diversification shifts in models of trait-dependent speciation and extinction. *Syst. Biol.* 65:583–601.

Beaulieu, J. M., B. C. O'Meara, and M. J. Donoghue. 2013. Identifying hidden rate changes in the evolution of a binary morphological character: The evolution of plant habit in campanulid angiosperms. *Syst. Biol.* 62:725–37.

Benítez-Alvarez, L., A. Maria Leal-Zanchet, A. Oceguera-Figueroa, R. Lopes Ferreira, D. de Medeiros Bento, J. Braccini, R. Sluys, and M. Riutort. 2020. Phylogeny and biogeography of the Cavernicola (Platyhelminthes: Tricladida): Relics of an epigean group sheltering in caves? *Mol. Phylogenet. Evol.* 145:106709.

Benun Sutton, F., and A. B. Wilson. 2019. Where are all the moms? External fertilization predicts the rise of male parental care in bony fishes. *Evolution* 73:2451–60.

Blackburn, D. G. 1982. Evolutionary origins of viviparity in the reptilia: I. Sauria. *Amphib-reptil.* 3:185–205.

Blomberg, S. P., T. Garland Jr., and A. R. Ives. 2003. Testing for phylogenetic signal in comparative data: Behavioral traits are more labile. *Evolution* 57:717–45.

Blomberg, S. P., J. G. Lefevre, J. A. Wells, and M. Waterhouse. 2012. Independent contrasts and PGLS regression estimators are equivalent. *Syst. Biol.* 61:382–91.

Boettiger, C., G. Coop, and P. Ralph. 2012. Is your phylogeny informative? Measuring the power of comparative methods. *Evolution* 66:2240–51.

Bokma, F. 2002. Detection of punctuated equilibrium from molecular phylogenies. *J. Evol. Biol.* 15:1048–56.

Bokma, F. 2008. Detection of "punctuated equilibrium" by Bayesian estimation of speciation and extinction rates, ancestral character states, and rates of anagenetic and cladogenetic evolution on a molecular phylogeny. *Evolution* 62:2718–26.

Brandley, M. C., J. P. Huelsenbeck, and J. J. Wiens. 2008. Rates and patterns in the evolution of snake-like body form in squamate reptiles: Evidence for repeated re-evolution of lost digits and long-term persistence of intermediate body forms. *Evolution* 62:2042–64.

Branstetter, M. G., A. Ješovnik, J. Sosa-Calvo, M. W. Lloyd, B. C. Faircloth, S. G. Brady, and T. R. Schultz. 2017. Dry habitats were crucibles of domestication in the evolution of agriculture in ants. *Proc. Biol. Sci.* 284:20170095.

Broeckhoven, C., G. Diedericks, C. Hui, B. G. Makhubo, and P. le Fras N. Mouton. 2016. Enemy at the gates: Rapid defensive trait diversification in an adaptive radiation of lizards. *Evolution* 70:2647–56.

Burnham, K. P., and D. R. Anderson. 2003. *Model selection and multimodel inference: A practical information theoretic approach.* New York: Springer Science & Business Media.

Butler, M. A., and A. A. King. 2004. Phylogenetic comparative analysis: A modeling approach for adaptive evolution. *Am. Nat.* 164:683–95.

Chang, J., D. L. Rabosky, and M. E. Alfaro. 2020. Estimating diversification rates on incompletely sampled phylogenies: Theoretical concerns and practical solutions. *Syst. Biol.* 69:602–11.

Collar, D. C., and P. C. Wainwright. 2006. Discordance between morphological and mechanical diversity in the feeding mechanism of centrarchid fishes. *Evolution* 60:2575–84.

Collar, D. C., P. C. Wainwright, M. E. Alfaro, L. J. Revell, and R. S. Mehta. 2014. Biting disrupts evolutionary integration of the skull in eels. *Nat. Commun.* 5:5505.

Cracraft, J. 1988. Deep-history biogeography: Retrieving the historical pattern of evolving continental biotas. *Syst. Biol.* 37:221–36.

Davis, A. M., P. J. Unmack, R. P. Vari, and R. Betancur-R. 2016. Herbivory promotes dental disparification and macroevolutionary dynamics in Grunters (Teleostei: Terapontidae), a freshwater adaptive radiation. *Am. Nat.* 187:320–33.

Eastman, J. M., M. E. Alfaro, P. Joyce, A. L. Hipp, and L. J. Harmon. 2011. A novel comparative method for identifying shifts in the rate of character evolution on trees. *Evolution* 65:3578–89.

Eldredge, N., and S. J. Gould. 1972. Punctuated equilibria: An alternative to phyletic gradualism. In: T. J. M. Schopf (Ed). *Models in Paleobiology.* Freeman, Cooper, & Co., San Francisco, CA, pp. 82–115.

Edwards, A. W. F. 1992. *Likelihood.* Baltimore: Johns Hopkins University Press.

Esquerré, D., D. Ramírez-Álvarez, C. J. Pavón-Vázquez, J. Troncoso-Palacios, C. F. Garín, J. S. Keogh, and A. D. Leaché. 2019. Speciation across mountains: Phylogenomics, species delimitation and taxonomy of the *Liolaemus leopardinus* clade (Squamata, Liolaemidae). *Mol. Phylogenet. Evol.* 139:106524.

Etienne, R. S., and B. Haegeman. 2012. A conceptual and statistical framework for adaptive radiations with a key role for diversity dependence. *Am. Nat.* 180:E75–89.

Etienne, R. S., and B. Haegeman. 2021. DDD: Diversity-Dependent Diversification. R package version 5.0. https://CRAN.R-project.org/package=DDD.

Etienne, R. S., B. Haegeman, T. Stadler, T. Aze, P. N. Pearson, A. Purvis, and A. B. Phillimore. 2012. Diversity-dependence brings molecular phylogenies closer to agreement with the fossil record. *Proc. Biol. Sci.* 279:1300–09.

Etienne, R. S., A. L. Pigot, and A. B. Phillimore. 2016. How reliably can we infer diversity-dependent diversification from phylogenies? *Methods Ecol. Evol.* 7:1092–99.

Evans, M. E. K., S. A. Smith, R. S. Flynn, and M. J. Donoghue. 2009. Climate, niche evolution, and diversification of the "Bird-Cage" evening primroses (Oenothera, sections Anogra and Kleinia). *Am. Nat.* 173:225–40.

Felsenstein, J. 1973. Maximum-likelihood estimation of evolutionary trees from continuous characters. *Am. J. Hum. Genet.* 25:471–92.

Felsenstein, J. 1985. Phylogenies and the comparative method. *Am. Nat.* 125:1–15.

Felsenstein, J. 1993. PHYLIP (phylogeny inference package), version 3.5c. Joseph Felsenstein.

Felsenstein, J. 2004. *Inferring phylogenies.* Sunderland, MA: Sinauer Associates.

Felsenstein, J. 2005. Using the quantitative genetic threshold model for inferences between and within species. *Philos. Trans. R. Soc. Lond. B Biol. Sci.* 360:1427–34.

Felsenstein, J. 2012. A comparative method for both discrete and continuous characters using the threshold model. *Am. Nat.* 179:145–56.

Fisher, R. A. 1922. On the mathematical foundations of theoretical statistics. *Phil. Trans. R. Soc. Lond. A* 222:309–68.

FitzJohn, R. G. 2010. Quantitative traits and diversification. *Syst. Biol.* 59:619–33.

FitzJohn, R. G. 2012. Diversitree: Comparative phylogenetic analyses of diversification in R. *Methods Ecol. Evol.* 3:1084–92.

FitzJohn, R. G., W. P. Maddison, and S. P. Otto. 2009. Estimating trait-dependent speciation and extinction rates from incompletely resolved phylogenies. *Syst. Biol.* 58:595–611.

Garamszegi, L. Z. 2014. *Modern phylogenetic comparative methods and their application in evolutionary biology: Concepts and practice.* New York: Springer.

Garland, T., P. H. Harvey, and A. R. Ives. 1992. Procedures for the analysis of comparative data using phylogenetically independent contrasts. *Syst. Biol.* 41:18–32.

Garland, T. Jr., A. W. Dickerman, C. M. Janis, and J. A. Jones. 1993. Phylogenetic analysis of covariance by computer simulation. *Syst. Biol.* 42:265–92.

Gibson, B., and A. Eyre-Walker. 2019. Investigating evolutionary rate variation in bacteria. *J. Mol. Evol.* 87:317–26.

Gilbert, P., and R. Varadhan. 2019. NumDeriv: Accurate numerical derivatives. R package version 2016.8-1.1.

Glazier, D. S. 2013. Log-transformation is useful for examining proportional relationships in allometric scaling. *J. Theor. Biol.* 334:200–3.

Glossip, D., and J. B. Losos. 1997. Ecological correlates of number of subdigital lamellae in anoles. *Herpetologica* 53:192–99.

Goldberg, E. E., and B. Igić. 2012. Tempo and mode in plant breeding system evolution. *Evolution* 66:3701–09.

Gonzalez-Voyer, A., R.-J. Den Tex, A. Castelló, and J. A. Leonard. 2013. Evolution of acoustic and visual signals in Asian barbets. *J. Evol. Biol.* 26:647–59.

Gradstein, F. M., J. G. Ogg, M. D. Schmitz, and G. M. Ogg. 2020. *The geologic time scale 2020.* Amsterdam: Elsevier Science.

Grafen, A. 1989. The phylogenetic regression. *Philos. Trans. R. Soc. Lond. B Biol. Sci.* 326:119–57.

Green, P. J. 1995. Reversible jump Markov chain Monte Carlo computation and Bayesian model determination. *Biometrika* 82:711–32.

Greenberg, D. A., and A. Ø. Mooers. 2017. Linking speciation to extinction: Diversification raises contemporary extinction risk in amphibians. *Evol. Lett.* 1:40–48.

Griffin, R. H. 2017. Primate activity pattern and orbit morphology: A phylogenetic analysis of data from Kirk & Kay (2004). https://www.randigriffin.com/2017/11/17/primate-orbit-size.html.

Halali, S., E. Bergen, C. J. Breuker, P. M. Brakefield, and O. Brattström. 2020. Seasonal environments drive convergent evolution of a faster pace-of-life in tropical butterflies. *Ecol. Lett.* 24:102–12.

Hamilton, H., N. Saarman, G. Short, A. B. Sellas, B. Moore, T. Hoang, C. L. Grace, M. Gomon, K. Crow, and W. B. Simison. 2017. Molecular phylogeny and patterns of diversification in syngnathid fishes. *Mol. Phylogenet. Evol.* 107:388–403.

Hansen, T. F. 1997. Stabilizing selection and the comparative analysis of adaptation. *Evolution* 51:1341–51.

Hardy, N. B., D. A. Peterson, and B. B. Normark. 2016. Nonadaptive radiation: Pervasive diet specialization by drift in scale insects? *Evolution* 70:2421–28.

Harmon, L. 2019. *Phylogenetic comparative methods: Learning from trees.* Moscow, Idaho: Ecoevorxiv.

Harmon, L. J., and S. Harrison. 2015. Species diversity is dynamic and unbounded at local and continental scales. *Am. Nat.* 185:584–93.

Harmon, L. J., J. B. Losos, T. Jonathan Davies, R. G. Gillespie, J. L. Gittleman, W. Bryan Jennings, K. H. Kozak, M. A. McPeek, F. Moreno-Roark, T. J. Near, A. Purvis, R. E. Ricklefs, D. Schluter, J. A. Schulte II, O. Seehausen, B. L. Sidlauskas, O. Torres-Carvajal, J. T. Weir, and A. Ø. Mooers. 2010. Early bursts of body size and shape evolution are rare in comparative data. *Evolution* 64:2385–96.

Harmon, L. J., J. A. Schulte, A. Larson, and J. B. Losos. 2003. Tempo and mode of evolutionary radiation in iguanian lizards. *Science* 301:961–64.

Harmon, L. J., J. T. Weir, C. D. Brock, R. E. Glor, and W. Challenger. 2008. GEIGER: Investigating evolutionary radiations. *Bioinformatics* 24:129–31.

Harvey, P. H., and M. D. Pagel. 1991. *The comparative method in evolutionary biology*. Oxford: Oxford University Press.

Hedges, S. B., J. Dudley, and S. Kumar. 2006. TimeTree: A public knowledge-base of divergence times among organisms. *Bioinformatics* 22:2971–72.

Hohenlohe, P. A., and S. J. Arnold. 2008. MIPoD: A hypothesis-testing framework for microevolutionary inference from patterns of divergence. *Am. Nat.* 171:366–85.

Huelsenbeck, J. P., R. Nielsen, and J. P. Bollback. 2003. Stochastic mapping of morphological characters. *Syst. Biol.* 52:131–58.

Huey, R. B., T. Garland Jr., and M. Turelli. 2019. Revisiting a key innovation in evolutionary biology: Felsenstein's "Phylogenies and the comparative method." *American Naturalist* 193:755–72.

Hunt, G. 2006. Fitting and comparing models of phyletic evolution: Random walks and beyond. *Paleobiology* 32:578–601.

Ikeda, H., M. Nishikawa, and T. Sota. 2012. Loss of flight promotes beetle diversification. *Nat. Commun.* 3:648.

Ingram, T., A. Harrison, D. L. Mahler, M. D. R. Castaneda, R. E. Glor, A. Herrel, Y. E. Stuart, and J. B. Losos. 2016. Comparative tests of the role of dewlap size in *Anolis* lizard speciation. *Proc. Biol. Sci.* 283:20162199.

Ives, A. R. 2019. R^2s for correlated data: Phylogenetic models, LMMs, and GLMMs. *Syst. Biol.* 68:234–51.

Jablonski, D. 2008. Species selection: Theory and data. *Ann. Rev. Ecol. Evol. Syst.* 39:501–24.

Kembel, S. W., P. D. Cowan, M. R. Helmus, W. K. Cornwell, H. Morlon, D. D. Ackerly, S. P. Blomberg, and C. O. Webb. 2010. Picante: R tools for integrating phylogenies and ecology. *Bioinformatics*. 26:1463–64.

Keogh, J. S., D. L. Edwards, R. N. Fisher, and P. S. Harlow. 2008. Molecular and morphological analysis of the critically endangered Fijian iguanas reveals cryptic diversity and a complex biogeographic history. *Philos. Trans. R. Soc. Lond. B Biol. Sci.* 363:3413–26.

Kirk, E. C., and R. F. Kay. 2004. The evolution of high visual acuity in the Anthropoidea. In: Ross, C. F., Kay R. F. (Eds). *Anthropoid Origins. Developments in Primatology: Progress and Prospects*. Springer, Boston, MA, pp. 539–602.

Klaus, K. V., and N. J. Matzke. 2020. Statistical comparison of trait-dependent biogeographical models indicates that podocarpaceae dispersal is influenced by both seed cone traits and geographical distance. *Syst. Biol.* 69:61–75.

Lewis, P. O. 2001. A likelihood approach to estimating phylogeny from discrete morphological character data. *Syst. Biol.* 50:913–25.

Losos, J. B. 1999. Uncertainty in the reconstruction of ancestral character states and limitations on the use of phylogenetic comparative methods. *Anim. Behav.* 58:1319–24.

Losos, J. B. 2009. *Lizards in an evolutionary tree: Ecology and adaptive radiation of anoles*. Berkeley: University of California Press.

Louca, S., and M. W. Pennell. 2020. Extant timetrees are consistent with a myriad of diversification histories. *Nature* 580:502–05.

Maddison, W. P., and R. G. FitzJohn. 2015. The unsolved challenge to phylogenetic correlation tests for categorical characters. *Syst. Biol.* 64:127–36.

Maddison, W. P., P. E. Midford, S. P. Otto, and T. Oakley. 2007. Estimating a binary character's effect on speciation and extinction. *Syst. Biol.* 56:701–10.

Mahler, D. L., L. J. Revell, R. E. Glor, and J. B. Losos. 2010. Ecological opportunity and the rate of morphological evolution in the diversification of greater antillean anoles. *Evolution* 64:2731–45.

Martins, E. P. 1999. Estimation of ancestral states of continuous characters: A computer simulation study. *Syst. Biol.* 48:642–50.

Martins, E. P., and T. F. Hansen. 1997. Phylogenies and the comparative method: A general approach to incorporating phylogenetic information in the analysis of interspecific data. *Am. Nat.* 149:646–67.

Marx, H. E., D. E. Giblin, P. W. Dunwiddie, and D. C. Tank. 2016. Deconstructing Darwin's naturalization conundrum in the San Juan islands using community phylogenetics and functional traits. *Divers. Distrib.* 22:318–31.

Matzke, N. J. 2012. Founder-event speciation in BioGeoBEARS package dramatically improves likelihoods and alters parameter inference in dispersal-extinction-cladogenesis (DEC) analyses. *Front. Biogeogr* 4:210.

Matzke, N. J. 2013. Probabilistic historical biogeography: New models for founder-event speciation, imperfect detection, and fossils allow improved accuracy and model-testing. *Front. Biogeogr.* 5:242–48.

Matzke, N. J. 2014. Model selection in historical biogeography reveals that founder-event speciation is a crucial process in island clades. *Syst. Biol.* 63:951–70.

Morlon, H., E. Lewitus, F. L. Condamine, M. Manceau, J. Clavel, and J. Drury. 2016. RPANDA: An R package for macroevolutionary analyses on phylogenetic trees. *Methods Ecol. Evol.* 7:589–97.

Morlon, H., T. L. Parsons, and J. B. Plotkin. 2011. Reconciling molecular phylogenies with the fossil record. *Proc. Natl. Acad. Sci. U. S. A.* 108:16327–32.

Morlon, H., M. D. Potts, and J. B. Plotkin. 2010. Inferring the dynamics of diversification: A coalescent approach. *PLoS Biol.* 8:e1000493.

Murdoch, D., and D. Adler. 2021. rgl: 3D visualization using OpenGL. R package version 0.107.14. https://CRAN.R-project.org/package=rgl.

Near, T. J., C. M. Bossu, G. S. Bradburd, R. L. Carlson, R. C. Harrington, P. R. Hollingsworth Jr., B. P. Keck, and D. A. Etnier. 2011. Phylogeny and temporal diversification of darters (Percidae: Etheostomatinae). *Syst. Biol.* 60:565–95.

Nee, S. 2006. Birth-death models in macroevolution. *Annu. Rev. Ecol. Evol. Syst.* 37:1–17.

Nee, S., E. C. Holmes, R. M. May, and P. H. Harvey. 1994. Extinction rates can be estimated from molecular phylogenies. *Philos. Trans. R. Soc. Lond. B Biol. Sci.* 344:77–82.

Nee, S., R. M. May, and P. H. Harvey. 1994. The reconstructed evolutionary process. *Philos. Trans. R. Soc. Lond. B Biol. Sci.* 344:305–11.

Nee, S., A. Ø. Mooers, and P. H. Harvey. 1992. Tempo and mode of evolution revealed from molecular phylogenies. *Proc. Natl. Acad. Sci. U. S. A.* 89:8322–26.

Neter, J., M. H. Kutner, C. J. Nachtsheim, and W. Wasserman. 1996. *Applied linear statistical models.* Chicago: Irwin.

Neuwirth, E. 2014. RColorBrewer: ColorBrewer Palettes. R package version 1.1-2. https://CRAN.R-project.org/package=RColorBrewer.

Nunn, C. L. 2011. *The comparative approach in evolutionary anthropology and biology.* Chicago: University of Chicago Press.

Oakley, T. H., and C. W. Cunningham. 2000. Independent contrasts succeed where ancestor reconstruction fails in a known bacteriophage phylogeny. *Evolution* 54:397–405.

O'Meara, B. C., C. Ané, M. J. Sanderson, and P. C. Wainwright. 2006. Testing for different rates of continuous trait evolution using likelihood. *Evolution* 60:922–33.

Orme, D., R. Freckleton, G. Thomas, T. Petzoldt, S. Fritz, N. Isaac, W. Pearse. 2018. Caper: Comparative analyses of phylogenetics and evolution in R. R package version 1.01:458.

Pagel, M. 1994. Detecting correlated evolution on phylogenies: A general method for the comparative analysis of discrete characters. *Proc. R. Soc. Lond. B Biol. Sci.* 255:37–45.

Pagel, M. 1999a. Inferring the historical patterns of biological evolution. *Nature* 401:877–84.

Pagel, M. 1999b. The maximum likelihood approach to reconstructing ancestral character states of discrete characters on phylogenies. *Syst. Biol.* 48:612–22.

Pagel, M., and A. Meade. 2013. *BayesTraits v. 2.0.* Reading: University of Reading.

Paradis, E., J. Claude, and K. Strimmer. 2004. APE: Analyses of phylogenetics and evolution in R language. *Bioinformatics* 20:289–90.

Paradis, E., and K. Schliep. 2019. Ape 5.0: An environment for modern phylogenetics and evolutionary analyses in R. *Bioinformatics* 35:526–28.

Pennell, M. W., J. M. Eastman, G. J. Slater, J. W. Brown, J. C. Uyeda, R. G. FitzJohn, M. E. Alfaro, and L. J. Harmon. 2014. Geiger v2. 0: An expanded suite of methods for fitting macroevolutionary models to phylogenetic trees. *Bioinformatics* 30:2216–18.

Pennell, M. W., and L. J. Harmon. 2013. An integrative view of phylogenetic comparative methods: Connections to population genetics, community ecology, and paleobiology. *Ann. N. Y. Acad. Sci.* 1289:90–105.

Pinheiro, J., D. Bates, S. DebRoy, D. Sarkar, and R Core Team. 2019. Nlme: Linear and nonlinear mixed effects models. R package version 3.1–149.

Plummer, M., N. Best, K. Cowles, and K. Vines. 2006. CODA: Convergence diagnosis and output analysis for mcmc. *R News* 6:7–11.

Price, S. A., J. J. Tavera, T. J. Near, and P. C. Wainwright. 2013. Elevated rates of morphological and functional diversification in reef-dwelling haemulid fishes. *Evolution* 67:417–28.

Pybus, O. G., and P. H. Harvey. 2000. Testing macro-evolutionary models using incomplete molecular phylogenies. *Proc. Biol. Sci.* 267:2267–72.

Rabosky, D. L. 2014. Automatic detection of key innovations, rate shifts, anddiversity-dependence on phylogenetic trees. *PLoS One* 9:e89543.

Rabosky, D. L., and E. E. Goldberg. 2015. Model inadequacy and mistaken inferences of trait-dependent speciation. *Syst. Biol.* 64:340–55.

Rabosky, D. L., and A. H. Hurlbert. 2015. Species richness at continental scales is dominated by ecological limits. *Am. Nat.* 185:572–83.

Rabosky, D. L., and A. R. McCune. 2010. Reinventing species selection with molecular phylogenies. *Trends Ecol. Evol.* 25:68–74.

Ramm, T., E. J. Roycroft, and J. Müller. 2020. Convergent evolution of tail spines in squamate reptiles driven by microhabitat use. *Biol. Lett.* 16:20190848.

R Development Core Team. 2020. *R: A language and environment for statistical computing*. Vienna: R Foundation for Statistical Computing.

Ree, R. H., B. R. Moore, C. O. Webb, and M. J. Donoghue. 2005. A likelihood framework for inferring the evolution of geographic range on phylogenetic trees. *Evolution* 59:2299–311.

Ree, R. H., and I. Sanmartin. 2018. Conceptual and statistical problems with the DEC+J model of founder-event speciation and its comparison with DEC via model selection. *J. Biogeogr.* 45: 741–49.

Ree, R. H., and S. A. Smith. 2008. Maximum likelihood inference of geographic range evolution by dispersal, local extinction, and cladogenesis. *Syst. Biol.* 57:4–14.

Revell, L. J., and L. J. Harmon. 2008. Testing quantitative genetic hypotheses about the evolutionary rate matrix for continuous characters. *Evol. Ecol. Res.* 10:311–31.

Revell, L. J., L. J. Harmon, and D. C. Collar. 2008. Phylogenetic signal, evolutionary process, and rate. *Syst. Biol.* 57:591–601.

Revell, L. J. 2009. Size-correction and principal components for interspecific comparative studies. *Evolution* 63:3258–68.

Revell, L. J., and D. C. Collar. 2009. Phylogenetic analysis of the evolutionary correlation using likelihood. *Evolution* 63:1090–100.

Revell, L. J. 2012. Phytools: An R package for phylogenetic comparative biology (and other things). *Methods Ecol. Evol.* 3:217–23.

Revell, L. J., D. L. Mahler, P. R. Peres-Neto, and B. D. Redelings. 2012. A new phylogenetic method for identifying exceptional phenotypic diversification. *Evolution* 66:135–46.

Revell, L. J., and S. A. Chamberlain. 2014. Rphylip: An R interface for PHYLIP. *Methods Ecol. Evol.* 5: 976–81.

Revell, L. J. 2014a. Ancestral character estimation under the threshold model from quantitative genetics. *Evolution* 68:743–59.

Revell, L. J. 2014b. Graphical methods for visualizing comparative data on phylogenies. In *Modern phylogenetic comparative methods and their application in evolutionary biology: Concepts and practice*, ed. L. Z. Garamszegi, 77–103. Berlin: Springer.

Revell, L. J., L. E. Gonzalez-Valenzuela, A. Alfonso, L. A. Castellanos-Garcia, C. E. Guarnizo, and A. J. Crawford. 2018. Comparing evolutionary rates between trees, clades and traits. *Methods Ecol. Evol.* 9:994–1005.

Revell, L. J. 2021. A variable-rate quantitative trait evolution model using penalized-likelihood. *PeerJ* 9:e11997.

Revell, L. J., K. P. Schliep, D. L. Mahler, and T. Ingram. 2021. Testing for heterogeneous rates of discrete character evolution on phylogenies. bioRxiv, doi: 10.1101/2021.09.14.460362.

RStudio Team. 2020. RStudio: Integrated development for R. RStudio, PBC. http://www.rstudio.com/.

Schliep, K. P. 2011. Phangorn: Phylogenetic analysis in R. *Bioinformatics* 27:592–93.

Schluter, D. 2000. *The ecology of adaptive radiation*. Oxford: Oxford University Press.

Sepkoski, J. J. 1981. A factor analytic description of the phanerozoic marine fossil record. *Paleobiology* 7:36–53.

Shine, R., and J. J. Bull. 1979. The evolution of live-bearing in lizards and snakes. *Am. Nat.* 113:905–23.

Sjaarda, C. P., N. Rustom, G. A. Evans, D. Huang, S. Perez-Patrigeon, M. L. Hudson, H. Wong, Z. Sun, T. H. Guan, M. Ayub, C. N. Soares, R. I. Colautti, and P. M. Sheth. 2021. Phylogenomics reveals viral sources, transmission, and potential superinfection in early-stage COVID-19 patients in Ontario, Canada. *Sci. Rep.* 11:3697

Slater, G. J., L. J. Harmon, and M. E. Alfaro. 2012. Integrating fossils with molecular phylogenies improves inference of trait evolution. *Evolution* 66:3931–44.

Slowinski, J. B., and C. Guyer. 1989. Testing the stochasticity of patterns of organismal diversity: An improved null model. *Am. Nat.* 134:907–21.

Stadler, T. 2011. Simulating trees with a fixed number of extant species. *Syst. Biol.* 60:676–84.

Stadler, T. 2013. How can we improve accuracy of macroevolutionary rate estimates? *Syst. Biol.* 62:321–29.

Stadler, T. 2015. TreePar: Estimating birth and death rates based on phylogenies. R package version 3.3. 564. http://CRAN.R-project.org/package=TreePar.

Stadler, T. 2019. TreeSim: Simulating phylogenetic trees. R package version 2.4. https://CRAN.R-project.org/package=TreeSim.

Stadler, T., and S. Bonhoeffer. 2013. Uncovering epidemiological dynamics in heterogeneous host populations using phylogenetic methods. *Philos. Trans. R. Soc. Lond. B Biol. Sci.* 368:20120198.

Sun, M., R. A. Folk, M. A. Gitzendanner, P. S. Soltis, Z. Chen, D. E. Soltis, and R. P. Guralnick. 2020. Estimating rates and patterns of diversification with incomplete sampling: A case study in the rosids. *Am. J. Bot.* 107:895–909.

Swenson, N. G. 2014. *Functional and phylogenetic ecology in R*. New York: Springer.

Tuffley, C., and M. Steel. 1997. Links between maximum likelihood and maximum parsimony under a simple model of site substitution. *Bull. Math.* 59:581–607.

Uyeda, J. C., and L. J. Harmon. 2014. A novel Bayesian method for inferring and interpreting the dynamics of adaptive landscapes from phylogenetic comparative data. *Syst. Biol.* 63:902–18.

Uyeda, J. C., L. J. Harmon, and C. E. Blank. 2016. A comprehensive study of cyanobacterial morphological and ecological evolutionary dynamics through deep geologic time. *PLoS One* 11:e0162539.

Wang, L., X. Didelot, J. Yang, G. Wong, Y. Shi, W. Liu, G. F. Gao, and Y. Bi. 2020. Inference of person-to-person transmission of COVID-19 reveals hidden super-spreading events during the early outbreak phase. *Nat. Commun.* 11:5006.

Webb, C. O., D. D. Ackerly, M. A. McPeek, and M. J. Donoghue. 2002. Phylogenies and community ecology. *Annu. Rev. Ecol. Syst.* 33:475–505.

Webster, A. J., and A. Purvis. 2002. Testing the accuracy of methods for reconstructing ancestral states of continuous characters. *Proc. Biol. Sci.* 269:143–49.

Wickham, H., J. Hester, and W. Chang. 2020. Devtools: Tools to make developing R packages easier. R package version 2.3.0.

Wilks, S. S. 1938. The large-sample distribution of the likelihood ratio for testing composite hypotheses. *Ann. Math. Stat.* 9:60–62.

Williams, J. H., M. L. Taylor, and B. C. O'Meara. 2014. Repeated evolution of tricellular (and bicellular) pollen. *Am. J. Bot.* 101:559–71.

Wright, S. 1934. The results of crosses between inbred strains of guinea pigs, differing in number of digits. *Genetics* 19:537–51.

Yang, Z. 2006. *Computational molecular evolution.* Oxford: Oxford University Press.

Yu, G. 2020. Using ggtree to visualize data on tree-like structures. *Curr. Protoc. Bioinformatics* 69:e96.

Yu, G., T. T.-Y. Lam, H. Zhu, and Y. Guan. 2018. Two methods for mapping and visualizing associated data on phylogeny using ggtree. *Mol. Biol. Evol.* 35:3041–43.

Yu, G., D. K. Smith, H. Zhu, Y. Guan, and T. T. Lam. 2017. Ggtree: An R package for visualization and annotation of phylogenetic trees with their covariates and other associated data. *Methods Ecol. Evol.* 8:28–36.

Zeileis, A., and T. Hothorn. 2002. Diagnostic checking in regression relationships. *R News* 2:7–10.

Index

Index of R functions

eeeeeeeeeeeeeeeeeeeeeeeeeeeeeeeeeeeeeeeeeeeeeeeeeeeeeeeeeee

Page numbers typically give the first use of a function, unless it is discussed in more detail later in the book.

grep, 18
grid, 47

head, 26, 43
help, 8
hessian, 184
hisse, 330
hist, 93
history.from.sim.discrete, 311
HPDinterval, 213

install_github, 136
install.packages, 6
is.null, 47

keep.tip, 23
keep.tip.contMap, 227

labelnodes, 13
lapply, 54–56
library, 7
lines, 42
lm, 40, 44
log, 40
lrtest, 161
ltt, 256

make.bd, 288
make.bd.split, 306
make.bd.t, 291
make.linear.x, 349
make.musse, 324
make.prior, 138
make.quasse, 349
make.simmap, 244
mapply, 54–56
MarginReconHiSSE, 334
matrix, 148
mccr, 262
mcmc, 289
mean, 93
mergeMappedStates, 186
mtext, 10
multi2di, 261

name.check, 27, 29
Ntip, 16

OUwie, 118

packageVersion, 6
paintBranches, 199
paintSubTree, 19–20
pairs, 82
par, 10
pd, 373

pd.test, 378
phenogram, 233
phylomorphospace, 31, 51, 128
phylosig, 90
phyl.pca, 30
pic, 43
plot_BioGeoBEARS_results, 364
plot.hisse.states, 335
plot.phylo, 9, 10, 15
plotMKmodel, 198
plotSimmap.mcmc, 142
plotTree, 13
plotTree.datamatrix, 169, 240
points, 37
priorOU, 139
priorSim, 139
profiles.plot, 290

R.version, 5
ratebytree, 108, 112
rateshift, 130
read.csv, 25
read.newick, 15
read.simmap, 113
read.tree, 7, 15
replicate, 54–55
rnorm, 76
run, 141

sapply, 54–56
scores, 30
ses.mntd, 380
ses.mpd, 378
ses.pd, 377
set.burnin, 141
setMap, 227
setNames, 43, 88
setwd, 15
simBMphylo, 98
sim.history, 148, 186
starting.point.bisse, 315
starting.point.musse, 324
starting.point.quasse, 347
str, 12
stripchart, 185
summary, 28, 41, 45

threshBayes, 212
to.matrix, 31
TransMatMakerHiSSE, 330
tree.bisse, 310

var, 79

while, 47
write.tree, 16, 24

Milton Keynes UK
Ingram Content Group UK Ltd.
UKHW020353170624
444191UK00005B/12